METHODS AND PROGRAMS FOR MATHEMATICAL FUNCTIONS

MATHEMATICS AND ITS APPLICATIONS

Series Editor: G. M. BELL, Professor of Mathematics, King's College London (KQC), University of London

NUMERICAL ANALYSIS, STATISTICS AND OPERATIONAL RESEARCH

Editor: B. W. CONOLLY, Emeritus Professor of Mathematics (Operational Research), Queen Mary College, University of London

Mathematics and its applications are now awe-inspiring in their scope, variety and depth. Not only is there rapid growth in pure mathematics and its applications to the traditional fields of the physical sciences, engineering and statistics, but new fields of application are emerging in biology, ecology and social organization. The user of mathematics must assimilate subtle new techniques and also learn to handle the great power of the computer efficiently and economically.

The need for clear, concise and authoritative texts is thus greater than ever and our series will endeavour to supply this need. It aims to be comprehensive and yet flexible. Works surveying recent research will introduce new areas and up-to-date mathematical methods. Undergraduate texts on established topics will stimulate student interest by including applications relevant at the present day. The series will also include selected volumes of lecture notes which will enable certain important topics to be presented earlier than would otherwise be possible.

In all these ways it is hoped to render a valuable service to those who learn, teach, develop and use mathematics.

Mathematics and its Applications

Series Editor: G. M. BELL, Professor of Mathematics, King's College London (KQC), University of London

Abaffy, J. & Spedicato, E. **ABS Projection Algorithms: Mathematical Techniques for Linear and Nonlinear Equations**
Anderson, I. **Combinatorial Designs**
Artmann, B. **Concept of Number: From Quaternions to Monads and Topological Fields**
Arczewski, K. & Pietrucha, J. **Mathematical Modelling in Discrete Mechanical Systems**
Arczewski, K. and Pietrucha, J. **Mathematical Modelling in Continuous Mechanical Systems**
Bainov, D.D. & Konstantinov, M. **The Averaging Method and its Applications**
Baker, A.C. & Porteous, H.L. **Linear Algebra and Differential Equations**
Balcerzyk, S. & Jösefiak, T. **Commutative Rings**
Balcerzyk, S. & Jösefiak, T. **Commutative Noetherian and Krull Rings**
Baldock, G.R. & Bridgeman, T. **Mathematical Theory of Wave Motion**
Ball, M.A. **Mathematics in the Social and Life Sciences: Theories, Models and Methods**
de Barra, G. **Measure Theory and Integration**
Bartak, J., Herrmann, L., Lovicar, V. & Vejvoda, D. **Partial Differential Equations of Evolution**
Bell, G.M. and Lavis, D.A. **Statistical Mechanics of Lattice Models, Vols. 1 & 2**
Berry, J.S., Burghes, D.N., Huntley, I.D., James, D.J.G. & Moscardini, A.O. **Mathematical Modelling Courses**
Berry, J.S., Burghes, D.N., Huntley, I.D., James, D.J.G. & Moscardini, A.O. **Mathematical Modelling Methodology, Models and Micros**
Berry, J.S., Burghes, D.N., Huntley, I.D., James, D.J.G. & Moscardini, A.O. **Teaching and Applying Mathematical Modelling**
Blum, W. **Applications and Modelling in Learning and Teaching Mathematics**
Brown, R. **Topology: A Geometric Account of General Topology, Homotopy Types and the Fundamental Groupoid**
Burghes, D.N. & Borrie, M. **Modelling with Differential Equations**
Burghes, D.N. & Downs, A.M. **Modern Introduction to Classical Mechanics and Control**
Burghes, D.N. & Graham, A. **Introduction to Control Theory, including Optimal Control**
Burghes, D.N., Huntley, I. & McDonald, J. **Applying Mathematics**
Burghes, D.N. & Wood, A.D. **Mathematical Models in the Social, Management and Life Sciences**
Butkovskiy, A.G. **Green's Functions and Transfer Functions Handbook**
Cartwright, M. **Fourier Methods: Applications in Mathematics, Engineering and Science**
Cerny, I. **Complex Domain Analysis**
Chorlton, F. **Textbook of Dynamics, 2nd Edition**
Chorlton, F. **Vector and Tensor Methods**
Cohen, D.E. **Computability and Logic**
Cordier, J.-M. & Porter, T. **Shape Theory: Categorical Methods of Approximation**
Crapper, G.D. **Introduction to Water Waves**
Cross, M. & Moscardini, A.O. **Learning the Art of Mathematical Modelling**
Cullen, M.R. **Linear Models in Biology**
Dunning-Davies, J. **Mathematical Methods for Mathematicians, Physical Scientists and Engineers**
Eason, G., Coles, C.W. & Gettinby, G. **Mathematics and Statistics for the Biosciences**
El Jai, A. & Pritchard, A.J. **Sensors and Controls in the Analysis of Distributed Systems**
Exton, H. **Multiple Hypergeometric Functions and Applications**
Exton, H. **Handbook of Hypergeometric Integrals**
Exton, H. ***q*-Hypergeometric Functions and Applications**
Faux, I.D. & Pratt, M.J. **Computational Geometry for Design and Manufacture**
Firby, P.A. & Gardiner, C.F. **Surface Topology**
Gardiner, C.F. **Modern Algebra**

Series continued at back of book

METHODS AND PROGRAMS FOR MATHEMATICAL FUNCTIONS

STEPHEN LLOYD BALUK MOSHIER

ELLIS HORWOOD LIMITED
Publishers · Chichester

Halsted Press: a division of
JOHN WILEY & SONS
New York · Chichester · Brisbane · Toronto

First published in 1989 by
ELLIS HORWOOD LIMITED
Market Cross House, Cooper Street,
Chichester, West Sussex, PO19 1EB, England
The publisher's colophon is reproduced from James Gillison's drawing of the ancient Market Cross, Chichester.

Distributors:
Australia and New Zealand:
JACARANDA WILEY LIMITED
GPO Box 859, Brisbane, Queensland 4001, Australia
Canada:
JOHN WILEY & SONS CANADA LIMITED
22 Worcester Road, Rexdale, Ontario, Canada
Europe and Africa:
JOHN WILEY & SONS LIMITED
Baffins Lane, Chichester, West Sussex, England
North and South America and the rest of the world:
Halsted Press: a division of
JOHN WILEY & SONS
605 Third Avenue, New York, NY 10158, USA
South-East Asia
JOHN WILEY & SONS (SEA) PTE LIMITED
37 Jalan Pemimpin # 05–04
Block B, Union Industrial Building, Singapore 2057
Indian Subcontinent
WILEY EASTERN LIMITED
4835/24 Ansari Road
Daryaganj, New Delhi 110002, India

British Library Cataloguing in Publication Data
Moshier, Stephen L. B.
Methods and programs for mathematical functions.
1. Mathematics. Applications of computer systems.
I. Title II. Series
510'.28'5

Library of Congress CIP data available

ISBN 0–7458–0289–3 (Ellis Horwood Limited)
ISBN 0–470–21609–3 (Halsted Press)

Printed in Great Britain by Hartnolls, Bodmin

Contents

Preface

This book provides a working collection of mathematical software for computing various elementary and higher functions. It also supplies tutorial information of a practical nature; the purpose of this is to assist in constructing numerical programs for the reader's special applications.

Though some of the main analytical techniques for deriving functional expansions are described, the emphasis is on computing; so there has been no attempt to incorporate or supplant the many books on functional and numerical analysis that are available.

Nearly all of the example programs are designed for double precision (16 or 17 decimal place) floating point arithmetic. Rational approximations for lower precision routines can be calculated as needed using the development program that is supplied. Some single precision expansions are included for the elementary functions.

Numerical approximation coefficients presented herein are very close to the correct coefficients of the true minimax rational approximations. Leveling of the error peaks was carried out to the maximum extent possible in a 144 bit (43 decimal place) arithmetic. For some of the higher functions, a 336 bit arithmetic was used throughout, so that reference function values could be calculated to the desired 144 bit accuracy. Of course, knowing the correct values of the mathematical constants may satisfy nothing more than idle curiosity. There is no guarantee that these numbers are any better in practice than others derived by less rigorous error leveling. Nevertheless, some of them may be found to give improved results.

The computer programs in this book are copyrighted 1984 - 1988 by the author; they may not be copied in any form for resale or distribution without permission. Computer readable versions of the programs can be obtained from the publisher, Ellis Horwood, Ltd., or from Oasys, 230 Second Avenue, Waltham, Massachusetts 02154, USA. Versions of the programs for elementary functions are distributed in connection with the GNU project by the Free Software Foundation, 675 Massachusetts Avenue, Cambridge, Massachusetts 02139, USA.

1

Floating Point Arithmetic

1.1 Numeric Data Structures

You have undoubtedly studied arithmetic before, at some earlier time in your career. In that study you learned that numerical errors were always the fault of the pupil. Error did not reside in the subject matter itself. It may please you to find that this book is devoted to an arithmetic in which the numbers are often wrong though the pupil is right.

Of what use is an arithmetic that gives incorrect answers? Alas, sometimes, none at all. But nearly every digital computer extant has such an arithmetic, so one must make the best of it. This arithmetic, called **floating point**, expresses quantities by a fixed number of significant figures together with an exponent that has a fixed number of other figures. For example, the number 999 might be represented in floating point by the expression $.9990 \cdot 10^{03}$, meaning .9990 multiplied by $10^3 = 1000$. Here there are two digits allotted to the **exponent** (03) and four to the **significand** (.9990). This representation is similar to scientific notation, in which the same number would be written $9.990 \cdot 10^2$. A vital semantic difference is that the trailing 0 in the 9.990 of scientific notation is significant — it is included only if the writer intends to indicate that degree of precision — whereas a floating point computer number always indicates a certain number of significant digits, even if all of them are wrong.

A quantitative way to denote the accuracy with which a floating point data structure can represent real numbers is to specify the **precision** of a number system. The **relative precision** is measured by the number of digits in the significand; neither the exponent nor the actual value of the significand has a bearing. With four decimal digits the relative precision might be given as 4S. The **absolute precision** of a number is the unit value of its least significant digit; in this measure the exponent is important. For example, the absolute precision of 0.0999 might be denoted 4D.

The accuracy of a specific number is measured by its **absolute error**.

This is the arithmetic difference between it and whatever is considered to be its correct value. A number could be very erroneous or inaccurate, even if it is given to a high precision. Ordinarily the precision of an arithmetic system does tell something about the accuracy of its basic operations such as addition and multiplication; but after a long sequence of calculations the result may have considerably lower accuracy. **Relative error** is the the absolute error divided arithmetically by the correct value; it has the character of a percentage.

Since the precision of a floating point system is fixed and finite, it is easy to think of numbers that do not fit into it. For example, there is no item equal to $1/3$; the closest you can get in the example system is $.3333 \cdot 10^{00}$. The dynamic range of the system (an engineering term for the ratio between its largest and smallest elements) is also limited. With a two digit exponent there is no item larger than $.9999 \cdot 10^{99}$, and no normal item smaller than $.1000 \cdot 10^{-99}$ except for zero.

Having understood that the number of significant figures in a particular system is four, you may be led to the false impression that arithmetic done with these items is accurate to four figures. To see that this is not so, consider the arithmetic problem

$$\frac{1001}{3000} - \frac{1}{3} = \frac{1001}{3000} - \frac{1000}{3000} = \frac{1}{3000} .$$

The closest representation of $1/3000$ is $.3333 \cdot 10^{-4}$. That is indeed accurate to four figures, but the computer does not know how to combine fractions (that would be called **rational arithmetic**, not floating point arithmetic). The computer finds instead

$$\begin{aligned} \frac{1001}{3000} &= .3337 \cdot 10^{0} \\ \frac{1}{3} &= .3333 \cdot 10^{0} \\ \text{answer} &= .0004 \cdot 10^{0} \\ &= .4000 \cdot 10^{-4} \end{aligned}$$

— a result whose significand contains no correct figures at all because the leading digits cancelled to zero. This effect, a loss of significance due to cancellation, is known as **cancellation error**. It is the bane of all those who do numerical work in floating point arithmetic.

Observe, however, that if the answer were left in the form $.0004 \cdot 10^{0}$ where the three leading zeros of .0004 are regarded as significant digits, then the answer is as accurate as you could expect. The **absolute error** of the computed result is $.0004 - .0003333\ldots$, or less than .0001. The error is reasonable from this point of view. The **relative error** of the computed result is the ratio

$$\frac{.0004 - .0003333\ldots}{.0003333\ldots} = 0.2$$

of the absolute error to the correct answer. This amounts to 20%, which could be regarded as unreasonable compared to the precision 10^{-4} of the number system.

Note that the above problem could be sidestepped by restructuring the arithmetic to read

$$\frac{1001}{3000} - \frac{1}{3} = \frac{1001 - 1000}{3000} = \frac{1}{3000} .$$

In floating point, $1001 - 1000$ is 1 exactly, so the expression $(1001 - 1000)/3000$ is computed satisfactorily to four figure accuracy. In effect, the programmer has filled in the instructions to perform rational arithmetic for this specific calculation. It may not be obvious, either to the computer or to the computer programmer, that the original problem needed restructuring.

To put these matters a little more formally, a computer represents the real number x by a three-part data structure comprising a significand s, an exponent e and a sign S such that

$$\begin{aligned} x &= S \cdot s \cdot b^e, \text{ where} \\ 1/b &\leq s < 1. \end{aligned}$$

Typically, the **base** b of the arithmetic is

for **binary arithmetic**,	base $b = 2$
for **octal arithmetic**,	base $b = 8$
for **decimal arithmetic**,	base $b = 10$
for **hexadecimal arithmetic**,	base $b = 16$.

Desktop and larger computers typically use binary arithmetic for floating point operations. The programs in this book are optimized for this application. Pocket calculators, cash registers, and business oriented software languages often have decimal arithmetic.

Any given floating point system has a largest number that will be denoted `MAXNUM`. An attempt to represent a larger number is called **overflow**. The system also has a smallest nonzero number. An attempt to represent a smaller number is **underflow**. Underflow can be gradual if the restriction $1/b \leq s$ is lifted when the exponent reaches a minimum value; the computer number is then called **denormalized**. For example, $.1000 \cdot 10^{-99}$ is normal, because the most significant digit is nonzero; but $.0100 \cdot 10^{-99} = 10^{-101}$ is denormal. If denormal numbers are allowed, then the smallest number in this system is $.0001 \cdot 10^{-99} = 10^{-103}$.

Computers vary widely in their reaction to underflow and overflow conditions. Some react by terminating the program. Others may produce a number of fixed magnitude, either a special overflow value called infinity or else a maximum saturation value that is a valid, though incorrect, number. Underflow may produce a result of zero. Some systems allow you to specify what will happen under these anomalous conditions. A great many computers, including one or two whose names are familiar to you, will sometimes do other things, exhibiting completely faulty underflow or overflow behavior. Since these effects may be built into the hardware, there is no way to deal with this other than try to ensure that the conditions seldom occur, by including tests and error escapes in your computer programs.

The number of base b digits in the significand determines the relative precision with which a number can be represented by the computer. When the number system is binary ($b = 2$), the most significant bit of the significand is 1 unless the number is denormalized. This bit is usually omitted in the computer data structure. The space left by the omitted bit is used to include an additional bit of significance at the small end of the significand. For normalized binary numbers in which the most significant bit is omitted, the actual precision in binary bits is one more than the number of significand bits that are physically present.

Let P denote the precision in base b digits. The value of the least significant digit is smaller by a factor of exactly b^{1-P} than the value of the (possibly omitted) most significant digit of the significand. Regard full scale in the P-digit significand to be a count of b^P, representing a signficand of $s = 1$. Then the numerical absolute precision is one part in b^P of the full scale $1 \cdot b^e$ determined by the value of the exponent.

A standard promulgated by the Institute of Electrical and Electronics Engineers (IEEE)[1] specifies several consistent floating point formats and conditions that, it is hoped, will come into universal practice. In IEEE double precision arithmetic the base $b = 2$ and the precision $P = 53$, so the least significant bit of the significand has a value 2^{-52} times the value of the most significant bit. The next IEEE number after $1.0 = 0.5 \cdot 2^1$ is $1.0+2^{-52} = 1.0+2.22\ldots\cdot 10^{-16}$. However the last IEEE number before 1.0 is $1.0 - 2^{-53} = 1.0 - 1.11 \cdot 10^{-16}$. The reason for this is that the exponent of the slightly smaller number is one less than the exponent of 1.0, so its absolute precision is finer by a factor of two. An analogous situation in decimal arithmetic is the sequence

$$\begin{aligned} 0.9998 &= .9998 \cdot 10^0 \\ 0.9999 &= .9999 \cdot 10^0 \\ 1.000 &= .1000 \cdot 10^1 \\ 1.001 &= .1001 \cdot 10^1 \\ 1.002 &= .1002 \cdot 10^1 \end{aligned}$$

of successive four-digit floating point numbers. The absolute precision decreases by a factor of ten as the decimal exponent changes from 0 to 1.

[1] ANSI/IEEE Std 754-1985, published by the Institute of Electrical and Electronics Engineers, Inc., 345 East 47th Street, New York, NY 10017, USA.

1.2 Rounding

Roundoff error refers to the difference between a real number and the nearest machine representation of that number. In magnitude this error cannot exceed half the value of the least significant bit or digit of the nearest machine number. Except for the special case $s = 1/b$ (where the bound from below is actually smaller), this is a function of the exponent e and is independent of the value of the significand s. In binary systems this upper bound ϵ on the absolute error magnitude is

$$\epsilon = 2^{e-P-1}$$

and is ordinarily referred to as the **roundoff error** of the arithmetic. It is also equal to half the precision of the number system. In the computer programs that will appear later, this value is denoted by the symbol `MACHEP`. Roundoff error limits the accuracy of numerical computations. However, it is often *not* the most important source of arithmetic error; as we have seen, cancellation error can be far more disastrous than rounding error.

A **rounded** quantity is derived from an unrounded quantity by deleting least signficant digits from the significand in such a way that the remaining number is as close as possible to the original. The rules for strict rounding, known as *round to nearest or even*, are as follows in a base b system:

1. If the first digit to be dropped is less than $b/2$, *round down* by simply truncating it and all less significant digits, leaving unchanged all the places to be retained.

2. If the first digit to be dropped is greater than $b/2$, then *round up* by adding 1 at the least significant place to be retained, carrying out to higher places as required.

3. If the first digit to be dropped is equal to $b/2$ and any less significant digit, down to infinite precision, is nonzero, then *round up* as above.

4. If the first digit to be dropped is equal to $b/2$ and all less significant digits, down to infinite precision, are zero, then examine the least significant digit to be retained. If it is odd, then *round up*. If it is even, then *round down*.

The last item will be referred to as the "even" clause of the "round to nearest or even" rule, since under it the least significant digit of the rounded result will be an even number. In order to obey the strict rounding rules during an arithmetic operation, it is necessary to know enough about the correct result to behave as if it were infinitely precise. The computer programs presented later in this chapter illustrate the correct, strict rounding rules.

The majority of commercial arithmetic software packages and even hardware arithmetic chips available today do not obey the strict rounding rules, in part because their decisions are not based on the infinite precision results of operations. This indictment includes some products that allegedly implement IEEE arithmetic. It may be argued that this is not very important, but it does mean that calculations carried out by one computer or program may not give exactly the same results as another computer even though the arithmetic is supposed to be the same.

A **chopped** number is one that has had some of its least significant digits deleted. Chopped arithmetic may simply chop the infinite precision results to its working precision or, worse, it may omit intermediate calculations such as least significant partial products. Its basic operations may therefore have a relative error that is more than twice as large as the relative error of rounded arithmetic.

1.3 Addition and Subtraction

To compute

$$x_1 + x_2$$

the computer attempts to find the nearest floating point representation of the sum assuming that both of the arguments are exact. By the associative law, for $x_1 \geq x_2$

$$\begin{aligned} x_1 + x_2 &= s_1\, b^{e_1} + s_2\, b^{e_2} \\ &= \left(s_1 + s_2\, b^{e_2 - e_1}\right) b^{e_1} \ . \end{aligned}$$

In the computer, this means that when $e_1 > e_2$ the significand of the smaller number must be shifted $e_1 - e_2$ places to the right (or down) before it is added to the larger number.

To perform chopped arithmetic the adder circuit or program requires exactly P digits of precision; but additional information is required to find a correctly rounded sum. To know whether or not to round up by incrementing at the least signficant digit b^{-P}, it is essential that at least one additional digit, located at the b^{-P-1} position, be computed. This extra digit is known as a guard digit. To conform to strict rounding rules the adder must also be able to tell whether any digits below the guard digit position are nonzero. It may accomplish this simply by remembering that one or more nonzero digits below the guard digit were shifted out and thrown away.

If the digit at the guard position is equal to $b/2$ then it is necessary to invoke tests for rounding up. When nonzero digits have been shifted out, there are two rounding cases.

1. If x_1 and x_2 have the same sign, then any nonzero digits below the guard position indicate that the sum is larger than indicated. In this case the sum should be incremented.

2. If the arguments have opposite sign, then the existence of lost nonzero digits indicates that the sum is smaller in magnitude than indicated. In this case the sum should be rounded down (i.e., left alone).

As discussed in the last section, the strict rounding rule is "round to nearest or even." In the critical case, the guard digit at b^{-P-1} equals $b/2$ and all less significant digits are zero. The two rounding choices are equally near, so the rule becomes "round to even." For this case if the least significant digit, at b^{-P}, is odd then the sum is incremented; else it is left alone.

1.4 Multiplication

Consider two floating point numbers

$$\begin{aligned} x_1 &= s_1\, b^{e_1} \\ x_2 &= s_2\, b^{e_2}\ . \end{aligned}$$

By the commutative law, their product is

$$\begin{aligned} x_1\, x_2 &= s_1\, b^{e_1}\, s_2\, b^{e_2} \\ &= s_1\, s_2\, b^{e_2+e_1}\ . \end{aligned}$$

Since the product of the two significands may range from 0.25 (if $b = 2$) to nearly 1.0, it is necessary to check the normalization of the product. Normalizing may involve shifting the significand up or down 1 bit, so the multiplication must be carried to 1 extra bit of precision to ensure that the product will always contain at least P significant bits.

Commonly used techniques for multiplying the two significands depend on the fact that any fractional number s decomposes into

$$s = \sum_{k=0}^{P-1} \sigma_k\, c^{k-P}$$

where, for example, σ_0 is the least significant bit or digit, σ_1 is the next more significant digit, and so on up the the most significant digit σ_{P-1}. If the decomposition is into binary bits, then $c = 2$. The decomposition may be in terms of multiple digits. For example, in a computer with 16-bit words c might be equal to 65536. In terms of this decomposition, the product of s_1 and s_2 is

$$s_1\, s_2 = \sum_{k=1}^{P} \sigma_k\, c^{k-P} s_2\ .$$

1.4.1 Long Multiplication in Binary Radix

In binary arithmetic the multiplication of two significands can be carried out by a long shift and add algorithm, as follows. Note that this is nothing more than a particular implementation of the equation stated above.

1. Provide an accumulator with at least $P+2$ bits of precision. Initialize the accumulator to contain all zeros.

2. If the least significant bit of x_1 is a 1, then add x_2 to the partial product accumulator.

3. Shift both the accumulator and x_1 down 1 bit.

4. Repeat steps 2 and 3 until all the bits of x_1 have been shifted out.

At each step, an accumulator bit is shifted out and lost. This bit is a correct bit of the product since no partial products of that magnitude will be generated by the succeeding steps. Therefore at the last step all of the bits retained in the accumulator are correct.

Note that this algorithm computes all the bits of the exact product even though the least significant bit is thrown away at each step. Strictly correct rounding of the result is therefore made possible merely by noting whether any nonzero bits were discarded.

An improvement in average speed may be achieved by stopping when all the remaining bits of the multiplier are zero; but the benefit of doing this depends on the length of time it takes for the computer to test the bits.

1.4.2 Multiplication in Word Integer Radix

Many computers contain hardware circuits that can multiply two unsigned integers quickly. If the hardware can multiply two 16 bit integers, for example, then the decomposition of a 64 bit significand into 16-bit pieces would be

$$s = \sum_{k=0}^{3} \sigma_k 65536^{k-4}$$

in which σ_k ranges from 0 to 65535. The product rs of two such decomposed numbers r and s is the sum

$$rs = \sum_{j=0}^{3} \sum_{k=0}^{3} \rho_j \sigma_k 65536^{j+k-8}$$

of all the partial products. These are most conveniently accumulated from least to most signficant, so that the need for multiple precision arithmetic carries is minimized. This technique is usually much faster than the shift and add method of long multiplication. It is very similar to the long multiplication procedure taught in grammar school. The main difference is that the computer method accumulates each product as it is formed, and does not retain several rows of partial product values to be summed later.

There is an important practical difference between this method and binary long multiplication. It lies in the need to carry out from the product to the next more significant word integer place. Since there can be arithmetic carries from the very least significant word product, all the partial products must actually be computed and accumulated to assure a correct strictly rounded result. This is actually done also in the binary case, but the fact is not so obvious from the way it is implemented. Rounding of the product follows the same rules given above in the discussion of addition, except that the special case noted for subtraction does not occur. In the critical rounding case, nonzero bits below the guard position always mean that the number is larger in magnitude than indicated.

If the size, in digits, of the significand is the same as n word integers, a full precision product takes n^2 integer multiplications. However if strict rounding rules are not required, then the least significant partial products may be ignored. By truncating at the same precision as the desired result, the number of multiplications is reduced to $n(n+1)/2$. The error is the sum of all the omitted terms, which could amount to several times the value of the least significant digit.

Some computers provide a *signed* integer multiply instruction but not an *unsigned* multiply. An unsigned integer is regarded as a positive number, even if its most significant bit is a one. In a signed integer, the most significant bit is the sign. To use the signed multiply instruction for extended precision multiplication, each integer has to be regarded as the sum of a number containing only the most significant bit and a number containing the remaining bits. A partial product then takes the form of a double precision multiplication itself comprising four partial products. Alternatively, the significand can be transformed into an array of integers each of which has a zero in its most significant bit. Despite the inconvenience it is usually better to use the signed multiply than to resort to the shift and add method. The computer program for this may be long and ugly, however.

1.4.3 Fast Multiplication

Multiplication methods have been developed that are asymptotically faster than the ones described here. They approach order n operations for an n-word by n-word product. These are very useful for extremely high precision arithmetic — thousands or millions of decimal places. Unfortunately they have a minimum processing overhead that usually makes them inapplicable at medium and low precision. Several such methods have been discussed and reviewed by Knuth.[2]

[2]Knuth, Donald E., *The Art of Computer Programming*, Vol. 2, *Seminumerical Algorithms*, 2nd ed., p. 278 *ff*; Addison-Wesley, 1981.

1.5 Division

By rearrangement of the factors, it is evident that division of two floating point numbers is a matter of subtracting the exponents and finding the quotient of the significands:

$$\begin{aligned}\frac{x_1}{x_2} &= \frac{s_1 b^{e_1}}{s_2\ b^{e_2}} \\ &= \frac{s_1}{s_2}\ b^{e_1 - e_2}\ .\end{aligned}$$

Since the quotient s_1/s_2 of the significands has nearly a four to one range, it is necessary to check for the need to renormalize the quotient and adjust its exponent.

1.5.1 Long Division

In base 2 arithmetic, a binary shift and subtract method can be used to find the quotient of the significands. At each step the next quotient bit q is given implicitly by the expression

$$s_1 = q\ s_2 + (s_1 - q\ s_2)$$

where q is either 0 or 1 and the term in parentheses is the nonnegative remainder less than s_2. The above equation may be rephrased as

$$\frac{s_1}{s_2} = q\ + \frac{1}{2}\ \frac{2(s_2 - q\ s_2)}{s_2}\ ,$$

so the problem for the next step is to find the remainder divided by s_2. Thus the algorithm is as follows.

1. Test whether the remainder is greater than or equal to s_2.

2. If it is, then shift a one bit up into the least significant bit of the quotient and subtract s_2 from the remainder to form the new remainder for the next step.

3. If it is not, then shift a zero bit up into the quotient and do not subtract s_2.

4. Shift the new remainder up 1 bit to end the step.

5. Repeat until the desired number of quotient bits have been computed.

Since the quotient of the significands may have zero as its most significant bit, as many as $P + 2$ steps may be required to find P significant quotient bits and a guard bit.

Rounding follows the same procedure as multiplication. Since the exact remainder is known, strict rounding rules can be obeyed without difficulty.

1.5.2 Division by Taylor Series

Many integer oriented computers are ill-equipped for floating point division, and usually take much longer to divide than to multiply. Often, however, there are viable alternatives to the shift and subtract approach to division. If the computer has both divide and multiply hardware capability for integers, then a faster method can be constructed from the Taylor series expansion

$$\frac{s_1}{s_2} = \frac{s_1}{a+x} = \frac{s_1}{a}\left[1 - \frac{x}{a} + \left(\frac{x}{a}\right)^2 - \left(\frac{x}{a}\right)^3 + \ldots\right] .$$

Here $s_2 = a + x$ is partitioned into a comprising its topmost bits and x having the topmost bits set equal to zero. The integer divide facility is used to divide by a; the integer multiply is used to form the powers of x/a accurate to the desired number of significant digits. The number of integer multiplies required to form $(x/a)^n$ from the previously computed powers $< n$ decreases with n, since the arithmetic is essentially fixed point at this stage. This method is suitable for modestly extended precision computations though it is simple only when the significand has less than twice the number of bits in an integer. For instance, to calculate a single precision result having 24 bits of significand using 16 bit arithmetic, only the terms up to x/a of the expansion are required.

1.5.3 Newton-Raphson Division

An even faster division algorithm can be constructed from the Newton-Raphson iteration

$$\begin{aligned} y_{i+1} &= y_i(2 - x\, y_i) \\ &= 2y_i - x\, y_i^2 \end{aligned}$$

where y_i is the ith approximant to $1/x$. Convergence is very rapid, but each step requires two multiplications. This formula is derived in the section on iterative methods. The method is superior to the Taylor series approach for extended precision arithmetic since the number of correct digits doubles on each iteration.

An example will illustrate the precision required for each stage of the calculation. To compute $1/\pi$, the first approximation, to one digit accuracy, is $y_1 = 1/3 = 0.3$. This might be found by using the computer's integer divide instruction on the first digit or word of the numerator (1.0) and denominator (π) significands. The second approximation requires arithmetic carried to two digits or words:

$$y_2 = 2 \cdot 0.3 - 3.1 \cdot 0.3^2 = 0.6 - 0.28 = 0.32 .$$

The next iteration may be carried to four digits:

$$y_3 = 2 \cdot 0.32 - 3.141 \cdot 0.32^2 = 0.64 - .3216 = 0.3184 .$$

The next is

$$y_4 = 2 \cdot 0.3184 - 3.1415926 \cdot 0.3184^2 = 0.31830987$$

which has an error of 2 units in the last decimal place. There is no need to calculate the early iterations in full precision arithmetic. This can make for a substantial saving in the total number of integer multiplications required to achieve an extended precision result. Furthermore, in the iteration formula one of the multiplications is a square (y_i^2). If the significand is the size of n integers, the square can be calculated in $(n/2+1)^2$ integer multiplications (n an even number) instead of the n^2 operations that would be required for a normal extended precision multiply.

Strictly correct rounding is difficult to achieve with either the Taylor series or the iterative method since the result nearly always has a nonzero error. Extra, costly iterations are required to detect the critical rounding case. However, the speed advantage of these methods is so great that a small rounding discrepancy may be considered tolerable in certain applications.

1.6 C Language

The next section contains a listing of a computer program, so a comment on the software nomenclature is in order. All computer programs in this book are written in C language.[3] In an effort to improve the appearance and legibility of the source code, the names of function subroutines and global variables are set in bold face type, while the names of local variables are in italics. Program statements are in a typewriter style type font, while comments are set in the same type style as the text but in smaller letters. In C, comments are supposed to be set off by the delimiters `/* */`; for the most part these have been removed. All numeric constants in the book and all the source code listings were typeset by computer processing of the operationally tested program source files. Except for some editing of comment lines, the programs that appear here should be identical to the ones that were tested.

Fortunately, most high level programming languages have nearly the same syntax of algebraic expression; but for those not familiar with the oddities of C, the following Rosetta Stone may help to clarify the source program listings.

FORTRAN	BASIC	C
.EQ.	=	==
.NE.	<>	!=
.GT.	>	>

[3] Kernighan, Brian W. and Dennis M. Ritchie, *The C Programming Language*, Prentice-Hall, 1978.

```
.LT.            <               <
.GE.            >=              >=
.LE.            <=              <=

X = X + Y       X = X + Y       X += Y;
X = X - Y       X = X - Y       X -= Y;
X = X * Y       X = X * Y       X *= Y;
X = X / Y       X = X / Y       X /= Y;
I = I + 1       I = I + 1       I++;
X = A(1)        X = A(1)        X = A[0];
DO 3 I=1,10     FOR I=1 TO 10   for(I=1; I<=10; I++){
3 CONTINUE      NEXT I          }

CALL FUNC       GOSUB 4         FUNC();

GOTO 5          GO TO 5         goto five;
5               5               five:
```

Statements and blocks of statements in C are separated by visible punctuation marks. Generally, little attention is paid to the ends of lines or to absolute position within a line. In C, program flow control statements such as

```
if( expression is true ){
     do this;
     }
```

either execute the immediately following code enclosed in braces { } (if the condition tested is true), or else skip around it. An expression whose value is nonzero is regarded as true; zero is interpreted as false. If there is only one following statement to be executed when the `if()` condition is true, then the braces can be (and often are) omitted. Conversely, if the braces are omitted, the compiler puts implicit braces around that one statement. C distinguishes between a logical AND of true or false expressions (operator `&&`) and a bitwise AND of two integers (operator `&`). The same holds for the logical (`||`) and bitwise (`|`) OR operators. Indirect addressing in C is done explicitly by statements such as

p = &x; Load p with the memory address of x
y = *p; Load y with the item whose memory address is p

1.7 An Extended Double Arithmetic: ieee.c

This section presents computer programs for a floating point arithmetic that satisfies the IEEE standards for rounding and other matters. A sufficient number of features has been included to achieve an "excellent!" rating from the PARANOIA test program (see Chapter 3). Some points of the IEEE specification, such as settable rounding rules and signaling, are not shown; so this is *not* a complete implementation of the standard. The programs are slow, since they are written entirely in C language using binary shift-and-add algorithms. They provide a model or a development base from which faster and more suitable machine language programs might be derived.

The numeric format has a 64 bit significand and a 15 bit exponent. Its data structure is the same as the one used by the Intel 8087 or Motorola 68881 coprocessor chip. Numbers are stored in C language as arrays of 16-bit unsigned short integers. The arguments of the routines are pointers to the arrays. Table 1.1 gives the structure of the number in its external bit format. Table 1.2 gives the exploded structure of the internal format, which includes high and low guard words around the significand.

These programs implement overflow to a maximum saturation value that is an ordinary, legal number. A special value for infinity might be preferred; this would require tests for that value and implementation of special rules for arithmetic operations involving inifinity.

Denormalized tiny numbers are implemented, both in the 64 bit format and in conversions to and from IEEE double precision. Table 1.3 lists the subroutines that have the external number format as an argument; Table 1.4 lists the subroutines that operate on the internal data format. The programs `efloor.c`, `efrexp.c`, `eldexp.c` are in Section 3.4. A square root program written for this arithmetic is given in Section 2.10.5.

The test programs for all of the mathematical functions treated in this book were implemented in an extended precision (144 or 336 bit) arithmetic that used essentially identical versions of these routines. The only important differences were that the bit shifts and arithmetic on significands were programmed in machine language; the machine programs had algorithms using integer multiply and divide instructions where possible, as suggested in the preceding sections.

To save some space, the adjustments required for variable rounding precision have been removed from the listings. The conversions to and from 24 bit precision are not shown. Also, the 53 bit conversions show only the format for Intel (IBM PC) double precision data structure.

Number of 16 bit words in external number format

```
#define NE 5
```

Number of 16 bit words in internal format

```
#define NI (NE+3)
```

Array offset to exponent in internal format

```
#define E 1
```

Array offset to high guard word

```
#define M 2
```

Number of bits of precision

Word	Contents
e[0] to e[NX-2]	Significand: least significant word first, most significant bit normally set
e[NX-1]	Biased exponent: value = EXONE for 1.0, top bit is the sign of the entire number.

Table 1.1: Data structure of external format number

Word	Contents
ei[0]	Sign word: 0 for positive, `0xffff` for negative
ei[1]	Biased exponent: value EXONE for the number 1.0
ei[2]	High guard word, always zero after normalization
ei[3] to ei[NI-2]	NI-4 significand words, most significant word first, most significant bit is normally set.
ei[NI-1]	Low guard word: top bit is rounding place

Table 1.2: Data structure of internal format number

asctoe(string, e)	ASCII string to e type
e24toe(&d, e)	IEEE single precision to e type
e53toe(&d, e)	IEEE double precision to e type
eabs(e)	absolute value
eadd(a, b, c)	c = b + a
eclear(e)	e = 0
ecmp(a, b)	compare a to b, return 1, 0, or –1
ediv(a, b, c)	c = b / a
efloor(a, b)	truncate to integer, toward –infinity
efrexp(a, exp, s)	extract exponent and significand
eifrac(e, &l, frac)	e to long integer plus e type fraction
einfin(e)	set e to infinity, leaving its sign alone
eldexp(a, n, b)	multiply by 2^n
emov(a, b)	b = a
emul(a, b, c)	c = b times a
eneg(e)	e = −e
eremain(a, b, c)	c = remainder of b/a
eround(a, b)	b = nearest integer value to a
esub(a, b, c)	c = b − a
etoasc(e, str, n)	e to ASCII string, n digits after decimal
etoe24(e, &d)	convert e type to IEEE single precision
etoe53(e, &d)	convert e type to IEEE double precision
ltoe(&l, e)	long (32 bit) integer to e type

Table 1.3: Arithmetic routines for external format numbers

eaddm(ai, bi)	add significands, bi = bi + ai
ecleaz(ei)	ei = 0
ecmpm(ai, bi)	compare significands, return 1, 0, or −1
edivm(ai, bi)	divide significands, bi = bi / ai
ernorm(ai,l,s,exp,r)	normalize and round off
emovi(a, ai)	convert external a to internal ai
emovo(ai, a)	convert internal ai to external a
emovz(ai, bi)	bi = ai, low guard word of bi = 0
emulm(ai, bi)	multiply significands, bi = bi * ai
enormlz(ei)	left-justify the significand
eshdn1(ai)	shift significand and guards down 1 bit
eshdn8(ai)	shift down 8 bits
eshdn16(ai)	shift down 16 bits
eshift(ai, n)	shift ai n bits up (or down if n < 0)
eshup1(ai)	shift significand and guards up 1 bit
eshup8(ai)	shift up 8 bits
eshup16(ai)	shift up 16 bits
esubm(ai, bi)	subtract significands, bi = bi − ai

Table 1.4: Routines for internal format numbers

```
#define NBITS ((NI-4)*16)
```
Maximum number of decimal digits in ASCII conversion = NBITS*log10(2)
```
#define NDEC (NBITS*8/27)
```
The biased exponent of the number 1.0
```
#define EXONE (0x3fff)
```
References to constants stored in external number format
```
extern unsigned short ezero[], ehalf[], eone[];
```

Program to set external format number x to 0.0.

```
void eclear( x )
register unsigned short *x;
    {
    register int i;

    for( i=0; i<NE; i++ )
        *x++ = 0;
    }
```

Move external format number from a to b.

```
void emov( a, b )
register unsigned short *a, *b;
    {
    register int i;

    for( i=0; i<NE; i++ )
        *b++ = *a++;
    }
```

Take absolute value of external format number.
The sign is the top bit of the last word.

```
void eabs(x)
unsigned short x[];
    {
    x[NE-1] &= 0x7fff;
    }
```

Negate external format number.

```
void eneg(x)
unsigned short x[];
    {
    x[NE-1] += 0x8000;
    }
```

Set $x = \infty$ or largest possible number.

```
void einfin(x)
register unsigned short *x;
    {
    register int i;

    for( i=0; i<NE-1; i++ )
        *x++ = 0xffff;
    *x |= 0x7fff;
    }
```

Move in external format number,
converting it to internal format.

```
void emovi( a, b )
unsigned short *a, *b;
    {
    register unsigned short *p, *q;

    q = b;
```

Point to last word of external number.

```
    p = a + (NE-1);
```

Get the sign bit.

```
    if( *p & 0x8000 )
        *q++ = 0xffff;
    else
        *q++ = 0;
```

Get the exponent.

```
    *q = *p--;
```

Delete the sign bit.

```
	*q++ &= 0x7fff;
```
Clear high guard word.
```
	*q++ = 0;
```
Move in the significand.
```
	*q++ = *p--;
	*q++ = *p--;
	*q++ = *p--;
	*q++ = *p--;
```
Clear low guard word.
```
	*q = 0;
	}
```

Move internal format number out,
converting it to external format.

```
void emovo( a, b )
unsigned short *a, *b;
	{
	register unsigned short *p, *q;
	unsigned short i;

	p = a;
```
Point to output exponent.
```
	q = b + (NE-1);
```
Combine sign and exponent.
```
	i = *p++;
	if( i )
		*q-- = *p++ | 0x8000;
	else
		*q-- = *p++;
```
Copy significand, omitting guard words.
```
	++p;
	*q-- = *p++;
	*q-- = *p++;
	*q-- = *p++;
	*q-- = *p++;
	}
```

Clear out internal format number.

```
void ecleaz( xi )
register unsigned short *xi;
	{
	register int i;

	for( i=0; i<NI; i++ )
		*xi++ = 0;
	}
```

Move internal format number from a to b.

```
void emovz( a, b )
register unsigned short *a, *b;
	{
	register int i;

	for( i=0; i<NI-1; i++ )
		*b++ = *a++;
```

Clear the low guard word of the destination.

```
	*b = 0;
	}
```

Compare significands of numbers in internal format.
Returns +1 if $a > b$, 0 if $a == b$, -1 if $a < b$.

```
ecmpm( a, b )
register unsigned short *a, *b;
	{
	int i;

	a += M;
	b += M;
	for( i=M; i<NI; i++ )
		{
		if( *a++ != *b++ )
			goto difrnt;
		}
	return(0);
difrnt:
	if( *(--a) > *(--b) )
		return(1);
```

```
    else
        return(-1);
    }
```

Shift significand down by 1 bit.

```
void eshdn1(x)
register unsigned short *x;
    {
    register unsigned short bits;
    int i;

    x += M;
    bits = 0;
    for( i=M; i<NI; i++ )
        {
        if( *x & 1 )
            bits |= 1;
        *x >>= 1;
        if( bits & 2 )
            *x |= 0x8000;
        bits <<= 1;
        ++x;
        }
    }
```

Shift significand up by 1 bit.

```
void eshup1(x)
register unsigned short *x;
    {
    register unsigned short bits;
    int i;

    x += NI-1;
    bits = 0;

    for( i=M; i<NI; i++ )
        {
        if( *x & 0x8000 )
            bits |= 1;
```

```
		*x <<= 1;
		if( bits & 2 )
			*x |= 1;
		bits <<= 1;
		--x;
		}
	}
```

Shift significand down by 8 bits.

```
void eshdn8(x)
register unsigned short *x;
	{
	register unsigned short newbyt, oldbyt;
	int i;

	x += M;
	oldbyt = 0;
	for( i=M; i<NI; i++ )
		{
		newbyt = *x << 8;
		*x >>= 8;
		*x |= oldbyt;
		oldbyt = newbyt;
		++x;
		}
	}
```

Shift significand up by 8 bits.

```
void eshup8(x)
register unsigned short *x;
	{
	int i;
	register unsigned short newbyt, oldbyt;

	x += NI-1;
	oldbyt = 0;

	for( i=M; i<NI; i++ )
		{
```

```
		newbyt = *x >> 8;
		*x <<= 8;
		*x |= oldbyt;
		oldbyt = newbyt;
		--x;
		}
	}
```

Shift significand up by 16 bits.

```
void eshup16(x)
register unsigned short *x;
	{
	int i;
	register unsigned short *p;

	p = x + M;
	x += M + 1;
	for( i=M; i<NI-1; i++ )
		*p++ = *x++;
	*p = 0;
	}
```

Shift significand down by 16 bits.

```
void eshdn16(x)
register unsigned short *x;
	{
	int i;
	register unsigned short *p;

	x += NI-1;
	p = x + 1;
	for( i=M; i<NI-1; i++ )
		*(--p) = *(--x);
	*(--p) = 0;
	}
```

Add significands: $x + y$ replaces y.

```
void eaddm( x, y )
unsigned short *x, *y;
    {
    register unsigned long a;
    int i;
    unsigned int carry;

    x += NI-1;
    y += NI-1;
    carry = 0;
    for( i=M; i<NI; i++ )
        {
        a = (unsigned long )(*x)
            + (unsigned long )(*y) + carry;
        if( a & 0x10000 )
            carry = 1;
        else
            carry = 0;
        *y = (unsigned short )a;
        --x;
        --y;
        }
    }
```

Subtract significands: $y - x$ replaces y.

```
void esubm( x, y )
unsigned short *x, *y;
    {
    unsigned long a;
    int i;
    unsigned int carry;

    x += NI-1;
    y += NI-1;
    carry = 0;
    for( i=M; i<NI; i++ )
        {
        a = (unsigned long )(*y)
            - (unsigned long )(*x) - carry;
        if( a & 0x10000 )
            carry = 1;
```

```
        else
            carry = 0;
        *y = (unsigned short )a;
        --x;
        --y;
        }
    }
```

Divide significands.

```
static unsigned short equot[NI] = {0};

int edivm( den, num )
unsigned short den[], num[];
    {
    int i, j;
    register unsigned short *p, *q;

    p = &equot[0];
    *p++ = num[0];
    *p++ = num[1];

    for( i=M; i<NI; i++ )
        {
        *p++ = 0;
        }
```

Use faster compare and subtraction if denominator has only 15 bits of significance.

```
    p = &den[M+2];
    if( *p++ == 0 )
        {
        for( i=M+3; i<NI; i++ )
            {
            if( *p++ != 0 )
                goto fulldiv;
            }
        if( (den[M+1] & 1) != 0 )
            goto fulldiv;
        eshdn1(num);
        eshdn1(den);

        p = &den[M+1];
        q = &num[M+1];
```

```
        for( i=0; i<NBITS+2; i++ )
             {
             if( *p <= *q )
                  {
                  *q -= *p;
                  j = 1;
                  }
             else
                  {
                  j = 0;
                  }
             eshup1(equot);
             equot[NI-2] |= j;
             eshup1(num);
             }
        goto divdon;
        }
```

The number of quotient bits to calculate is
NBITS plus 1 scaling guard bit plus 1 roundoff bit.

```
fulldiv:
     p = &equot[NI-2];
     for( i=0; i<NBITS+2; i++ )
          {
          if( ecmpm(den,num) <= 0 )
               {
               esubm(den, num);
```

Quotient bit = 1

```
               j = 1;
               }
          else
               j = 0;
          eshup1(equot);
          *p |= j;
          eshup1(num);
          }
divdon:
     eshdn1( equot );
     eshdn1( equot );
```

Test for nonzero remainder after roundoff bit.

```
     p = &num[M];
     j = 0;
     for( i=M; i<NI; i++ )
          {
          j |= *p++;
```

```
	}
if( j )
	j = 1;
for( i=0; i<NI; i++ )
	num[i] = equot[i];
return(j);
}
```

Multiply significands.

```
int emulm( a, b )
unsigned short a[], b[];
	{
	unsigned short *p, *q;
	int i, j, k;

	equot[0] = b[0];
	equot[1] = b[1];
	for( i=M; i<NI; i++ )
		equot[i] = 0;
	p = &a[NI-2];
	k = NBITS;
```

Calling program checks that significand is not zero.

```
	while( *p == 0 )
		{
		eshdn16(a);
		k -= 16;
		}
	if( (*p & 0xff) == 0 )
		{
		eshdn8(a);
		k -= 8;
		}
	q = &equot[NI-1];
	j = 0;
	for( i=0; i<k; i++ )
		{
		if( *p & 1 )
			eaddm(b, equot);
```

Remember if there were any nonzero bits shifted out.

```
		if( *q & 1 )
			j |= 1;
		eshdn1(a);
```

```
        eshdn1(equot);
        }

    for( i=0; i<NI; i++ )
        b[i] = equot[i];
```

Return flag for lost nonzero bits.

```
    return(j);
    }
```

```
static unsigned short rbit[NI] = {0,0,0,0,0,0,0,0};
```

Normalize and round off.

The internal format number to be rounded is "s."
Input *lost* indicates whether or not the number is exact.
This is the so-called sticky bit.
Input *subflg* indicates whether the number was obtained
by a subtraction operation. In that case if *lost* is nonzero
then the number is slightly smaller than indicated.
Input *exp* is the biased exponent, which may be negative.
The exponent field of *s* is ignored but is replaced by
exp as adjusted by normalization and rounding.

```
void ernorm( s, lost, subflg, exp, rcntrl )
unsigned short s[];
int lost;
int subflg;
long exp;
int rcntrl;
    {
    int i, j;
```

Normalize

```
    j = enormlz( s );
    if( j > NBITS )
        {
        ecleaz( s );
        return;
        }
    exp -= j;
    if( exp > 32767L )
        goto overf;
```

```
	if( exp < 0L )
		{
		if( exp > (long )(-NBITS-1) )
			{
			j = (int )exp;
			i = eshift( s, j );
			if( i )
				lost = 1;
			}
		else
			{
			ecleaz( s );
			return;
			}
		}
Round off
	if( ((s[NI-1] & 0x8000) != 0) && (rcntrl != 0) )
		{
		if( s[NI-1] == 0x8000 )
			{
			if( lost == 0 )
				{
Round to even
				if( (s[NI-2] & 1) == 0 )
				goto mddone;
				}
			else
				{
				if( subflg != 0 )
					goto mddone;
				}
			}
		rbit[NI-2] = 1;
		eaddm( rbit, s );
		if( s[2] != 0 )
			{
Overflow on roundoff
			eshdn1(s);
			exp += 1;
			}
		}
mddone:
	if( exp > 32767 )
		{
overf:
```

```
        s[1] = 32767;
        s[2] = 0;
        for( i=M+1; i<NI-1; i++ )
            s[i] = 0xffff;
        s[NI-1] = 0;
        return;
        }

    if( exp < 0 )
        s[1] = 0;
    else
        s[1] = (unsigned short )exp;
    s[NI-1] = 0;
    }
```

Subtract external format numbers.

```
static unsigned short subflg = 0;

void esub( a, b, c )
unsigned short *a, *b, *c;
    {
    void eadd1();

    subflg = 1;
    eadd1( a, b, c );
    }
```

Add external format numbers.

```
void eadd( a, b, c )
unsigned short *a, *b, *c;
    {

    subflg = 0;
    eadd1( a, b, c );
    }

void eadd1( a, b, c )
unsigned short *a, *b, *c;
```

```
{
unsigned short ai[NI], bi[NI], ci[NI];
int i, lost, j, k;
long lt, lta, ltb;

emovi( a, ai );
emovi( b, bi );
if( subflg )
      ai[0] =    ai[0];
```

Compare exponents.

```
lta = ai[E];
ltb = bi[E];
lt = lta - ltb;
if( lt > 0L )
      {
```

Put the larger number in *bi*.

```
      emovz( bi, ci );
      emovz( ai, bi );
      emovz( ci, ai );
      ltb = bi[E];
      lt = -lt;
      }
lost = 0;
if( lt != 0L )
      {
      if( lt < (long )(-NBITS-1) )
            goto done;
      k = (int )lt;
      lost = eshift( ai, k );
      }
else
      {
```

Exponents were the same, so must compare significands.

```
      i = ecmpm( ai, bi );
      if( i == 0 )
            {
```

If same magnitude but different signs, result is zero.

```
            if( ai[0] != bi[0] )
                  {
                  eclear(c);
                  return;
                  }
```

If same magnitude and same sign, result is double.
Special handling to double a denomalized tiny number:

```
            if( (bi[E] == 0) && ((bi[3] & 0x8000) == 0) )
```

```
                {
                eshup1( bi );
                goto done;
                }
```

Add 1 to the exponent, unless both numbers are zero.

```
            for( j=1; j<NI-1; j++ )
                {
                if( bi[j] != 0 )
                    {
                    ltb += 1;
                    if( ltb > 32767 )
                        {
                        einfin( c );
                        return;
                        }
                    break;
                    }
                }
            bi[E] = (unsigned short )ltb;
            goto done;
            }
        if( i > 0 )
            {
```

Put the larger number in *bi*

```
            emovz( bi, ci );
            emovz( ai, bi );
            emovz( ci, ai );
            }
        }
    if( ai[0] == bi[0] )
        {
        eaddm( ai, bi );
        subflg = 0;
        }
    else
        {
        esubm( ai, bi );
        subflg = 1;
        }
    ernorm( bi, lost, subflg, ltb, 64 );
done:
    emovo( bi, c );
    }
```

Divide external format numbers.

```
void ediv( a, b, c )
unsigned short *a, *b, *c;
     {
     unsigned short ai[NI], bi[NI];
     int i;
     long lt, lta, ltb;

     emovi( a, ai );
     emovi( b, bi );
     lta = ai[E];
     ltb = bi[E];
```
See if numerator is zero.
```
     if( bi[E] == 0 )
          {
          for( i=1; i<NI-1; i++ )
               {
               if( bi[i] != 0 )
                    {
                    ltb -= enormlz( bi );
                    goto dnzro1;
                    }
               }
          eclear(c);
          return;
          }
dnzro1:
```
See if denominator is zero.
```
     if( ai[E] == 0 )
          {
          for( i=1; i<NI-1; i++ )
               {
               if( ai[i] != 0 )
                    {
                    lta -= enormlz( ai );
                    goto dnzro2;
                    }
               }
          einfin(c);
          mtherr( "ediv", SING );
          return;
          }
dnzro2:
```

```
    i = edivm( ai, bi );
```
Calculate exponent of quotient.
```
    lt = ltb - lta + EXONE;
    ernorm( bi, i, 0, lt, 64 );
```
Set the sign.
```
    if( ai[0] == bi[0] )
        bi[0] = 0;
    else
        bi[0] = 0Xffff;
    emovo( bi, c );
    }
```

Multiply external format numbers.

```
void emul( a, b, c )
unsigned short *a, *b, *c;
    {
    unsigned short ai[NI], bi[NI];
    int i, j;
    long lt, lta, ltb;

    emovi( a, ai );
    emovi( b, bi );
    lta = ai[E];
    ltb = bi[E];
    if( ai[E] == 0 )
        {
        for( i=1; i<NI-1; i++ )
            {
            if( ai[i] != 0 )
                {
                lta -= enormlz( ai );
                goto mnzer1;
                }
            }
        eclear(c);
        return;
        }
mnzer1:
    if( bi[E] == 0 )
        {
        for( i=1; i<NI-1; i++ )
            {
```

```
                if( bi[i] != 0 )
                    {
                    ltb -= enormlz( bi );
                    goto mnzer2;
                    }
                }
            eclear(c);
            return;
            }
mnzer2:
```
Multiply significands.
```
    j = emulm( ai, bi );
```
Calculate exponent of product.
```
    lt = lta + ltb - (EXONE - 1);
    ernorm( bi, j, 0, lt, 64 );
```
Calculate sign of product.
```
    if( ai[0] == bi[0] )
        bi[0] = 0;
    else
        bi[0] = 0xffff;
    emovo( bi, c );
    }
```

Convert IEEE double precision to *e* type

```
void e53toe( e, y )
unsigned short *e, *y;
    {
    register unsigned short r;
    register unsigned short *p;
    unsigned short yy[NI];
    int denorm, k;
```

Flag to indicate denormalized number.
```
    denorm = 0;
    ecleaz(yy);
    e += 3;
    r = *e;
    yy[0] = 0;
    if( r & 0x8000 )
        yy[0] = 0xffff;
    yy[M] = (r & 0x0f) | 0x10;
```
Strip sign and 4 significand bits.

```
	r &= ~ 0x800f;
	r >>= 4;
```
If exponent is zero, then the significand is denormalized;
so, take back the understood high significand bit.
```
	if( r == 0 )
		{
		denorm = 1;
		yy[M] &= ~ 0x10;
		}
	r += EXONE - 01777;
	yy[E] = r;
	p = &yy[M+1];
	*p++ = *(--e);
	*p++ = *(--e);
	*p++ = *(--e);
	eshift( yy, -5 );
	if( denorm )
		{
```
If exponent was zero, then normalize the significand.
```
		if( (k = enormlz(yy)) > NBITS )
			ecleaz(yy);
		else
			yy[E] -= k-1;
		}
	emovo( yy, y );
	}
```

Convert *e* type to IEEE double precision

```
void etoe53( x, e )
unsigned short *x, *e;
	{
	unsigned short xi[NI], ri[NI];
	int i, k;
	register unsigned short *p;

	e += 3;
	*e = 0;
	p = &xi[0];
	emovi( x, p );
	if( *p++ )
		*e = 0x8000;
```
Round off to nearest or even.

```
    i = xi[M+4];
    if( (i & 0x400) != 0 )
        {
        if( (i & 0x07ff) == 0x400 )
            {
            if( (i & 0x800) == 0 )
                goto nornd;
            }
        ecleaz( ri );
        ri[M+4] = 0x400;
        eaddm( ri, xi );
        k = enormlz( xi );
        *p -= k;
        }
nornd:
```

Note, $*p$ = exponent of xi.

```
    if( *p < (EXONE - 1081) )
        goto ozero;
    if( *p > (EXONE + 1023) )
        {
```

Saturate at largest legal number.

```
        *e |= 0x7fef;
        *(--e) = 0xffff;
        *(--e) = 0xffff;
        *(--e) = 0xffff;
        return;
        }
    i = (int )*p++ -(EXONE - 01777);
```

Handle denormalized small numbers.

```
    if( i <= 0 )
        {
        if( i > -53 )
            {
            eshift( xi, i-1 );
            i = 0;
            }
        else
            {
ozero:
            *(--e) = 0;
            *(--e) = 0;
            *(--e) = 0;
            return;
            }
        }
```

```
    i <<= 4;
    eshift( xi, 5 );
```
Note, $*p = xi$[M].
```
    i |= *p++ & 0x0f;
    *e |= (unsigned short )i;
    *(--e) = *p++;
    *(--e) = *p++;
    *(--e) = *p;
    }
```

Compare two e type numbers.
Important note: If you do not implement denormal numbers, you should arrange for $a == b$ to imply that $a - b == 0$.

```
int ecmp( a, b )
unsigned short *a, *b;
    {
    unsigned short ai[NI], bi[NI];
    register unsigned short *p, *q;
    register int i;
    int msign;

    emovi( a, ai );
    p = ai;
    emovi( b, bi );
    q = bi;
    if( *p != *q )
        {
```
The signs are different, but ensure that -0 equals $+0$.
```
        for( i=1; i<NI-1; i++ )
            {
            if( ai[i] != 0 )
                goto nzro;
            if( bi[i] != 0 )
                goto nzro;
            }
        return(0);
nzro:
        if( *p == 0 )
            return( 1 );
        else
            return( -1 );
        }
```

Both numbers have the same sign.

```
	if( *p == 0 )
		msign = 1;
	else
		msign = -1;
	i = NI-1;
	do
		{
		if( *p++ != *q++ )
			{
			goto diff;
			}
		}
	while( --i > 0 );
```

The numbers are equal.

```
	return(0);
diff:
	if( *(--p) > *(--q) )
```

p is bigger

```
		return( msign );
	else
```

p is littler

```
		return( -msign );
	}
```

Convert long (32-bit) integer to *e* type.

```
void ltoe( lp, y )
long *lp;
unsigned short *y;
	{
	unsigned short yi[NI];
	unsigned long ll;
	int k;

	ecleaz( yi );
	if( *lp < 0 )
		{
		ll = -(*lp);
		yi[0] = 0xffff;
		}
	else
		{
```

```
        ll = *lp;
        }
```
Move the long integer to *yi* significand area.
```
    yi[M] = (unsigned short )(ll >> 16);
    yi[M+1] = (unsigned short )ll;
```

Normalize the significand.
```
    yi[E] = EXONE + 15;
    if( (k = enormlz( yi )) > NBITS )
        ecleaz( yi );
    else
        yi[E] -= k;
    emovo( yi, y );
    }
```

Find long integer and fractional part of *e* type number.

```
void eifrac( x, i, frac )
unsigned short *x;
long *i;
unsigned short *frac;
    {
    unsigned short xi[NI];
    int k;

    emovi( x, xi );
    k = (int )xi[E] - (EXONE - 1);
    if( k <= 0 )
        {
```
If exponent ≤ 0, integer $= 0$ and argument is fraction.
```
        *i = 0L;
        emovo( xi, frac );
        return;
        }
    if( k > 31 )
        {
```
Long integer overflow: output large integer and correct fraction.
```
        *i = 0x7fffffff;
        eshift( xi, k );
        goto lab10;
        }

    if( k > 16 )
```

```
		{
		k -= 16;
		eshift( xi, k );
		*i = (long )(((unsigned long )xi[M] << 16)
			| (unsigned short )xi[M+1]);
		eshup16( xi );
		goto lab10;
		}

	eshift( xi, k );
	*i = (long )xi[M] & 0xffff;
lab10:
	if( xi[0] )
		*i = -(*i);
	xi[0] = 0;
	xi[E] = EXONE - 1;
	xi[M] = 0;
	if( (k = enormlz( xi )) > NBITS )
		ecleaz( xi );
	else
		xi[E] -= k;

	emovo( xi, frac );
	}
```

Shift significand of internal format number by n bits.

```
int eshift( x, sc )
unsigned short *x;
int sc;
	{
	int lost;
	unsigned short *p;

	if( sc == 0 )
		return( 0 );

	lost = 0;
	p = x + NI-1;
	if( sc < 0 )
		{
		sc = -sc;
		while( sc >= 16 )
```

```
			{
Remember lost bits.
			lost |= *p;
			eshdn16(x);
			sc -= 16;
			}

		while( sc >= 8 )
			{
			lost |= *p & 0xff;
			eshdn8(x);
			sc -= 8;
			}

		while( sc > 0 )
			{
			lost |= *p & 1;
			eshdn1(x);
			sc -= 1;
			}
		}
	else
		{
		while( sc >= 16 )
			{
			eshup16(x);
			sc -= 16;
			}

		while( sc >= 8 )
			{
			eshup8(x);
			sc -= 8;
			}

		while( sc > 0 )
			{
			eshup1(x);
			sc -= 1;
			}
		}
	if( lost )
		lost = 1;
	return( lost );
	}
```

Normalize significand of internal format number.

```
int enormlz(x)
unsigned short x[];
    {
    register unsigned short *p;
    int sc;

    sc = 0;
    p = &x[M];
    if( *p != 0 )
        goto normdn;
    ++p;
    if( *p & 0x8000 )
        return( 0 );
    while( *p == 0 )
        {
        eshup16(x);
        sc += 16;
```

With guard word, there are NBITS+16 bits available.

```
        if( sc > NBITS )
            return( sc );
        }
```

See if high byte is zero.

```
    while( (*p & 0xff00) == 0 )
        {
        eshup8(x);
        sc += 8;
        }
```

Now shift 1 bit at a time.

```
    while( (*p & 0x8000) == 0)
        {
        eshup1(x);
        sc += 1;
        if( sc > NBITS )
            {
            mtherr( "enormlz", UNDERFLOW );
            return( sc );
            }
        }
    return( sc );
```

Normalize by shifting down out of the high guard word.

```
normdn:
	if( *p & 0xff00 )
		{
		eshdn8(x);
		sc -= 8;
		}
	while( *p != 0 )
		{
		eshdn1(x);
		sc -= 1;

		if( sc < -NBITS )
			{
			mtherr( "enormlz", OVERFLOW );
			return( sc );
			}
		}
	return( sc );
	}
```

c = remainder after dividing b by a
Least significant integer quotient bits left in equot[].

```
int eremain( a, b, c )
unsigned short a[], b[], c[];
	{
	unsigned short den[NI], num[NI];

	if( ecmp(a,ezero) == 0 )
		{
		mtherr( "eremain", SING );
		eclear( c );
		return;
		}
	emovi( a, den );
	emovi( b, num );
	eiremain( den, num );
```

Sign of remainder = sign of quotient

```
	if( a[0] == b[0] )
		num[0] = 0;
	else
```

```
        num[0] = 0xffff;
    emovo( num, c );
    }
eiremain( den, num )
unsigned short den[], num[];
    {
    long ld, ln;
    int j;

    ld = den[E];
    ld -= enormlz( den );
    ln = num[E];
    ln -= enormlz( num );
    ecleaz( equot );
    while( ln >= ld )
        {
        if( ecmpm(den,num) <= 0 )
            {
            esubm(den, num);
            j = 1;
            }
        else
            {
            j = 0;
            }
        eshup1(equot);
        equot[NI-1] |= j;
        eshup1(num);
        ln -= 1;
        }
```

Normalize, but do not round off.

```
    emdnorm( num, 0, 0, ln, 0 );
    }
```

1.8 Binary — Decimal Conversion

An adequate facility to convert between decimal numbers and binary floating point numbers demands arithmetic of more than the normal working precision. Otherwise it would not always be possible to convert a binary number to a decimal string and back to the same binary value. The two programs that follow use the 64 bit arithmetic of Section 1.7, which is sufficient to meet the IEEE specification for 17 decimal double precision conversion.

Correctly rounded conversions between radix 2 and radix 10 notations are possible for numbers that can be expressed exactly in the binary computer format as an integer multiplied by a value equal to a power of 10. Both the integer and the power of 10 must be exact in their floating point representation, so that the only possible error incurred is the roundoff in forming their product or quotient. If either of these conditions fails, then the conversion may incur more than one rounding error.

To convert a number from decimal to binary, one first converts the string of significant decimal digits to a binary integer using an accumulator having as large a capacity as possible (64 bits, in this case). The algorithm is to multiply the previous value in the accumulator by 10 and add in the value of the next decimal digit. This is repeated for all digits in the decimal string, until the accumulator overflows. The decimal point, if any, is ignored except to count the number of digits that follow it. The result should be strictly rounded to the nearest or even integer, using the information from any excess decimals that may have been given.

The input decimal string may specify a power of 10 as a decimal exponent. The integer found above must be multiplied by this power of 10, diminished by the number of significant digits appearing after the decimal point. Since the net power of 10 may be too large to be represented exactly by a computer number, the integer accumulator should be padded with trailing zeros so as to minimize the power. If the accumulator did not overflow, and the power of 10 is exact, the only error in the procedure is possible rounding on multiplying (or dividing) by the power of 10. The program `asctoe.c`, listed below, illustrates the details.

To convert a number x from binary to decimal one finds the largest power of 10, say 10^n, that is less than x. On dividing x by 10^n, the integer part of the quotient is the leading decimal digit. The remainder r is less than 10^n; it is multiplied by 10. The quotient on dividing $10r$ by 10^n is the second decimal digit, etc. This procedure gives an exact decimal result, provided that 10^n is an exact computer number. To minimize n, trailing zeros should be stripped from or padded onto the significand, which may be regarded as an integer for this purpose. The details are shown below in `asctoe.c`.

1.8.1 etoasc.c

Convert e type number to decimal format ASCII string.
The constants are for 64 bit precision. Larger tables could be used. For example, computing 10^n from a table containing $10^1, 10^2, \ldots, 10^7, 10^8, 10^{16}, 10^{32}, \cdots, 10^{112}, 10^{128}, 10^{256}, \cdots$ would require fewer operations but seems no more accurate.

```
#define NTEN 12
#define MAXP 4096
```

```
static unsigned short etens[NTEN+1][NE] = {
        {0x979b,0x8a20,0x5202,0xc460,0x7525,},  10^4096
        {0x5de5,0xc53d,0x3b5d,0x9e8b,0x5a92,},  10^2048
        {0x0c17,0x8175,0x7586,0xc976,0x4d48,},
        {0x91c7,0xa60e,0xa0ae,0xe319,0x46a3,},
        {0xde8e,0x9df9,0xebfb,0xaa7e,0x4351,},
        {0x8ce0,0x80e9,0x47c9,0x93ba,0x41a8,},
        {0xa6d5,0xffcf,0x1f49,0xc278,0x40d3,},
        {0xb59e,0x2b70,0xada8,0x9dc5,0x4069,},
        {0x0000,0x0400,0xc9bf,0x8e1b,0x4034,},
        {0x0000,0x0000,0x2000,0xbebc,0x4019,},
        {0x0000,0x0000,0x0000,0x9c40,0x400c,},
        {0x0000,0x0000,0x0000,0xc800,0x4005,},
        {0x0000,0x0000,0x0000,0xa000,0x4002,},  10^1
    };

static unsigned short emtens[NTEN+1][NE] = {
        {0x9fde,0xd2ce,0x04c8,0xa6dd,0x0ad8,},  10^-4096
        {0x2de4,0x3436,0x534f,0xceae,0x256b,},
        {0xc0be,0xda57,0x82a5,0xa2a6,0x32b5,},
        {0xd21c,0xdb23,0xee32,0x9049,0x395a,},
        {0x193a,0x637a,0x4325,0xc031,0x3cac,},
        {0xe4a1,0x64bc,0x467c,0xddd0,0x3e55,},
        {0xe9a5,0xa539,0xea27,0xa87f,0x3f2a,},
        {0x94ba,0x4539,0x1ead,0xcfb1,0x3f94,},
        {0xe15b,0xc44d,0x94be,0xe695,0x3fc9,},
        {0xcefd,0x8461,0x7711,0xabcc,0x3fe4,},
        {0x652c,0xe219,0x1758,0xd1b7,0x3ff1,},
        {0xd70a,0x70a3,0x0a3d,0xa3d7,0x3ff8,},
        {0xcccd,0xcccc,0xcccc,0xcccc,0x3ffb,},  10^-1
    };

void etoasc( x, string, ndigs )
unsigned short x[];
char *string;
int ndigs;
    {
    long digit;
    unsigned short y[NI], t[NI], u[NI], w[NI];
    unsigned short *p, *r, *ten, *tenth;
    unsigned short sign;
    int i, j, k, expon;
    char *s, *ss;

    emov( x, y );  Retain external number format.
```

```
    if( y[NE-1] & 0x8000 )
        {
        sign = 0xffff;
        y[NE-1] &= 0x7fff;
        }
    else
        {
        sign = 0;
        }
    expon = 0;
    ten = &etens[NTEN][0];
    emov( eone, t );
```

Test for zero exponent

```
    if( y[NE-1] == 0 )
        {
        for( k=0; k<NE-1; k++ )
            {
            if( y[k] != 0 )
                goto tnzro; Denormalized number
            }
        goto isone; Legal all zeros
        }
tnzro:
```

Test for exponent nonzero but significand denormalized.
This is an error condition.

```
    if( (y[NE-1] != 0) && ((y[NE-2] & 0x8000) == 0) )
        {
        printf( "NaN " );
        for( i=0; i<NE; i++ )
        printf( "%04x ", x[i] );
        printf( "\n" );
        return;
        }
```

Compare to 1.0

```
    i = ecmp( eone, y );
    if( i == 0 )
        goto isone;

    if( i < 0 )
        { Number is greater than 1
```

Convert significand to an integer and strip trailing decimal zeros.

```
        emov( y, u );
        u[NE-1] = EXONE + NBITS - 1;
```

```
            p = &etens[NTEN-4][0];
            k = 16;
            do
                  {
                  ediv( p, u, t );
                  efloor( t, w );
                  for( j=0; j<NE-1; j++ )
                        {
                        if( t[j] != w[j] )
                              goto noint;
                        }
                  emov( t, u );
                  expon += k;
noint:
                  p += NE;
                  k >>= 1;
                  }
            while( k > 0 );
Rescale from integer significand
            k = (EXONE + NBITS - 1) - u[NE-1];
            k = y[NE-1] - k;
            u[NE-1] = k;
            emov( u, y );
Find power of 10
            emov( eone, t );
            k = MAXP;
            p = &etens[0][0];
            while( ecmp( ten, u ) <= 0 )
                  {
                  if( ecmp( p, u ) <= 0 )
                        {
                        ediv( p, u, u );
                        emul( p, t, t );
                        expon += k;
                        }
                  k >>= 1;
                  if( k == 0 )
                        break;
                  p += NE;
                  }
            }
      else
            { Number is less than 1.0
Pad significand with trailing decimal zeros.
```

```
if( y[NE-1] == 0 )
     {
     while( (y[NE-2] & 0x8000) == 0 )
          {
          emul( ten, y, y );
          expon -= 1;
          }
     }
else
     {
     emovi( y, w );
     for( i=0; i<NDEC+1; i++ )
          {
          if( (w[NI-1] & 0x7) != 0 )
               break;
multiply by 10
          emovz( w, u );
          eshdn1( u );
          eshdn1( u );
          eaddm( w, u );
          u[1] += 3;
          while( u[2] != 0 )
               {
               eshdn1(u);
               u[1] += 1;
               }
          if( u[NI-1] != 0 )
               break;
          if( eone[NE-1] <= u[1] )
               break;
          emovz( u, w );
          expon -= 1;
          }
     emovo( w, y );
     }
k = -MAXP;
p = &emtens[0][0];
r = &etens[0][0];
emov( y, w );
emov( eone, t );
while( ecmp( eone, w ) > 0 )
     {
     if( ecmp( p, w ) >= 0 )
          {
          emul( r, w, w );
```

```
                    emul( r, t, t );
                    expon += k;
                    }
                k /= 2;
                if( k == 0 )
                    break;
                p += NE;
                r += NE;
                }
            ediv( t, eone, t );
            }
isone:
```
Find the first (leading) digit.
```
        emovi( t, w );
        emovz( w, t );
        emovi( y, w );
        emovz( w, y );
        eiremain( t, y, y );
        digit = equot[NI-1];
        while( (digit == 0) && (ecmp(y,ezero) != 0) )
            {
            eshup1( y );
            emovz( y, u );
            eshup1( u );
            eshup1( u );
            eaddm( u, y );
            eiremain( t, y, y );
            digit = equot[NI-1];
            expon -= 1;
            }
        s = string;
        if( sign )
            *s++ = '-';
        else
            *s++ = ' ';
        *s++ = (char )digit + '0';
        *s++ = '.';
```
Examine number of digits requested by caller.
```
        if( ndigs < 0 )
            ndigs = 0;
        if( ndigs > NDEC )
            ndigs = NDEC;
```
Generate digits after the decimal point.
```
        for( k=0; k<=ndigs; k++ )
            {
```

Multiply current number by 10, without normalizing.

```
        eshup1( y );
        emovz( y, u );
        eshup1( u );
        eshup1( u );
        eaddm( u, y );
        eiremain( t, y, y );
        *s++ = (char )equot[NI-1] + '0';
        }
    digit = equot[NI-1];
    --s;
    ss = s;
```

Round off the ASCII string.

```
    if( digit > 4 )
        {
```

Test for critical rounding case in ASCII output.

```
        if( digit == 5 )
            {
            emovo( y, t );
            if( ecmp(t,ezero) != 0 )
                goto roun; round to nearest
            if( (*(s-1) & 1) == 0 )
                goto doexp; round to even
            }
```

Round up and propagate carry-outs.

```
roun:
        --s;
        k = *s & 0x7f;
```

Carry out to most significant digit?

```
        if( k == '.' )
            {
            --s;
            k = *s;
            k += 1;
            *s = (char )k;
```

Most significant digit carries to 10?

```
            if( k > '9' )
                {
                expon += 1;
                *s = '1';
                }
            goto doexp;
            }
```

Round up and carry out from less significant digits.

```
        k += 1;
```

```
          *s = (char )k;
          if( k > '9' )
               {
               *s = '0';
               goto roun;
               }
          }
doexp:
     sprintf( ss, "E%d\0", expon );
     }
```

1.8.2 asctoe.c

This program converts a decimal ASCII string to an *e* type number as defined in Section 1.7. It uses the table of powers of 10 from the previous section.

```
void asctoe( s, y )
char *s;
unsigned short *y;
     {
     unsigned short yy[NI], xt[NI];
     int esign, decflg, sgnflg, nexp, exp, prec, lost, k, trail, c;
     long lexp;
     unsigned short nsign, *p;
     char *sp;

     lost = 0;
     nsign = 0;
     decflg = 0;
     sgnflg = 0;
     nexp = 0;
     exp = 0;
     prec = 0;
     ecleaz( yy );
     trail = 0;

nxtcom:
     k = *s - '0';
     if( (k >= 0) && (k <= 9) )
          {
```

Ignore leading zeros.

```
          if( (prec == 0) && (decflg == 0) && (k == 0) )
```

```
            goto donchr;
```
Identify and strip trailing zeros after the decimal point.
```
        if( (trail == 0) && (decflg != 0) )
            {
            sp = s;
            while( (*sp >= '0') && (*sp <= '9') )
                ++sp;
```
Check for syntax error.
```
            c = *sp & 0x7f;
            if( (c != 'e') && (c != 'E') && (c != '\0')
                && (c != '\n') && (c != '\r')
                && (c != ' ') )
                goto error;
            --sp;
            while( *sp == '0' )
                *sp-- = ',';
            trail = 1;
            if( *s == ',' )
                goto donchr;
            }
```
If enough digits were given to more than fill up the yy register, continuing until overflow into the high guard word $yy[2]$ guarantees that there will be a roundoff bit at the top of the low guard word after normalization.
```
        if( yy[2] == 0 )
            {
            if( decflg )
                nexp += 1;  count digits after decimal point
            eshup1( yy );  multiply current number by 10
            emovz( yy, xt );
            eshup1( xt );
            eshup1( xt );
            eaddm( xt, yy );
            ecleaz( xt );
            xt[NI-2] = k;
            eaddm( xt, yy );
            }
        else
            {
            lost |= k;
            }
        prec += 1;
        goto donchr;
        }
    switch( *s )
```

```
        {
        case ' ':
        case ',':
            break;
        case 'E':
        case 'e':
            goto expnt;
        case '.':
            if( decflg )
                goto error;
            ++decflg;
            break;
        case '-':
            nsign = 0xffff;
        case '+':
            if( sgnflg )
                goto error;
            ++sgnflg;
            break;
        case '\0':
        case '\n':
        case '\r':
            goto daldone;
        default:
        error:
            mtherr( "asctoe", DOMAIN );
            eclear(y);
            return;
        }
donchr:
    ++s;
    goto nxtcom;
```

Exponent interpretation

```
expnt:

    esign = 1;
    exp = 0;
    ++s;
```

check for + or -

```
    if( *s == '-' )
        {
        esign = -1;
        ++s;
        }
```

```
	if( *s == '+' )
		++s;
	while( (*s >= '0') && (*s <= '9') )
		{
		exp *= 10;
		exp += *s++ - '0';
		}
	if( esign < 0 )
		exp = -exp;

daldone:
	nexp = exp - nexp;
```

Pad trailing zeros to minimize power of 10, per IEEE spec.

```
	while( (nexp > 0) && (yy[2] == 0) )
		{
		emovz( yy, xt );
		eshup1( xt );
		eshup1( xt );
		eaddm( yy, xt );
		eshup1( xt );
		if( xt[2] != 0 )
			break;
		nexp -= 1;
		emovz( xt, yy );
		}
	if( (k = enormlz(yy)) > NBITS )
		{
		eclear(y);
		return;
		}
	lexp = (EXONE - 1 + NBITS) - k;
	ernorm( yy, lost, 0, lexp, 64 );
```

Convert to external format.

```
	yy[0] = nsign;
	emovo( yy, y );
```

Multiply by 10**nexp. If precision is 64 bits,
the maximum relative error incurred in forming 10^n
for $0 \le n \le 324$ is $8.2 \cdot 10^{-20}$, at 10^{180}.
For $0 \le n \le 999$, the peak relative error is $1.4 \cdot 10^{-19}$ at 10^{947}.
For $0 \ge n \ge -999$, it is $-1.55 \cdot 10^{-19}$ at 10^{-435}.

```
	if( nexp == 0 )
		goto aexit;
	esign = 1;
	if( nexp < 0 )
		{
```

```
        nexp = -nexp;
        esign = -1;
        }
    p = &etens[NTEN][0];
    emov( eone, xt );
    exp = 1;
    do
        {
        if( exp & nexp )
            emul( p, xt, xt );
        p -= NE;
        exp <<= 1;
        }
    while( exp <= MAXP );
Notice! If less than 64 bit precision output is required, you must
provide a way to round directly to that precision in the following.
Otherwise there will be double rounding.
    if( esign < 0 )
        ediv( xt, y, y );
    else
        emul( xt, y, y );
aexit:
    return;
    }
```

1.9 Analysis of Error

1.9.1 Roundoff and Cancellation

Computer arithmetic is designed to produce the closest possible sum, product, or quotient under the assumption that the input numbers are exact. Under this assumption the relative error of addition, multiplication or division will not exceed the roundoff error. However, a machine number may be either an exact or an inexact representation of an intended real number. The assumption that all numbers in a computer calculation are exact is seldom valid. If the numbers are not exact, then the relative error of the arithmetic result may be much larger than the roundoff error.

An example will illustrate what can happen in the worst case. Let ϵ be the roundoff error of the arithmetic. The computer attempts the following simple calculation:

$$\begin{aligned} a &= 1 + 5\epsilon \\ b &= 1 + 3\epsilon \\ a - b &= 2\epsilon. \end{aligned}$$

It finds the answer 0 instead of the correct answer 2ϵ. In the first line, the computer has no number equal to $1 + 5\epsilon$. The sum rounds down to $1 + 4\epsilon$ by the "even" clause of the "round to nearest or even" rule. In the second line, $1 + 3\epsilon$ rounds up to $1 + 4\epsilon$ by the same clause. Since the machine numbers a and b have the same exponent, the difference $a - b = 0$ of the rounded values is calculated with no error; but the damage has already been accomplished by rounding the first two sums.

In this example the *absolute* error of the computed $a - b$ is 2ϵ. The *relative error* is $(0 - 2\epsilon)/2\epsilon = -1.0$, which signifies a total loss of relative precision. The four input values 1, 5ϵ, 1, and 3ϵ are all exact, yet the computed arithmetic result has no correct significant digits whatsoever.

As mentioned earlier, the effect illustrated by the above computation is called **cancellation error**. This type of error happens much more frequently than one would like. It could be argued that cancellation is by far the most important source of error in numerical programs.

For the effect to occur, at least one of the numbers must be inaccurate. If both were exact, then the result of subtraction would be exactly correct since the equality of the exponents means that no significant bits are shifted out of either significand.

There is no cancellation effect in multiplication or division, nor in adding numbers of the same sign. Cancellation error can occur only when adding inaccurate numbers that have the same exponent value and opposite signs.

Consider

$$(2 + 6\epsilon) - (1 + 5\epsilon) = 1 + 1\epsilon$$

which becomes, in the computer,

$$(2 + 8\epsilon) - (1 + 4\epsilon) = 1 + 4\epsilon.$$

The absolute error is 3ϵ. In relative terms this is a much smaller error than in the first example, since it is about $3\epsilon/2$ times the larger term summed. The two inexact numbers being subtracted differ by nearly a factor of two, so it is impossible for the difference to be near zero.

When there is no cancellation, the effect of errors in the arguments of an operation can be analyzed as follows. Suppose that two computer numbers have errors, so that in terms of the correct values x and y the values actually represented are $x + \epsilon_x$ and $y + \epsilon_y$ respectively. If these two numbers are multiplied, the maximum relative error of the product will be ϵ, so the computed product in terms of the correct arguments is

$$[x(1 \pm \epsilon_x)]\ [y(1 \pm \epsilon_y)](1 \pm \epsilon) = xy(1 \pm \epsilon \pm \epsilon_x \pm \epsilon_y + \ldots)$$

where the terms left out are of order ϵ^2. The same analysis holds for division and addition.

If there is cancellation, then the exponents of $x(1 + \epsilon_x)$, $y(1 + \epsilon_y)$ must be the same and there is no error in computing their difference. However, the computed difference between x and y is

$$x(1 \pm \epsilon_x) - y(1 \pm \epsilon_y) = x - y \pm x\epsilon_x \pm y\epsilon_y \ .$$

The relative error of the computed difference is

$$\frac{\pm x\epsilon_x \pm y\epsilon_y}{x - y}$$

which may be unbounded as $x - y$ approaches zero.

In terms of the exponents e_x, e_y, e_{x+y} of the arguments and the sum, the absolute error of addition or subtraction is bounded by

$$\begin{aligned} \epsilon_{x+y} &= \epsilon\ 2^{e_m}, \text{ where} \\ e_m &= \max\{e_x, e_y, e_{x+y}\} \end{aligned}$$

when the arguments are exact. If they are in error, then the computed sum is

$$x(1 \pm \epsilon_x) + y(1 \pm \epsilon_y) \pm \epsilon\ 2^{e_{m'}} = x + y \pm x\epsilon_x \pm y\epsilon_y \pm \epsilon\ 2^{e_{m'}}$$

where $e_{m'}$ refers to the exponents of the inaccurate x and y representations and of their computed difference. The relative error of addition is

$$\frac{\pm x\epsilon_x \pm y\epsilon_y \pm \epsilon\ 2^{e_{m'}}}{x + y} \ .$$

The absolute error of the sum of two inexact numbers x, y is bounded by the expression

$$2\epsilon \max\{|x|, |y|, |x + y|\}$$

where ϵ is the roundoff error of x and y.

When the arguments of arithmetic are *exact*, the foregoing analysis leads to a much more optimistic conclusion in many cases. For example, over the wide range $1 \leq x \leq 1/\epsilon$, the subtraction $x - 1.0$ gives an exact result because no nonzero bit is shifted out of the significand at any point in the computation. This fact can be useful in the design of recurrence algorithms. In the case of multiplication, suppose that more than half of the least significant bits of the arguments are all zero; then the product is exact. This fact will be used later in reducing x modulo $\pi/4$ for trigonometric functions.

1.9.2 Error Propagation

Some mathematical functions intrinsically amplify any error that appears in their arguments or parameters. For example, the function

$$f(x) = \frac{1}{x - b}$$

varies more rapidly with x when x is near b than when x is far from b. This effect can be studied analytically by examining the derivative of the function with respect to the variable of interest. For example, the relative effect of an error in b for the above function is

$$\begin{aligned} \frac{b}{f}\frac{\partial f}{\partial b} &= b(x-b)\,\frac{1}{(x-b)^2} \\ &= \frac{b}{x-b}\,. \end{aligned}$$

This measures the relative change in f due to a relative change in b. Note if $x = 0$ there is no *relative* error amplification though there is amplification of the *absolute* error.

1.9.3 Error as a Random Variable

Consider an inexact number between 1 and 2. It may have an error up to the roundoff error ϵ of the arithmetic. For some purposes the actual value e of the error may be treated as an unknown random number that is described by a probability function. The distribution of e may be modeled by a uniform probability density function of amplitude $1/2\epsilon$ extending from $-\epsilon$ to $+\epsilon$. The probability of an error larger than ϵ is zero. For present purposes this description will be taken as a definition of the term **inexact number**. For the absolute value $|\epsilon|$ of the error the probability density function is a rectangle of height $1/\epsilon$ extending from 0 to $+\epsilon$. From this geometric picture it is evident that the expected or average error magnitude is $\epsilon/2$. If two such inexact numbers are added or subtracted, the graph of of the probability density of the signed absolute error e of the sum extends from -2ϵ to $+2\epsilon$ and has the shape of an isoceles triangle:

$$p_e(e) = \left(1 - \frac{|e|}{2\epsilon}\right)\,\frac{1}{2\epsilon}\,.$$

This formula assumes that the errors of the two numbers are statistically independent, so that the distribution of the sum is the convolution integral of the individual distributions. The root mean square (**rms**) error of the sum is $\epsilon\sqrt{2/3} = 0.82\epsilon$. The rms error of the sum of n independent inexact numbers of approximately unit magnitude is $\epsilon\sqrt{n/3}$. As n increases, the probability density of the error approaches a Gaussian shape. Even if the errors are not independent, the maximum possible absolute error of the sum is $n\epsilon$.

The analysis demonstrates that the maximum error of a calculation may be much larger than the average error. Though requirements vary, most mathematical function routines are designed to have the smallest feasible maximum (or *worst case*) error. Even if a specification calls for a certain average or rms error, it would be unwise to ignore special conditions that produce a very bad worst case. In reporting the accuracy of a function routine it is good practice to give both the rms error and the maximum error.

Often it is very unjustified to assume that the inexact terms have statistically independent errors. For example, the terms in the product $x(x+1)(x+2)$ might have highly correlated errors. Failure of the statistical independence assumption may make the probability analysis much more complicated.

1.9.4 Order of Summation

Note that computer arithmetic does not obey the associative law of real arithmetic:

$$(\epsilon + \epsilon) + 1 = 1 + 2\epsilon$$

but

$$\epsilon + (\epsilon + 1) = \epsilon + 1 = 1 \, .$$

Since the order of summation can be significant, the accuracy of critical calculations can be improved (on average) by arranging summations to proceed from smaller to larger terms.

The effect of rearrangement is usually very minor, unless the calculation involves a great many terms. Order of summation thus tends to be an overrated issue. From the probability analysis given earlier, the typical error of a calculation should not exceed the roundoff error multiplied by the number of terms in the calculation. If the error does exceed this estimate, then cancellation error should be suspected.

It is sometimes possible to remove cancellation error by rearranging the arithmetic. In the example given earlier of complete cancellation, the sum

$$\begin{aligned} a - b &= (1 + 5\epsilon) - (1 + 3\epsilon) \\ &= 0 \end{aligned}$$

would be computed correctly if it could be rearranged to read

$$\begin{aligned} a - b &= (5\epsilon - 3\epsilon) + (1 - 1) \\ &= 2\epsilon \, . \end{aligned}$$

Unfortunately this is usually difficult to accomplish simply by inspecting the arithmetic. More often, it is necessary to study the function analytically and derive an alternative theoretical expression that is less subject to cancellation error.

1.10 Complex Arithmetic

In languages that do not contain a complex number data type, the computer programmer must adopt a convention about how the real and imaginary parts of a complex number are to be stored and referenced. Subroutines to perform complex arithmetic must then be written and invoked, sometimes rather awkwardly, as function calls in place of the arithmetic operators +, -, *, /. For example, in structured languages such as C the following structure declaration and subroutine usages might be adopted.

Define the structure of a complex number.

```
typedef struct {
double r;
double i;
}cmplx;
```

Declare a, b, c to be complex numbers

```
cmplx a, b, c;
```

Typical subroutine invocations. Subroutine arguments are memory addresses of the structures.

cadd(&a, &b, &c);	$c = b + a$
csub(&a, &b, &c);	$c = b - a$
cmul(&a, &b, &c);	$c = b * a$
cdiv(&a, &b, &c);	$c = b/a$
cneg(&c);	$c = -c$
cmov(&b, &c);	$c = b$
cabs(&a);	$\sqrt{Re(a)^2 + Im(a)^2}$

It may also be convenient to provide globally accessible complex constants **czero** $= (0.0, 0.0)$ and **cone** $= (1.0, 0.0)$.

Let

$$z_1 = x_1 + iy_1$$
$$z_2 = x_2 + iy_2$$
$$z_3 = x_3 + iy_3$$

be complex numbers. Ignoring computational problems, the definitions of complex arithmetic are

Addition,

$$\begin{aligned} z_3 &= z_2 + z_1 \\ x_3 &= x_2 + x_1 \\ y_3 &= y_2 + y_1 \end{aligned}$$

Subtraction,

$$\begin{aligned} z_3 &= z_2 - z_1 \\ x_3 &= x_2 - x_1 \\ y_3 &= y_2 - y_1 \end{aligned}$$

Multiplication,

$$\begin{aligned} z_3 &= z_2 z_1 \\ x_3 &= x_2 x_1 - y_2 y_1 \\ y_3 &= x_2 y_1 + y_2 x_1 \end{aligned}$$

Division,

$$\begin{aligned} z_3 &= z_2 / z_1 \\ d &= x_1^2 + y_1^2 \\ x_3 &= (x_2 x_1 + y_2 y_1)/d \\ y_3 &= y_2 x_1 - x_2 y_1 \end{aligned}$$

Absolute value,

$$|z| = \sqrt{x^2 + y^2}$$

These computations can be made with the typical relative accuracies shown in Table 1.5. Functions were tested in the rectangle $[-10, +10]$. In some applications, detailed checking for potential overflow conditions is required so that the program, rather than the computer operating system, can always remain in control. In the programs that follow, this has been done partially for division and for the absolute value function.

1.10.1 cmplx.c

Complex arithmetic routines.

```
extern double MAXNUM;

typedef struct
    {
    double r;
    double i;
    }cmplx;

cmplx czero = {0.0, 0.0};
extern cmplx czero;
```

Function	Arithmetic	Trials	Peak	RMS
cadd	IEEE	100000	$1.1 \cdot 10^{-16}$	$2.7 \cdot 10^{-17}$
csub	IEEE	100000	$1.1 \cdot 10^{-16}$	$3.4 \cdot 10^{-17}$
cmul	IEEE	100000	$2.1 \cdot 10^{-16}$	$6.9 \cdot 10^{-17}$
cdiv	IEEE	100000	$3.7 \cdot 10^{-16}$	$1.1 \cdot 10^{-16}$
cabs	IEEE	100000	$2.7 \cdot 10^{-16}$	$6.9 \cdot 10^{-17}$
cadd	DEC	10000	$1.4 \cdot 10^{-17}$	$3.4 \cdot 10^{-18}$
csub	DEC	10000	$1.4 \cdot 10^{-17}$	$4.5 \cdot 10^{-18}$
cmul	DEC	3000	$2.3 \cdot 10^{-17}$	$8.7 \cdot 10^{-18}$
cdiv	DEC	18000	$4.9 \cdot 10^{-17}$	$1.3 \cdot 10^{-17}$
cabs	DEC	30000	$3.2 \cdot 10^{-17}$	$9.2 \cdot 10^{-18}$

Table 1.5: Accuracy of Complex Arithmetic Programs

```
cmplx cone = {1.0, 0.0};
extern cmplx cone;
```

$c = b + a$

```
cadd( a, b, c )
register cmplx *a, *b;
cmplx *c;
    {

    c->r = b->r + a->r;
    c->i = b->i + a->i;
    }
```

$c = b - a$

```
csub( a, b, c )
register cmplx *a, *b;
cmplx *c;
    {

    c->r = b->r - a->r;
```

```
    c->i = b->i - a->i;
    }
```

$c = b \cdot a$

```
cmul( a, b, c )
register cmplx *a, *b;
cmplx *c;
    {
    double y;

    y = b->r * a->r - b->i * a->i;
    c->i = b->r * a->i + b->i * a->r;
    c->r = y;
    }
```

$c = b/a$

```
cdiv( a, b, c )
register cmplx *a, *b;
cmplx *c;
    {
    double y, p, q, w;
    double fabs();


    y = a->r * a->r + a->i * a->i;
    p = b->r * a->r + b->i * a->i;
    q = b->i * a->r - b->r * a->i;

    if( y < 1.0 )
        {
        w = MAXNUM * y;
        if( (fabs(p) > w) || (fabs(q) > w) || (y == 0.0) )
            {
            c->r = MAXNUM;
            c->i = MAXNUM;
            mtherr( "cdiv", OVERFLOW );
            return;
            }
        }
    c->r = p/y;
```

```
	c->i = q/y;
	}
```

$b = a$

```
cmov( a, b )
register short *a, *b;
	{
	int i;

	i = 8;
	do
		*b++ = *a++;
	while( --i );
	}
```

$a = -a$

```
cneg( a )
register cmplx *a;
	{

	a->r = -a->r;
	a->i = -a->i;
	}
```

1.10.2 Absolute Value: cabs.c

Absolute value of complex number, also known as **hypot**().

Unnecessary arithmetic overflow in the complex absolute value function should be avoided. There are at least several ways to do this. One is to compute

$$w = \sqrt{1 + (b/a)^2}$$

where a is the larger of x and y in $z = x + iy$, and b is the smaller of x and y. Then

$$\sqrt{a^2 + b^2} = a\ w\ ,$$

but w is used in an overflow test prior to attempting the operation a times w. If $a >$ MAXNUM$/w$ there will be overflow.

An alternative method is shown in the program that follows. It uses error-free integer operations on the exponents of x and y to rescale them so that overflow is avoided.

```
#include "mconf.h"
typedef struct
     {
     double r;
     double i;
     }cmplx;
#ifdef UNK
#define PREC 27
#define MAXEXP 1024
#endif
#ifdef DEC
#define PREC 29
#define MAXEXP 128
#endif
#ifdef IBMPC
#define PREC 27
#define MAXEXP 1024
#endif
#ifdef MIEEE
#define PREC 27
#define MAXEXP 1024
#endif

extern double MAXNUM, MACHEP, PI, PIO2;

double cabs( z )
register cmplx *z;
     {
     double x, y, b, re, im;
     int ex, ey, e, f;
     double fabs(), sqrt(), frexp(), ldexp();

     re = fabs( z->r );
     im = fabs( z->i );
     if( re == 0.0 )
          {
          return( im );
          }
     if( im == 0.0 )
          {
          return( re );
          }
```

Get the exponents of the numbers.

```
     x = frexp( re, &ex );
     y = frexp( im, &ey );
```

Check if one number is tiny compared to the other.

```
        e = ex - ey;
        if( e > PREC )
                return( re );
        if( e < -PREC )
                return( im );
```

Find approximate exponent e of the geometric mean.

```
        e = (ex + ey) >> 1;
```

Rescale so mean is about 1.

```
        x = ldexp( re, -e );
        y = ldexp( im, -e );
```

Hypotenuse of the right triangle

```
        b = sqrt( x * x + y * y );
```

Compute the exponent of the answer.

```
        y = frexp( b, &ey );
        ey = e + ey;
        if( ey > MAXEXP )
                {
                mtherr( "cabs", OVERFLOW );
                return( MAXNUM );
                }
        if( ey < -MAXEXP )
                return(0.0);
```

Undo the scaling.

```
        b = ldexp( b, e );
        return( b );
        }
```

1.11 Rational Arithmetic

Occasionally the rounding and cancellation problems of floating point arithmetic are so troublesome that one must resort to some method of exact calculation. This situation often arises while carrying out algorithms to generate expansion coefficients such as Bernoulli numbers. For these problems a rational arithmetic implemented with the available floating point functions can be helpful. The programs that follow provide a serviceable arithmetic of structures, or arrays, each of which contains the integer valued numerator and denominator of a rational number. The largest integer that can be handled in IEEE double precision is

$$\frac{1}{\mathtt{MACHEP}} - 1 = 2^{53} - 1 = 9007199254740991$$

or about $9 \cdot 10^{15}$. The programs make use of Euclid's algorithm for reducing a fraction to lowest terms. Table 1.6 shows how the routines are invoked.

euclid(&*n*, &*d*)	Reduce n/d to lowest terms
radd(&*a*, &*b*, &*c*)	$c = b + a$
rsub(&*a*, &*b*, &*c*)	$c = b - a$
rmul(&*a*, &*b*, &*c*)	$c = b * a$
rdiv(&*a*, &*b*, &*c*)	$c = b/a$

Table 1.6: Rational Arithmetic Routines

1.11.1 euclid.c

Data structure of a rational number,
essentially an array of two double precision reals.
Note: The numerator, n, and the denominator, d,
are assumed without checking to be integer valued.

```
typedef struct
	{
	double n;
	double d;
	}fract;
```

Overflow threshold for integer valued doubles:

```
extern double MACHEP;
#define BIG (1.0 / MACHEP)
```

Include file defines error codes for **mtherr**().

```
#include "mconf.h"
```

Euclidean algorithm
reduces fraction to lowest terms,
returns greatest common divisor.

```
double euclid( num, den )
double *num, *den;
	{
	double n, d, q, r;
	double floor();

	n = *num;
	d = *den;
```

Make numbers positive, locally.

```
    if( n < 0.0 )
        n = -n;
    if( d < 0.0 )
        d = -d;
```

Abort if numbers are too big for integer arithmetic.

```
    if( (n >= BIG) || (d >= BIG) )
        {
        mtherr( "euclid", OVERFLOW );
        return(1.0);
        }
```

Divide by zero, gcd = 1.

```
    if(d == 0.0)
        return( 1.0 );
```

Zero. Return 0/1, gcd = denominator.

```
    if(n == 0.0)
        {
        *den = 1.0;
        return( d );
        }
```

Begin Euclid's algorithm.

```
    while( d > 0.5 )
        {
```

Find integer part of n divided by d.

```
        q = floor( n/d );
```

Find remainder after dividing n by d.

```
        r = n - d * q;
```

The next fraction is d/r.

```
        n = d;
        d = r;
        }
    if( n < 0.0 )
        printf( "euclid error, gcd = %.15e\n", n );
    *num /= n;
    *den /= n;
    return( n );
    }
```

Add fractions.

```
radd( f1, f2, f3 )
fract *f1, *f2, *f3;
    {
    double gcd, d1, d2, gcn, n1, n2;
```

```
    n1 = f1->n;
    d1 = f1->d;
    n2 = f2->n;
    d2 = f2->d;
    if( n1 == 0.0 )
        {
        f3->n = n2;
        f3->d = d2;
        return;
        }
    if( n2 == 0.0 )
        {
        f3->n = n1;
        f3->d = d1;
        return;
        }
```

Common divisors of denominators

```
    gcd = euclid( &d1, &d2 );
```

Common divisors of numerators

```
    gcn = euclid( &n1, &n2 );
```

Note, factoring the numerators makes overflow slightly less likely.

```
    f3->n = ( n1 * d2 + n2 * d1) * gcn;
    f3->d = d1 * d2 * gcd;
    euclid( &f3->n, &f3->d );
    }
```

Subtract fractions.

```
rsub( f1, f2, f3 )
fract *f1, *f2, *f3;
    {
    double gcd, d1, d2, gcn, n1, n2;

    n1 = f1->n;
    d1 = f1->d;
    n2 = f2->n;
    d2 = f2->d;
    if( n1 == 0.0 )
        {
        f3->n = n2;
        f3->d = d2;
        return;
        }
```

```
	if( n2 == 0.0 )
		{
		f3->n = -n1;
		f3->d = d1;
		return;
		}

	gcd = euclid( &d1, &d2 );
	gcn = euclid( &n1, &n2 );
	f3->n = (n2 * d1 - n1 * d2) * gcn;
	f3->d = d1 * d2 * gcd;
	euclid( &f3->n, &f3->d );
	}
```

Multiply fractions.

```
rmul( f1, f2, f3 )
fract *f1, *f2, *f3;
	{
	double d1, d2, n1, n2;
	double fabs();

	n1 = f1->n;
	d1 = f1->d;
	n2 = f2->n;
	d2 = f2->d;

	if( (n1 == 0.0) || (n2 == 0.0) )
		{
		f3->n = 0.0;
		f3->d = 1.0;
		return;
		}
```

Cross cancel common divisors.

```
	euclid( &n1, &d2 );
	euclid( &n2, &d1 );
	f3->n = n1 * n2;
	f3->d = d1 * d2;
```

Report overflow.

```
	if( (fabs(f3->n) >= BIG) || (fabs(f3->d) >= BIG) )
		{
		mtherr( "rmul", OVERFLOW );
		return(1.0);
		}
```

```
	}
```

Divide fractions.

```
rdiv( f1, f2, f3 )
fract *f1, *f2, *f3;
	{
	double d1, d2, n1, n2;
	double fabs();
```

Invert f1, then multiply

```
	n1 = f1->d;
	d1 = f1->n;
```

Keep denominator positive.

```
	if( d1 < 0.0 )
		{
		n1 = -n1;
		d1 = -d1;
		}
	n2 = f2->n;
	d2 = f2->d;
	if( (n1 == 0.0) || (n2 == 0.0) )
		{
		f3->n = 0.0;
		f3->d = 1.0;
		return;
		}
```

Cross cancel any common divisors.

```
	euclid( &n1, &d2 );
	euclid( &n2, &d1 );
	f3->n = n1 * n2;
	f3->d = d1 * d2;
```

Report overflow.

```
	if( (fabs(f3->n) >= BIG) || (fabs(f3->d) >= BIG) )
		{
		mtherr( "rdiv", OVERFLOW );
		return(1.0);
		}
	}
```

2

Approximation Methods

2.1 Power Series

A function whose derivatives exist to all orders in the neighborhood of a point a can be expanded in a power series

$$f(a+x) = \sum_{k=0}^{\infty} c_k x^k \ .$$

One way of finding the coefficients is to take the kth derivative of both sides, observing that the kth derivative of x^n is

$$\begin{aligned} \left.\frac{d^k}{dx^k}\, x^n\right|_{x=0} &= \left. n(n-1)\cdots(n-k+1)x^{n-k}\right|_{x=0} \\ &= \begin{cases} n!, & k=n \\ 0, & k \neq n \ . \end{cases} \end{aligned}$$

Denoting

$$f^{(k)}(a) \equiv \left.\frac{d^k}{dx^k} f(a+x)\right|_{x=0} ,$$

the observation implies that

$$f^{(k)}(a) = k!\; c_k \ .$$

Hence

$$f(a+x) = \sum_{k=0}^{\infty} \frac{f^{(k)}(a)}{k!} x^k$$

which is the Taylor series expansion of f about the point a. It can be shown that[1] the remainder after the nth term of the series is

$$\begin{aligned} R_{n+1} &= \sum_{k=n+1}^{\infty} \frac{f^{(k)}(a)}{k!} x^k \\ &= \frac{x^{n+1}}{(n+1)!} f^{(n+1)}(a+\theta x) \end{aligned}$$

for some θ between 0 and 1.

Although the expansion in this form is unique, other useful power series expansions can be often be found by change of variables. For example,

$$\ln(1+x) = x - \frac{1}{2}x^2 + \frac{1}{3}x^3 - \ldots$$

is a fundamental power series for the logarithm on the interval $-1 < x \leq 1$. But also

$$\ln x = 2\left(\frac{x-1}{x+1}\right) + \frac{2}{3}\left(\frac{x-1}{x+1}\right)^3 + \frac{2}{5}\left(\frac{x-1}{x+1}\right)^5 + \ldots$$

is a series that converges more rapidly and for all $x > 0$. Either series might be used for computation, the choice depending on the relative importance assigned to speed versus accuracy. While the latter series is faster, its first term has a rounding error in $x+1$ and another rounding error from division. The former series has no rounding error in its first term, and only the error in computing x^2 in its second term.

2.2 Chebyshev Expansions

Expansions in Chebyshev polynomials are used quite frequently in numerical work. They are often near-best polynomial approximations, their evaluation is numerically stable, and their coefficients are easy to compute.

With x varying from -1 to $+1$, the sequence of Chebyshev polynomials is defined by

$$\begin{aligned} T_0(x) &= 1 \\ T_1(x) &= x \\ T_2(x) &= 2x^2 - 1 \\ T_3(x) &= 4x^3 - 3x \\ T_{n+1}(x) &= 2x\, T_n(x) - T_{n-1}(x) \,. \end{aligned}$$

[1]See any advanced calculus text. For our purposes, computational error will almost always be greater than the theoretical error, so we will not have much occasion to be concerned with remainders of convergent series. Divergent series are another matter, however.

Chebyshev expansions are closely related to Fourier series. This stems from the fact that the kth Chebyshev polynomial is equal to

$$T_k(x) = \cos(k \cos^{-1} x) \ .$$

To see this, check that

$$\begin{aligned} \cos(0 \cos^{-1} x) &= \cos 0 = 1 \\ \cos(1 \cos^{-1} x) &= x \end{aligned}$$

so the formula is true for $k = 0$ and $k = 1$. By induction, the statement is true for all k if it satisfies the recurrence relation given above for T_{n+1}. Substituting gives

$$\cos((n+1)\cos^{-1} x) = 2x\cos(n\cos^{-1} x) - \cos((n-1)\cos^{-1} x)$$

which is readily verified to be a trigonometric identity.

Any function $f(x)$ that is continuous and of finite variation in the interval from -1 to $+1$ has an expansion

$$\begin{aligned} f(x) &= \tfrac{1}{2}a_0 + a_1T_1(x) + a_2T_2(x) + \ldots \\ &= \sum_{k=0}^{\infty}{}' a_k T_k(x) \ . \end{aligned}$$

Note the prime over the summation sign; it denotes that the first term is to be multiplied by $\frac{1}{2}$. The coefficients in the expansion are given by

$$\begin{aligned} a_k &= \frac{2}{\pi}\int_{-1}^{1} f(x)\ T_k(x)\ (1-x^2)^{-\frac{1}{2}}\, dx \\ &= \frac{2}{\pi}\int_{0}^{\pi} f(\cos\theta)\ \cos k\theta\ d\theta \\ &\approx \frac{2}{n}\sum_{j=0}^{n}{}'' f\left(\cos\frac{\pi j}{n}\right)\ \cos\frac{\pi k j}{n} \end{aligned}$$

for sufficiently large n. The double prime over the summation indicates that the first and last terms are to be multiplied by $\frac{1}{2}$. Chebyshev coefficients are thus Fourier cosine transform coefficients of the function evaluated at nonuniformly spaced points.

If the Chebyshev series

$$f(x) \approx \sum_{k=0}^{n}{}' a_k T_k(x)$$

to n terms converges fairly rapidly, so that the $n+2$nd term is small compared to the $n+1$st term, then the curve of the error will be approximately equal to $a_{n+1}T_{n+1}$. This is significant because of the shape of the polynomial T_{n+1}. Referring to the trigonometric representation, in general T_k is the cosine of an angle that varies continuously from $-k\pi/2$ to $+k\pi/2$. Therefore the graph of $T_k(x)$ has $k+1$ maxima and k zeros in the interval $-1 \le x \le 1$. In other words, T_{n+1} has $n+2$ points of alternation (see the secton on the Remes algorithm). Furthermore, the value of T_k at each maximum is either $+1$ or -1. Hence if convergence is rapid, as supposed, then the Chebyshev expansion is not far from the least maximum polynomial approximation to $f(x)$.

The discrete Fourier transform as shown above may be used in practice[2] to compute the Chebyshev coefficients. The error in a_k caused by using the discrete sum in place of the integral is on the order of a_{2n-k}. To be fairly sure that the accuracy of the calculated coefficients is adequate, the chosen value of n should be at least twice the highest required value of k. One should then also verify that

$$\sum_{i=k+1}^{n} |a_i|$$

is less than the desired maximum error. The Chebyshev coefficients presented in this book were calculated by the discrete Fourier transform with $n = 64$.

To compute a function whose Chebyshev coefficients a_k are given, use the following recurrence. First set

$$\begin{aligned} b_{n+2} &= 0 \\ b_{n+1} &= 0 \,. \end{aligned}$$

Starting with $k = n$, calculate

$$b_k = 2xb_{k+1} - b_{k+2} + a_k$$

repeating until $k = 0$. Then

$$f(x) = \tfrac{1}{2}(b_0 - b_2) \,.$$

Since the interval over which an approximation is desired usually is not $[-1, +1]$ it is necessary to transform the variable from the desired interval $[u, v]$ to the interval on which the Chebyshev polynomials are defined. This can be accomplished by a linear transformation of either x or $1/x$. A suitable transformation is

$$w = \frac{2x - v - u}{v - u}$$

[2] C. W. Clenshaw, *Mathematical Tables, Volume 5, Chebyshev Series for Mathematical Functions.* National Physical Laboratory, Her Majesty's Stationery Office, London 1962.

which maps $[u, v]$ to $[-1, 1]$. When the Chebyshev coefficients have been calculated for an inverted interval, that is, for the function $f(1/x)$ in the interval $[1/v, 1/u]$, the transformation is

$$w = \frac{\frac{2uv}{x} - v - u}{v - u}$$

which maps $[1/v, 1/u]$ to $[-1, 1]$. As v approaches ∞ this becomes

$$w = \frac{2u}{x} - 1\ .$$

To compute the Chebyshev coefficients a_k for these transformed intervals the argument x of $f(x)$ in the discrete Fourier transform must be changed. Where the Fourier transform calls for $f(x)$, compute instead $f(t)$ where

$$t = \tfrac{1}{2}(v - u)\ x + (v + u)$$

or, for the inverted interval,

$$t = \frac{2uv}{(v - u)\ x + (v + u)}\ .$$

With either transformation, to compute $f(x)$ insert the corresponding value of w into the recurrence.

2.2.1 chbevl.c

This program evaluates the Chebyshev expansion of a function at argument $x/2$, given its Chebyshev coefficients in *array*. Chebyshev coefficients must be stored in reverse order, with the zeroth order term is last in the array. Note n is the number of coefficients, not the order.

```
double chbevl( x, array, n )
double x;
double array[];
int n;
     {
     double b0, b1, b2, *p;
     int i;
     p = array;
     b0 = *p++;
     b1 = 0.0;
     i = n - 1;
     do
          {
          b2 = b1;
```

```
        b1 = b0;
        b0 = x * b1 - b2 + *p++;
        }
    while( --i );
    return( 0.5*(b0-b2) );
    }
```

2.3 Padé Approximations

A rational function

$$R(x) = \frac{P(x)}{Q(x)} = \frac{\sum_{i=0}^{n} p_i x^i}{\sum_{j=0}^{m} q_j x^j}$$

is the ratio of two polynomials. Padé approximations are rational functions that can be found by a closed form procedure from the derivatives of the function to be approximated. Suppose that the function $f(x)$ can be written

$$f(a+x)\sum_{j=0}^{\infty} q_j x^j = \sum_{i=0}^{\infty} p_i x^i \ .$$

The plan is to take the kth derivative of both sides evaluated at $x = 0$, and form a system of equations for $k = 0, 1, \ldots, n$. The required kth derivatives are

$$\begin{aligned} f^{(k)} &\equiv \left.\frac{d^k}{dx^k} f(a+x)\right|_{x=0} \\ f^{(0)} &= f(a) \\ \left.\frac{d^k}{dx^k}\sum_{i=0}^{\infty} q_i x^i\right|_{x=0} &= k!\ q_k \\ \left.\frac{d^k}{dx^k}\sum_{i=0}^{\infty} p_i x^i\right|_{x=0} &= k!\ p_k \ . \end{aligned}$$

While finding the kth derivative of

$$f(a+x)\sum_{i=0}^{\infty} q_i x^i \ ,$$

the rule for differentiating a product leads to a sequence of expressions involving binomial coefficients (Pascal's triangle). The resulting equations for $k = 0, 1, 2, \ldots$ are

$$\begin{array}{lllllllll}
f^{(0)}q_0 & & & & & & & = & 0!p_0 \\
f^{(1)}q_0 & + & f^{(0)}q_1 & & & & & = & 1!p_1 \\
f^{(2)}q_0 & + & 2f^{(1)}q_1 & + & f^{(0)}2!q_2 & & & = & 2!p_2 \\
f^{(3)}q_0 & + & 3f^{(2)}q_1 & + & 3f^{(1)}2!q_2 & + & f^{(0)}3!q_3 & = & 3!p_3
\end{array}$$

and so on.

To find an approximation of degree n, set all coefficients of degree $k > n$ and derivatives of order $m > 2n$ to zero:

$$f^{(m)} = q_k = p_k = 0,\ \ k > n,\ \ m > 2n\ .$$

The last $n + 1$ nonzero equations then consist of a square matrix on the left starting with the first row in the above array that contains q_n. The matrix proceeds downward a total of $n + 1$ rows. On the right hand side, p_n is set arbitrarily to 1. This system of linear equations can be solved for the q's, which are the coefficients of the denominator polynomial of the approximation. The n remaining p's can be determined by substituting the q's into the first n rows of the array of equations.

When this method is applied to the exponential function interesting approximations such as

$$e^x \approx \frac{665280 + 332640x + 75600x^2 + 10080x^3 + 840x^4 + 42x^5 + x^6}{665280 - 332640x + 75600x^2 - 10080x^3 + 840x^4 - 42x^5 + x^6}$$

are produced, in which the numerator and denominator terms are the same except for sign. These may be rearranged into the form

$$e^x \approx \frac{P(x^2) + xQ(x^2)}{P(x^2) - xQ(x^2)}$$

where, in this example,

$$\begin{aligned}
P(t) &= 665280 + 75600t + 840t^2 + t^3 \\
Q(t) &= 332640 + 10080t + 42t^2\ .
\end{aligned}$$

By substituting x^2 for t, you can see that the two expressions for e^x are the same. The example has relative accuracy of $2 \cdot 10^{-13}$ for $-1 \le x \le +1$ and $2 \cdot 10^{-17}$ for $-.5 \le x \le +.5$.

Another example of a Padé approximation is

$$\sin x \approx \frac{166320x - 22260x^3 + 551x^5}{166320 + 5460x^2 + 75x^4}$$

which has relative accuracy $2 \cdot 10^{-9}$ for $|x| \le \pi/4$.

2.4 Least Maximum Approximations

If the Taylor series expansion of a function $f(x)$ is truncated at the nth term of the series, the result is a polynomial of degree n,

$$f(x) \approx P(x) = c_0 + c_1x + c_2x^2 + \ldots + c_nx^n .$$

This polynomial approximates $f(x)$ with an error that can be estimated analytically. However, $P(x)$ is unlikely to be the best polynomial of degree n that could be used to approximate $f(x)$. Analytical expansions often suggest a good approximating form, but the coefficients of the analytical expansion are not the ones that you would actually use in an efficient computer routine. The process of adjusting the coefficients so as to improve the accuracy of the approximation is the subject of this section.

If the term "best" means having the smallest worst case error on a specified interval $a \leq x \leq b$, then a theorem of Chebyshev[3] states the characteristics of the best polynomial.

1. The graph of the best polynomial moves back and forth across the graph of $f(x)$, being first greater, then less, then greater, etc., than $f(x)$. On each excursion, the error reaches a maximum value.

2. The best polynomial of degree n makes at least $n+2$ such excursions, with $n+2$ corresponding points of maximum error. These are called *points of alternation.*

3. For the best polynomial, the errors at the points of alternation all have exactly the same magnitude.

2.4.1 Best Polynomial Approximations

A polynomial $P(x)$ that satisfies the first two of the above criteria but not the third is called a nearly least maximum, or near-best, approximation. It is relatively easy to find a polynomial of this type by setting up and solving an array of linear simultaneous equations for the coefficients. Each equation in this array is formed by setting x equal to some arbitrary constant value in the interval of approximation $[a, b]$, while considering the coefficients c_i of the polynomial to be the unknowns. The error, or deviation,

$$d(x) = P(x) - f(x)$$

can be included in the equations to ensure that the polynomial has points of alternation. Since the best polynomial has the magnitude of d equal at all the points of alternation, the equations may contain a single variable d,

[3]For a proof and other discussion of fundamentals, see Rivlin, Theodore J., *An Introduction to the Approximation of Functions*, Blaisdell, 1969.

regarded as an additional unknown and having a coefficient of either +1 or −1. Each equation thus has the form

$$c_0 \;+\; xc_1 \;+\; x^2c_2 \;+\; \ldots \;+\; x^nc_n \;\pm\; d \;=\; f(x)\;.$$

This is simply a rearrangement of the expression for the error of $P(x)$ at argument x. The solution will come close to satisfying all three of the criteria for a least maximum approximation. However, there is nothing in the equations to say that d is the maximum of anything, so the result may not satisfy the conditions completely. With this proviso, the $n+2$ equations in $n+2$ unknowns are

$$\begin{array}{l}
c_0 + x_1c_1 + x_1^2c_2 + \ldots + x_1^nc_n - d = f(x_1) \\
c_0 + x_2c_1 + x_2^2c_2 + \ldots + x_2^nc_n + d = f(x_2) \\
\qquad\ldots \\
c_0 + x_{n+2}c_1 + x_{n+2}^2c_2 + \ldots + x_{n+2}^nc_n \pm d = f(x_{n+2})\;.
\end{array}$$

These equations can be solved by standard techniques such as the Gauss elimination method.[4] The solution gives the coefficients c_i of a near-best polynomial approximation to $f(x)$.

To find the best approximation it is necessary to arrange for each x_i to be a point of alternation. Having solved the equations a first time, the curve of the error for the corresponding polynomial can be traced and the error peaks found. Then if each x_i is positioned so that $P(x_i)$ is at an error peak, the equations solved using these new values of x_i will yield a polynomial that is closer to the best approximation. After several iterations, the new set of x_i will be the same as the previous set to within the machine accuracy, and the process will have converged. This iterative solution procedure is known as the algorithm of Remes.

In the absence of better information, a reasonable set of starting points for the values x_i can be found by supposing that the Chebyshev polynomial of degree $n+1$ approximates the error curve for the best polynomial of degree n. This point was discussed earlier in the section on Chebyshev expansions. The maxima of the Chebyshev polynomial occur at the points

$$\begin{aligned}
r_k &= \tfrac{1}{2}[1-\cos(k\pi/n)] \\
x_k &= a+(b-a)\,r_k
\end{aligned}$$

where a and b are the ends of the approximation interval and k ranges from 0 through n.

[4]For an exposition of Gaussian elimination see Forsythe, George E. and Cleve B. Moler, *Computer Solution of Linear Algebraic Systems*, Prentice-Hall, 1967.

Having inserted these carefully reasoned x_k into the equations and determined the actual error peaks, one may find that the error is very much greater toward one end of the interval than it is near the other end. This means that more of the x_k need to be concentrated where the errors are highest. To produce a smooth distribution of the x_k concentrated toward the high end (toward b), replace r_k by its square root. To concentrate the points near the low end, replace r_k by $r_k^{3/2}$. With the initial guesses thus revised, the peak errors should be more nearly equal on the second attempt. The Remes algorithm may then begin on the third try, using the observed error peaks to determine the next set of x_k.

In most cases the Remes algorithm for polynomial approximations has second order convergence, and will converge to machine accuracy after relatively few iterations. If the function is difficult to compute, then the amount of work involved may be dominated by the number of function evaluations that have to be performed. If this is important, then the error peaks may be determined to within only a coarse approximation on the first iterations.

Searching for the error peaks involves starting at the previous estimate of x_k and testing the error at various points in its vicinity. One establishes a step size, moves x_k by the amount of one step, and calculates the error at that point. The steps continue until the error passes a maximum value or until the boundary of the search interval is reached. If a local maximum was found, its location can be improved by fitting a parabola to the three surrounding points; but this is an unnecessary refinement.

A workable strategy for setting the size of the search step is to make it proportional to the disparity between adjacent error peaks. As the peaks become more nearly equal in height, the step size then automatically decreases in proportion. A suitable formula for the step size Δ is

$$\Delta_k = \frac{1}{4}\left(\frac{|d_{k+1}|}{|d_k|} - 1\right)(x_{k+1} - x_k) \ ,$$

where d_k is the error at x_k.

The search must not be permitted to miss an error peak entirely. This can usually be accomplished by restricting the search interval for x_k to extend from x_{k-1} to x_{k+1}. A more conservative strategy is to stop if the error reverses sign, thus bounding the search to the interval between successive zeros of the error.

An approximation may be specified to have minimum absolute error; this is the case addressed by the equations shown. More often the goal is to minimize the relative error. In that case the error is

$$d(x) = (P(x) - f(x))/f(x) \ .$$

The corresponding equations to be solved have the form

$$P(x) \pm f(x)d(x) = -f(x)$$

which amounts merely to an adjustment of the coefficient of the unknown d. When searching for error peaks, the error reported must then be the relative error instead of the absolute error.

2.4.2 Best Rational Approximations

For a given computational labor, an approximating form that involves the ratio $P(x)/Q(x)$ of two polynomials is nearly always superior, in theory, to a best polynomial approximation of the same degree. Numerical instability occasionally renders the rational form less suitable in practice, however.

A procedure similar to the Remes algorithm can be used to find best rational approximations. The theorem of Chebyshev states that the best rational function

$$\frac{P_n(x)}{Q_m(x)} = \frac{p_0 + p_1 x + \ldots + p_n x^n}{q_0 + q_1 x + \ldots + q_m x^m}$$

of degree (n, m) has at least $m+n+2$ points of alternation. Equations for determining trial coefficients of P/Q can be set up as before. The error of the approximation is

$$d(x) = \frac{P(x)}{Q(x)} - f(x)$$

so the equations to be solved have the form

$$P(x) - (f(x) \pm d)Q(x) = 0 \; .$$

Note that the deviation d occurs multiplied by Q, so the equations are nonlinear if d is considered to be an unknown. The solution of the equations in this form can be approached as an eigenvalue problem or a linear programming problem.

The equations can be linearized, however, by choosing values x_i to make $d = 0$. This is the approach that will be taken here. When q_m is set equal to 1 to make the rational form irreducible, the equations become

$$\begin{array}{ccccccccccc}
p_0 & + & \ldots & + & x_1^n p_n & + & f(x_1)q_0 & + & \ldots & + & f(x_1)x_1^{m-1}q_{m-1} \\
 & & & & & & & & & = & -f(x_1)x_1^m \\
p_0 & + & \ldots & + & x_2^n p_n & + & f(x_2)q_0 & + & \ldots & + & f(x_2)x_2^{m-1}q_{m-1} \\
 & & & & & & & & & = & -f(x_2)x_2^m \\
 & & & & & & \ldots & & & & \\
p_0 & + & \ldots & + & x_r^n p_n & + & f(x_r)q_0 & + & \ldots & + & f(x_r)x_r^{m-1}q_{m-1} \\
 & & & & & & & & & = & -f(x_r)x_r^m
\end{array}$$

where $r = m + n + 1$ is the number of equations. The unknowns are the coefficients p_i and q_i of the polynomials P and Q. Since there are $m+n+2$ error peaks, the number of zeros between the peaks is $m + n + 1$ so the system is fully determined.

Supposing, with less reason than before, that the error curve for a rational function of total degree $(n+m)$ has the same shape as the Chebyshev polynomial of degree $m+n+1$, the initial guesses for x_i may be set to the points at which the value of this Chebyshev polynomial is equal to zero. These points are

$$\begin{aligned} r_k &= \tfrac{1}{2}[1 - cos(\ (2k-1)\pi/2N)] \\ x_k &= a + (b-a)\ r_k \end{aligned}$$

where a and b are the ends of the approximation interval, $N = m+n+1$, and k ranges from 1 through N. As before, these points can be squeezed toward the high or low end of the interval by replacing r with $r^{1/2}$ or $r^{3/2}$, respectively.

Solving the equations with the given points x_i yields the coefficients for a rational function P/Q whose error is zero at those points. To improve on this solution by leveling the error peaks it is necessary to search for the error maxima. The procedure for this is the same as described previously. If the starting x_i were the zeros of a Chebyshev polynomial, the search for the peaks on the first iteration may begin at the maxima of the same Chebyshev polynomial.

After finding the $m+n+2$ error peaks the x_i must be adjusted so as to improve the leveling of the peaks. The strategy is to move the x_i, which will be points of zero error on the next iteration, in the direction of the taller peaks. The amount to move each point can be made proportional to the disparity between adjacent peaks, and in fact the same strategy can be used for moving the zeros as was used in searching for the peaks. Moving x_k by an amount equal to Δ_k, as given above, is an acceptable choice:

$$\Delta_k = \frac{1}{4}\left(\frac{|d_{k+1}|}{|d_k|} - 1\right)(p_{k+1} - p_k)\ ,$$

where p_k and p_k are the values of x at the peaks on either side of x_k. Occasionally the scale factor $\frac{1}{4}$ is too large, with the result that the maximum error actually increases from one iteration to the next, instead of decreasing. It is then necessary to insert a smaller scale factor and try again. In the early iterations there is also a chance that this step is so big that x_k would move beyond the next error peak. If this happens, a compromise value should be chosen, such as half way between the peaks surrounding x_k.

This variant of the Remes algorithm has first order convergence after the guesses become sufficiently close to the final values. It therefore requires many more iterations than the second order convergence of the algorithm for polynomials. The functions presented in this book typically needed between 100 and 300 iterations to achieve error peaks level to 21 decimals. An automatic search strategy may fail if the initial guesses x_k are too far from their final values. This problem is likely to arise if the function has a cusp or is otherwise very unpolynomial-like. A strategy that covers all cases is to allow for easy manual intervention, so that the computer operator can insert his own guesses.

In this connection, it is well to understand that the computer routine that generates the reference function values has finite accuracy. The function actually computed may well have a cusp or other irregular behavior that interferes with orderly convergence of the algorithm. If the approximation is theoretically accurate to D decimal places and the error peaks are leveled to D decimal places relative to each other, the function must be computed to $2D$ decimals or better. Since many of the higher transcendental functions lose half the arithmetic precision due to cancellation, it could take a $4D$-decimal computer arithmetic to produce minimax approximation coefficients that are correct to D decimals. In the past, this amount of precision might have involved prohibitive computer costs; but today the job can be accomplished by a desk-top computer in a reasonable time.

2.4.3 Special Rational Forms

Often the analytical expansion of a function suggests a good approximation form that is not strictly the ratio of two arbitrary polynomials. For example, the Cody and Waite form

$$f(x) \approx x + x^3 \frac{P(x^2)}{Q(x^2)}$$

specifies in advance that certain coefficients of the equivalent general rational form are equal to zero or one. By a change of variables, the special form reduces to a rational form of lower total degree for which a minimax solution must exist. This solution may or may not satisfy the original problem. For example the above form readily transforms to

$$\begin{aligned} t &= x^2 \\ g(t) &= \frac{f(\sqrt{t}) - \sqrt{t}}{t^{3/2}} \approx \frac{P_1(t)}{Q_1(t)} . \end{aligned}$$

But, finding the best approximation to $g(t)$ is not the same as finding the best approximation to $f(x)$; they are different functions. A way around this problem is to incorporate the special form directly into the error equations. For the Cody and Waite form this yields equations of the form

$$x^3 P(x^2) - (f(x) - x \pm d)Q(x^2) = 0$$

which are solved by exactly the same technique as the ones for the simpler form. The only difference is a modification of the coefficients of p_i and q_i that appear in the array of linear equations.

A change of variable that does not affect the expression for the error requires no modification of the equations. For example,

$$f(x) \approx \frac{P(1/x)}{Q(1/x)}$$

can be handled by approximating $g(t)$, where

$$g(t) = f(1/t) \approx \frac{P(t)}{Q(t)} \ .$$

Here the reference function computed is the same as the original function; only the value of the independent variable is different. The interval of approximation also changes in accordance with the change of variable.

2.5 A Program to Find Best Approximations: remes.c

This section contains the complete listing of a computer program for generating least maximum rational approximations. The program implements the various forms of Remes algorithm discussed in the previous section. It includes the necessary adjustments for the forms $xR(x)$, $xR(x^2)$, $x^2R(x^2)$, the Cody and Waite form, and the Padé form of the type used in the exponential function. The program as it is displayed operates in double precision arithmetic. It can generate approximations with an accuracy of up to 8 or 10 decimals. The higher precision coefficients presented in this book were calculated by a substantially identical program that was configured to use extended precision arithmetic. The program prompts the user through the following steps:

1. Indicate whether error criterion is relative or absolute.
2. Accept or modify the default configuration for the form of approximation and the general positioning of initial trial solution points.
3. Indicate the degree of the numerator polynomial.
4. Indicate the degree of the denominator polynomial.
5. Give the start of the approximation interval.
6. Give the width of the approximation interval.
7. Accept or modify the initial guesses for the locations of zero and maximum error of the approximation.
8. Indicate number of iterations to perform.
9. After the iterations are done, enter a number of further iterations to do, or display the results, or write the results to a computer file.

`remesg.c`

Definitions and global symbol declarations for Remes program

`#define P 24` Maximum total degree of polynomials, + 2

`#define N 20` Number of items to tabulate for display

`extern int` **config**`;` Flag bits for type of approximation:

`#define PXSQ 1` $R(z^2)$

`#define XPX 2` $zR()$

`#define PADE 4` Pade form with denominator $Q(z) - zP(z)$

`#define CW 8` Cody & Waite form $x + x^2R(z)$

Note, if CW and degree of denominator = 0 then set ZER bit also.

`#define SQL 16` Squeeze toward low end of approximation interval

`#define ZER 32` Search for zeros, even if no denominator

`#define X2PX 64` $z^2R()$

`#define SQH 128` Squeeze toward high end of interval

`int` **IPS**`[P] = {0,};` simq() work vector

`double` **AA**`[P*P] = {0.0,};` coefficient matrix

`double` **BB**`[P] = {0.0,};` right hand side vector

`double` **param**`[P] = {0.0,};` solution vector

`double` **xx**`[P] = {0.0,};` points in approximation interval

`double` **mm**`[P] = {0.0};` points of maximum deviation

`double` **yy**`[P] = {0.0};` the error maxima

`double` **step**`[P] = {0.0};` step sizes for searching

`double` **qx** `= 0.0;` function argument

`double` **qy** `= 0.0;` accurate function value

`double` **qyaprx** `= 0.0;` approximate function value

`double` **dev** `= 0.0;` deviation at solution points

`double` **apstrt** `= 0.0;` start of approximation interval

`double` **apwidt** `= 0.0;` width of interval

`double` **apend** `= 0.0;` end of interval

`double` **xm** `= 0.0;` variables for search of maximum

`double` **xn** `= 0.0;`

`double` **ym** `= 0.0;`

`double` **yn** `= 0.0;`

`double` **delta** `= 0.0;`

`double` **eclose** `= 0.0;`

`double` **farther** `= 0.0;`

`double` **spread** `= 0.0;` error spread from last iteration

`int` **esign** `= 0;` sign of error returned by geterr

`int` **n** `= 0;` degree of numerator polynomial

`int` **d** `= 0;` degree of denominator polynomial

`int` **nd1** `= 0;` n + d + 1

`int` **neq** `= 0;` number of equations to solve

`int` **relerr** `= 0;` relative error criterion flag

`int` **search** `= 0;` flag for automatic search

`int` **iter** `= 0;` search-solve iteration number

```
int askitr = 0; iteration on which to stop
extern double PI; 3.14159...
```

Top level control of the program is in the module `remes.c`. This is the main program.

```
main()
     {
     int i;
     int chgflg; Indicates changes made in default values
     double x;
     char s[40];
     char *sp;

     printf(
     "\nRational Approximation by Remes Algorithm\n\n" );
START:
     remesa(); Get operator commands.
     goto showg; Jump to operator intervention point.

LOOP:

     iter += 1;
     printf( "Iteration %d\n", iter );

     if( search != 0 )
          {
          remess(); Search for error maxima.
          goto solveq;
          }
showg:
Display old values of guesses and let user change them if desired.
     chgflg = 0;
First get the step size if rational form
     if( (d > 0) && (search != 0) )
          {
There is a denominator polynomial.
          printf( "delta = %.4E ? ", delta );
          gets( s );
If input is not a null line, then decode the number.
          if( s[0] != '\0' )
               {
               chgflg = 1;
               sscanf( s, "%lf", &delta );
               }
```

```
        for( i=0; i<=neq; i++ )
            step[i] *= delta;
        }
```
Read in guesses for locations of solution.
```
    for( i=0; i<neq; i++ )
        {
        printf( "x[%d] = %.4E ? ", i, xx[i] );
        gets( s );
        if( s[0] != '\0' )
            {
            chgflg = 1;
            sscanf( s, "%lf", &xx[i] );
            if( (d == 0) && ((config & ZER) == 0) )
                mm[i] = xx[i];
            }
        }

    if( (d > 0) || ((config & ZER) != 0) )
        {
        for( i=0; i <=neq; i++ )
            {
            printf( "peak[%d] = %.4E ? ", i, mm[i] );
            gets( s );
            if( s[0] != '\0' )
                {
                chgflg = 1;
                sscanf( s, "%lf", &mm[i] );
                }
            }
        }
```
If there were any changes to the default values
then reinitialize the step size array.
```
    if( chgflg )
        stpini();
solveq:
    remese(); Solve equations.
    goto whtnxt;

ptabl:
    remesp(); Display the results

whtnxt:
```
Test solution against convergence criteria.
```
    if( (delta < 1.0e-15) || (spread < 1.0e-15) )
        askitr = iter;
```

```
	if( askitr > iter )
		goto LOOP;
```

Ask what to do next.

```
	printf(
	"Enter #, p(rint), w(rite), g(uess), x(it), or n(one))?"
	);
```

Get command line from operator

```
	sp = &s[0];
	gets( sp );
	if( *sp == 'w' )
		{
		remesw(); Write results to file
		goto whtnxt;
		}
	if( *sp == 'g' ) Modify the guesses
		goto showg;
	if( *sp == 'p' ) Display results
		goto ptabl;
	if( (*sp >= '1') && (*sp <= '9') )
		{
```

Numeric input is iteration count

```
		sscanf( sp, "%d", &askitr );
		askitr += iter;
		goto LOOP;
		}
	if( *sp == 'x' ) Close files and exit
		exit(0);
	else
		goto START;
	}
```

Module `remesa.c` interacts with the user to obtain parameters such as the degree of the approximation. It sets up the initial guesses for the locations of zero and maximum approximation error.

```
remesa.c
remesa()
	{
	char s[40];
	double a, b, c, q, r;
	int i, k, ncheb;
	double func(), cos(), sqrt();
```

Ask for error criterion

```
    printf( "Relative error (y or n) ?  " );
    gets( s );
    relerr = 0;
    if( s[0] == 'y' )
         relerr = 1;
    search = 1; Enable Automatic search.
getconf:
    printf( "\nConfiguration word = " );
    pconfig();
    printf( " ?  " );
    gets(s);
    if( s[0] != '\0' )
         {
Display a menu of configuration bits
         if( s[0] == '?'  )
              {
              k = config;
              config = 0xffff;
              pconfig();
              config = k;
              goto getconf;
              }
         else
              {
              sscanf( s, "%x", &config );
              pconfig();
              }
         printf( "\n" );
         }

    printf( "Degree of numerator polynomial?  " );
    gets( s );
    sscanf( s, "%d", &n );

    printf( "Degree of denominator polynomial?  " );
    gets( s );
    sscanf( s, "%d", &d );

    printf ( "Start of approximation interval ?  " );
    gets( s );
    sscanf( s, "%lf", &apstrt );
getwid:
    printf ( "Width of approximation interval ?  " );
    gets( s );
    sscanf( s, "%lf", &apwidt );
```

```
if( apwidt <= 0.0 )
     {
     puts( "?  must be > 0" );
     goto getwid;
     }
append = apstrt + apwidt;
nd1 = n + d + 1;
spread = 1.0e37;
iter = 0;
askitr = 1;
Supply initial guesses for solution points.
if( (d == 0) && ((config & ZER) == 0) )
     { There is no denominator polynomial.
     neq = n + 2;
     mm[neq] = apend;
     }
else
     {
     neq = nd1;
     }
ncheb = nd1;
Find ncheb+1 extrema of Chebyshev polynomial
a = ncheb;
mm[0] = apstrt;
for( i=1; i<ncheb; i++ )
     {
     r = 0.5 * (1.0 - cos( (PI * i) / a ) );
     if( config & SQL )
          r = r * sqrt(r);
     else if( config & SQH )
               r = sqrt(r);
     mm[i] = apstrt + r * apwidt;
     }
mm[ncheb] = apend;

if( (d == 0) && ((config & ZER) == 0) )
     { The operative locations are maxima.
     for( i=0; i<=neq; i++ )
          xx[i] = mm[i];
     }
else
     { Zeros of Chebyshev polynomial
     a = 2.0 * ncheb;
     for( i=0; i<=ncheb; i++ )
          {
```

```
                r = 0.5 * (1.0 - cos( PI * (2*i+1) / a ) );
```

Squeeze toward low end of approximation interval.

```
                if( config & SQL )
                        r = r * sqrt(r);
```

Squeeze toward high end.

```
                else if( config & SQH )
                                r = sqrt(r);
                xx[i] = apstrt + r * apwidt;
                }
```

Deviation at solution points

```
        dev = 0.0;
        }
```

If form is $xR(x)$, avoid $x = 0$.

```
    if( config & (XPX | X2PX) )
        {
        if( config & CW )
                qx = 1.0e-15 + xx[0]/2.0;
        else
                qx = 1.0e-15 + apstrt * (1.0 + 1.0e-15);
        mm[0] = qx;
        printf( "mm[0] = %.15E\n", qx );
        if( (d == 0) && ((config & ZER) == 0) )
                xx[0] = qx;
        }
    stpini(); Initialize step sizes
    }
```

Subroutine to initialize step sizes.

```
stpini()
    {
    int i;

    if( search )
        {
        xx[neq+1] = apend;
        delta = 0.25;
        if( (d > 0) || ((config & ZER) != 0) )
                {
                step[0] = xx[0] - apstrt;
                for( i=1; i<neq; i++ )
                        step[i] = xx[i] - xx[i-1];
                }
```

```
		else
			{
			step[0] = 0.0625 * (xx[1] - xx[0]);
			for( i=1; i<neq; i++ )
				step[i] = 0.0625 * (xx[i+1] - xx[i]);
			}
		step[neq] = step[neq-1];
		}
	}
```

Subroutine to display the configuration word

Table of bits in the configuration word

```
#define NCNFG 8
static char *cnfmsg[NCNFG] = {
	"PXSQ",
	"XPX",
	"PADE",
	"CW",
	"SQL",
	"ZER",
	"X2PX",
	"SQH"
	};

pconfig()
	{
	int i, k;

	k = 1;
	for( i=0; i<NCNFG; i++ )
		{
		if( k & config )
			printf( "%s ", cnfmsg[i] );
		k <<= 1;
		}
	printf( "hex value = %x ", config );
	}
```

Module `remese.c` sets up the array of simultaneous equations to be solved, based on the current estimates of the locations of zero or maximum error.

```
remese()
```

```
{
double x, y, z, gx, gxsq, pade;
int i, j, ip;
double gofx(), func();

if( (d == 0) && ((config & ZER) == 0) )
	{
```
There is no denominator.
In this case the deviation is an unknown variable adjoined to the other unknowns; it has a coefficient of plus or minus 1.
```
	dev = 1.0;
	}
else
	{
```
Otherwise dev = 0, since the solution sought is for the locations of zero error.
```
	dev = 0.0;
	}
```
Set up the equations for solution by simq()
```
for( i=0; i<neq; i++ )
	{
```
Offset to 1st element of this row of matrix
```
	ip = neq * i;
```
The guess for this row
```
	x = xx[i];
```
Right-hand-side vector
```
	y = func(x);
	gx = gofx(x);
	if( config & PXSQ )
		gxsq = gx * gx;
	if( config & PADE )
		{
		pade = 2.0 + y;
		}
```
Add the deviation, if there is a denominator.
```
	if( d > 0 )
		{
```
Relative error criterion
```
		if( relerr )
			y = y * (1.0+dev);
```
Absolute error criterion
```
		else
			y = y + dev;
		}
```

```
        if( config & CW )
            {
y(1 + dev) = z + z^2 P/Q
            y = (y - gx)/(gx*gx);
            }
```

Insert powers of x[i] for numerator coefficients.

```
        if( config & XPX )
            z = gx;
        else if( config & X2PX )
                z = gx * gx;
        else
            z = 1.0;
        for( j=0; j<=n; j++ )
            {
            if( config & PADE )
                AA[ip+j] = pade * z;
            else
                AA[ip+j] = z;
            if( config & PXSQ )
                z = z * gxsq;
            else
                z = z * gx;
            }
```

Insert denominator terms, if any.

```
        if( d > 0 )
            {
            z = 1.0;
            for( j=0; j<d; j++ )
                {
                AA[ip+n+1+j] = -y * z;
                if( config & PXSQ )
                    z = z * gxsq;
                else
                    z = z * gx;
                }
```

Right hand side vector

```
            BB[i] = y * z;
            }
        else
            {
```

Right hand side vector if no denominator

```
            BB[i] = y;
            z = dev;
            if( relerr )
                z = z * y;
```

```
                AA[ip+n+1] = z;
                }
```
Switch sign of deviation for next row of matrix.
```
            dev = -1.0 * dev;
            }
```
Solve the simultaneous linear equations.
```
        simq( &AA[0], &BB[0], &param[0], neq, 0, &IPS[0] );
        }
```

Equations as set up by `remese.c` are solved by the subroutine `simq.c`. The algorithm for Gaussian elimination is from *Computer Solution of Linear Algebraic Systems* by George E. Forsythe and Cleve B. Moler, Prentice-Hall, 1967, which contains detailed discussion and an equivalent Fortran program.

```
int simq( A, B, X, n, flag, IPS )
double A[], B[], X[];
int n, flag;
int IPS[];
        {
        int i, j, ij, ip, ipj, ipk, ipn;
        int idxpiv, iback;
        int k, kp, kp1, kpj, kpk, kpn;
        int nip, nkp, nm1;
        double em, q, rownrm, big, size, pivot, sum;
        double fabs();

        if( flag < 0 )
            goto solve;
```
Initialize IPS and X
```
        ij=0;
        for( i=0; i<n; i++ )
            {
            IPS[i] = i;
            rownrm = 0.0;
            for( j=0; j<n; j++ )
                {
                q = fabs( A[ij] );
                if( rownrm < q )
                    rownrm = q;
                ++ij;
                }
            if( rownrm == 0.0 )
                {
                puts("SIMQ ROWNRM=0");
```

```
                        return(1);
                        }
                X[i] = 1.0/rownrm;
                }
Gaussian elimination with partial pivoting
        nm1 = n-1;
        for( k=0; k<nm1; k++ )
                {
                big= 0.0;
                for( i=k; i<n; i++ )
                        {
                        ip = IPS[i];
                        ipk = n*ip + k;
                        size = fabs( A[ipk] ) * X[ip];
                        if( size > big )
                                {
                                big = size;
                                idxpiv = i;
                                }
                        }

                if( big == 0.0 )
                        {
                        puts( "SIMQ BIG=0" );
                        return(2);
                        }
                if( idxpiv != k )
                        {
                        j = IPS[k];
                        IPS[k] = IPS[idxpiv];
                        IPS[idxpiv] = j;
                        }
                kp = IPS[k];
                kpk = n*kp + k;
                pivot = A[kpk];
                kp1 = k+1;
                for( i=kp1; i<n; i++ )
                        {
                        ip = IPS[i];
                        ipk = n*ip + k;
                        em = -A[ipk]/pivot;
                        A[ipk] = -em;
                        nip = n*ip;
                        nkp = n*kp;
                        for( j=kp1; j<n; j++ )
```

```
                    {
                    ipj = nip + j;
                    A[ipj] = A[ipj] + em * A[nkp + j];
                    }
                }
            }
Last element of IPS[n] th row
    kpn = n * IPS[n-1] + n - 1;
    if( A[kpn] == 0.0 )
        {
        puts( "SIMQ A[kpn]=0");
        return(3);
        }
Back substitution
solve:
    ip = IPS[0];
    X[0] = B[ip];
    for( i=1; i<n; i++ )
        {
        ip = IPS[i];
        ipj = n * ip;
        sum = 0.0;
        for( j=0; j<i; j++ )
            {
            sum += A[ipj] * X[j];
            ++ipj;
            }
        X[i] = B[ip] - sum;
        }

    ipn = n * IPS[n-1] + n - 1;
    X[n-1] = X[n-1]/A[ipn];

    for( iback=1; iback<n; iback++ )
        {
i goes (n - 1),...,1
        i = nm1 - iback;
        ip = IPS[i];
        nip = n*ip;
        sum = 0.0;
        for( j=i+1; j<n; j++ )
        sum += A[nip+j] * X[j];
        X[i] = (X[i] - sum)/A[nip+i];
        }
    return(0);
```

```
    }
```

Module `remess.c` searches for new error maxima and adjusts the trial locations of zero or maximum error.

```
remess.c

remess()
    {
    double a, b, q, xm, ym, xn, yn, xx0, xx1;
    int i, meq, emsign, ensign, steps;
    double approx(), func(), geterr();

Search for maxima of error
    eclose = 1e30;
    farther = 0.0;
    meq = neq;
    if( (d > 0) || ((config & ZER) != 0) )
        meq += 1;
    xx0 = apstrt;

    for( i=0; i<meq; i++ )
        {
        steps = 0;
        if( (d > 0) || ((config & ZER) != 0) )
            xx1 = xx[i]; Next zero
        else
            xx1 = mm[i+1]; Next maximum
        if( i == meq-1 )
            xx1 = apend;
        xm = mm[i];
        ym = geterr(xm);
        emsign = esign; Sign of error
        q = step[i];
        xn = xm + q;
Cannot skip over adjacent boundaries
        if( xn < xx0 )
            goto revers;
        if( xn >= xx1 )
            goto revers;
        yn = geterr(xn);
        ensign = esign;
        if( yn < ym )
            {
revers:
```

```
                q = -q;
                xn = xm;
                yn = ym;
                ensign = emsign;
                }
March until error becomes smaller.
          while( yn >= ym )
                {
                if( ++steps > 10 )
                      goto tsear;
                ym = yn;
                xm = xn;
                emsign = ensign;
                a = xm + q;
                if( a == xm )
                      goto tsear;
Must not skip over the zeros either side.
                if( a <= xx0 )
                      goto tsear;
                if( a >= xx1 )
                      goto tsear;
                xn = a;
                yn = geterr(xn);
                ensign = esign;
                }
tsear:
          mm[i] = xm; Position of maximum
          yy[i] = ym; Value of maximum
          if( ym == 0.0 )
                printf( "yy[%d] = 0\n", i );
          if( eclose > ym )
                eclose = ym;
          if( farther < ym )
                farther = ym;
No denominator polynomial.
          if( (d == 0) && ((config & ZER) == 0) )
                xx[i] = xm;
Walk to next zero.
          if( (d > 0) || ((config & ZER) != 0) )
                xx0 = xx1;
          else
                xx0 = 0.5*(xm + xx1);
          } End of search loop.
Decrease step size if error spread increased.
     q = (farther - eclose);
```

```
if( eclose != 0.0 )
     q /= eclose;
if( q >= spread )
     {
```
Spread is increasing; decrease step size.
```
     delta *= 0.5;
     printf( "delta = %.4E\n", delta );
     }
spread = q;
printf(
     "peak error = %.4E, relative error spread = %.4E\n",
     farther, spread );
```
Calculate new step sizes
```
if( (d == 0) && ((config & ZER) == 0) )
     {
     if( spread < 0.25 )
          q = 0.0625;
     else
          q = 0.5;
     for( i=0; i<neq; i++ )
          step[i] *= q;
     }
else
     {
     for( i=0; i<neq; i++ )
          {
          q = yy[i+1];
          if( q != 0.0 )
               q = yy[i]/q - 1.0;
          else
               q = 0.0625;
          if( q > 0.25 )
               q = 0.25;
          q *= mm[i+1] - mm[i];
          step[i] = q * delta;
          }
     step[neq] = step[neq-1];
```
Insert new locations for the zeros.
```
     for( i=0; i<neq; i++ )
          {
          xm = xx[i] - step[i];
          if( xm <= apstrt )
               continue;
          if( xm >= apend )
               continue;
```

```
                if( xm <= mm[i] )
                        {
                        printf( "xx[%d] < mm\n", i );
                        xm = 0.5 * (mm[i] + xx[i]);
                        }
                if( xm >= mm[i+1] )
                        {
                        printf( "xx[%d] > mm\n", i );
                        xm = 0.5 * (mm[i+1] + xx[i]);
                        }
                xx[i] = xm;
                }
        }
sdone:  ;
        }
```

Module `remesf.c` is the interface to the function that is to be approximated. Subroutine **func()** must be modified as appropriate to compute the desired reference function.

`remesf.c`
Default flag bits for type of approximation
PXSQ — XPX — X2PX — SQL — SQH — PADE — CW, etc.

```
int config = SQH;
```

Insert here the function name and formulas for printout

```
char funnam[] = {
        "exp(x)"
        };
char znam[] = { "x" };
```

This subroutine computes the rational form P(x)/Q(x) using coefficients from the solution vector param[].

```
double approx(x)
double x;
        {
        double gx, z, yn, yd, q;
        double gofx();
        int i;

        gx = gofx(x);
        if( config & PXSQ )
                z = gx * gx;
        else
                z = gx;
```

Highest degree numerator coefficient

```
        yn = param[n];
```

Work backwards toward the constant term.

```
    for( i=n-1; i>=0; i-- )
        yn = z * yn + param[i];

    if( d > 0 )
        {
Highest degree denominator coefficient = 1.0
        yd = z + param[n+d];
        for( i=n+d-1; i>n; i-- )
            yd = z * yd + param[i];
        }
    else
        yd = 1.0; There is no denominator.
    if( config & XPX )
        yn = yn * gx;
    if( config & X2PX )
        yn = yn * gx * gx;
    if( config & PADE )
        { 2P/(Q - P)
        yd = yd - yn;
        yn = 2.0 * yn;
        }
    qyaprx = yn/yd;
    if( config & CW )
        qyaprx = gx + qyaprx * gx * gx;
    return( qyaprx );
    }
```

Subroutine to compute approximation error at x.

```
double geterr(x)
double x;
    {
    double e, f;
    double fabs(), approx(), func();

    f = func(x);
    e = approx(x) - f;
    if( relerr )
        {
        if( f != 0.0 )
            e /= f;
        }
    if( e < 0.0 )
        {
```

```
        esign = -1;
        e = -e;
        }
else
        esign = 1;
Here multiply e by x to weight the error by x.
return(e);
}
```

Subroutine for special argument transformations

```
double gofx(x)
double x;
     {

     return(x);
     }
```

func() is an accurate routine for the function to be approximated.

```
static int fflg = 0;
static double ff = 0.0;

double func(x)
double x;
     {
     double y, t;
     double exp();

     qx = x;

     if( fflg == 0 )
          {
          fflg = 1;
Use this space for initializing constants.
          }
     y = exp(x);
     qy = y;
     return( y );
     }
```

Module `remesp.c` formats and displays the coefficients of the solution polynomial or rational function. It then displays a table showing values of the function, the approximation, and the approximation error.

remesp.c

```
remesp()
	{
	int i, j, k, ip;
	double t, a, b, x, y, z, xm, ym;
	double approx(), func(), geterr();

	j = 0; Printout variable
	ip = 0; Solution vector counter
	printf( "Numerator coefficients:\n" );
	for( i=0; i<=n; i++, j++, ip++ )
		{
		if( j >= 3 )
			{
			printf( "\n" );
			j = 0;
			}
		printf( "%23.15E ", param[ip] );
		}
	if( d > 0 )
		{
		j = 0;
		printf( "\nDenominator coefficients:\n" );
		for( i=0; i<d; i++, j++, ip++ )
			{
			if( j >= 3 )
				{
				printf( "\n" );
				j = 0;
				}
			printf( "%23.15E ", param[ip] );
			}
		if( j >= 3 )
			printf( "\n" );
Leading denominator coefficient always = 1.
		printf( "%9.1E", 1.0 );
		}
	else
		printf( "\nDeviation:  %.4E", param[n+1] );

Display table of function and approximation error.
Note, sorting of the three interleaved sequences is imperfect
so the display is not strictly ordered by the value of x.
	printf(
```

```
	"\n\n x func approx error\n"
	);
	a = apwidt/N;
	b = apstrt;
	k = 0;
	j = 0;
	for( i=0; i<=N; i++ )
		{
		x = b + i * a;
		if( x >= mm[k] )
			{
			xm = mm[k];
			y = geterr(xm);
			z = qyaprx;
			ym = qy;
			printf( "%11.3E %11.3E %11.3E %11.3E*\n",
			xm, ym, z, y*esign );
			k += 1;
			}
```

Fill in the zeros also.

```
		if( (d > 0) && (x >= xx[j]) )
			{
			xm = xx[j];
			y = geterr(xm);
			z = qyaprx;
			ym = qy;
			printf( "%11.3E %11.3E %11.3E %11.3Eo\n",
			xm, ym, z, y*esign );
			j += 1;
			}
		y = geterr(x);
		z = qyaprx;
		ym = qy;
		printf( "%11.3E %11.3E %11.3E %11.3E\n",
		x, ym, z, y*esign );
		}
	}
```

Module `remesw.c` formats and writes the solution to a computer file.

`remesw.c`

```
#include <stdio.h>

remesw()
```

```
    {
    char s[40];
    int i;
    FILE *f, *fopen();
    double log10();

reopn:
    printf( "Log file name ?  " );
    gets(s);
    f = fopen( s, "w" );
    if( f == 0 )
        {
        printf( "Can't open %s\n", s );
        goto reopn;
        }
    fprintf( f, "\n%s = ", funnam );
    if( config & CW )
        fprintf( f, "x + x**2 * " );
    if( config & XPX )
        fprintf( f, "z " );
    if( config & X2PX )
        fprintf( f, "z**2 " );
    if( config & PXSQ )
        fprintf( f, "P(z**2)" );
    else
        fprintf( f, "P(z)" );
    if( d > 0 )
        {
        if( config & PXSQ )
            fprintf( f, "/Q(z**2)" );
        else
            fprintf( f, "/Q(z)" );
        }
    fprintf( f, "\nz(x) = %s\n", znam );
    if( relerr )
        fprintf( f, "Relative error\n" );
    else
        fprintf( f, "Absolute error\n" );
    fprintf( f, "n = %d, d = %d\n", n, d );
    fprintf( f, "precision = %23.15E ", -log10(farther) );
    fprintf( f, "error=%23.15E\n", farther );
    fprintf( f, "leveled=%23.4E\n", spread );
    fprintf( f, "\nNumerator:\n" );
    for( i=0; i<n+d+1; i++ )
        {
```

```
        if( i == n+1 )
            fprintf( f, "\nDenominator\n" );
        fprintf( f, "%23.15E\n", param[i] );
        }
    if( d > 0 )
        fprintf( f, "%23.15E\n", 1.0 );
    if( (d > 0) || ((config & ZER) != 0) )
        {
        fprintf( f, "\n Locations of zero error:\n" );
        for( i=0; i<=n+d; i++ )
        fprintf( f, "%23.15E\n", xx[i] );
        }
    fprintf( f, "\n Locations of peak error:\n" );
    for( i=0; i<=n+d+1; i++ )
    fprintf( f, "%23.15E\n", mm[i] );
    fclose(f);
    }
```

2.6 Forms of Approximation

Although the minimax polynomial or rational form may be a theoretically optimum approximation, it may not be so when the finite precision of the arithmetic is taken into account. Many functions, in fact, are best computed in practice by forms that are not analytically optimum.

One of the more important methods of reducing both roundoff and cancellation effects is to express the function as the sum of one term that can be computed very accurately and a second term that supplies a correction to the first. For example, the circular sine,

$$\sin(x) = x - \frac{x^3}{3!} + \frac{x^5}{5!} - \ldots$$

might be computed in the form

$$\sin(x) \approx x\; P_1(x^2)$$

where $P_1(x)$ is a minimax polynomial approximant. This, in IEEE arithmetic, yields a computer routine that has peak relative error $2.7 \cdot 10^{-16}$ and rms relative error $6.1 \cdot 10^{-17}$. But $P_1(t)$ is not exact, and it appears multiplied by the (assumed) exact value of x. Therefore its full relative error is reflected in the result. This error is made smaller by the Cody and Waite form

$$\sin(x) \approx x + x^3\; P_2(x^2)$$

which represents $\sin(x)$ as an exact function, x, plus a correction that is relatively small compared to x through most of the approximation interval.

A routine based on this form has a peak relative error of $2.1 \cdot 10^{-16}$ and rms relative error $5.5 \cdot 10^{-17}$.

The effect of this general technique is very striking when the function to be approximated has a singularity that can be subtracted or factored out. For example, the complete elliptic integrals have a logarithmic singularity that is well represented by Hastings' form

$$f(x) \approx R_1(x) \ln(x) + R_2(x) \ .$$

Functions of this sort can be approximated only very badly by purely polynomial or rational forms.

Functions that have a zero in the approximation interval may suffer from arithmetic cancellation near the zero. If possible, the zero should be transformed so that it has less effect on the approximation. This can sometimes be done by mapping the zero or other singularity onto the origin or to inifinity. For example, the form

$$f(x) = (x - r) \ R(x)$$

factors out a zero at $x = r$. The term $x - r$ may be computed in extra precision arithmetic, if desired, to maintain good relative accuracy even in the vicinity of the zero. A similar technique is to capitalize on an identity such as

$$\sin x = \sin(x - 2N\pi)$$

where N is an integer. This is used very effectively in trigonometric routines, where an accurate reduction modulo $\pi/4$ maps the zeros (or poles) to the origin.

Another example is the hyperbolic sine, which might be computed directly using the form of its definition

$$\sinh(x) = \tfrac{1}{2} \ (e^x - e^{-x}) \ .$$

But if x is near zero this is a good example of how *not* to compute a function. At $x = 0$, $\sinh(0) = 0$ but $e^0 = e^{-0} = 1$, so there is a complete loss of relative significance after subtracting e^{-x} from e^x. An alternative strategy for small x is to use a rational approximation based on the power series expansion about $x = 0$:

$$\begin{aligned} \sinh(x) &= x + \frac{x^3}{3!} + \frac{x^5}{5!} + \cdots \\ &\approx x + x^3 \ P(x^2)/Q(x^2) \ . \end{aligned}$$

2.7 Asymptotic Expansions

A basic technique for deriving asymptotic expansions[5] depends on integration by parts. For an illustration of the technique, consider the probability function

$$\operatorname{erfc}(x) = \frac{2}{\sqrt{\pi}} \int_x^\infty e^{-t^2} dt .$$

Using the formula for integration by parts,

$$\begin{aligned}
\int u \, dv &= uv - \int v \, du \\
v &= -\tfrac{1}{2} e^{-t^2} \\
dv &= t e^{-t^2} \\
u &= 1/t \\
du &= -1/t^2 \\
\int e^{-t^2} dt &= -\frac{1}{2t} e^{-t^2} - \int \frac{1}{2t^2} e^{-t^2} dt .
\end{aligned}$$

This yields the first term of the asymptotic expansion, plus an integral which may be regarded as the error term of the expansion:

$$\operatorname{erfc}(x) = \frac{1}{\sqrt{\pi}\, x} e^{-x^2} + \frac{2}{\sqrt{\pi}} \int_x^\infty \frac{1}{2t^2} e^{-t^2} dt .$$

Repeating the procedure on this integral produces the second term of the expansion:

$$\begin{aligned}
v &= -\tfrac{1}{2} e^{-t^2} \\
dv &= t e^{-t^2} \\
u &= 1/2t^3 \\
du &= -3/2t^4 \\
\int \frac{1}{2t^2} e^{-t^2} dt &= -\frac{1}{4t^3} e^{-t^2} - \int \frac{1 \cdot 3}{4t^4} e^{-t^2} dt .
\end{aligned}$$

Continuing n times and inserting the limits of integration results in the formula

$$\begin{aligned}
\operatorname{erfc}(x) &= \frac{1}{\sqrt{\pi}\, x} e^{-x^2} \left(1 + \sum_{k=1}^{n} \frac{(-1)^k 1 \cdot 3 \cdot 5 \cdots (2k-1)}{(2x^2)^k} \right) \\
&\quad + \frac{2}{\sqrt{\pi}} \int_x^\infty \frac{(-1)^{n+1} 1 \cdot 3 \cdot 5 \cdots (2n+1)}{(2x^2)^{n+1}} e^{-t^2} dt .
\end{aligned}$$

[5]For a thorough and readable exposition, see Norman Bleistein and Richard A. Handelsman, *Asymptotic Expansions of Integrals*, Holt, Rinehart and Winston, 1975, or Dover, 1986.

A useful bound on the error can be derived simply by noting that the error term (the integral on the second line above) changes sign when n in the summation increases to $n+1$. The sum as n increases is first larger, then smaller, etc., than the true function value. This means that the $(n+1)$st term in the summation must be larger in magnitude than the error after n terms — otherwise, the inclusion of the $(n+1)$st term would not change the sign of the error. Hence the error can be characterized as being smaller than the first omitted term of the series. This is a common (but not universal) feature of asymptotic expansions.

A second feature of asymptotic series is that the terms in the sum initially grow smaller, but eventually begin to increase without limit. This is the opposite of the behavior of a convergent power series. If n is allowed to increase indefinitely, then the sum of the asymptotic series diverges. Therefore it is necessary to stop at a value of n such that the error term is a minimum. If the error is smaller than the first omitted term, a good place to stop is just before the terms begin to grow in magnitude. The theoretical error at this point will not be zero. However, since the error term is a decreasing function of x, it can be said that if the error is adequately small for a given value of x, then it will be even smaller for all larger values of x. Furthermore, the error tends to zero as x tends to infinity.

Many of the higher transcendental functions are computed by supplying two expansions. The first is ordinarily derived from the ascending Taylor series expansion of the function. It will be accurate for small x but will suffer to an increasing degree from roundoff and cancellation effects as x increases. The second expansion is ordinarily an asymptotic series whose error is small for sufficiently large x. For each given arithmetic precision there is a crossover point at which the arithmetic error in calculating the ascending series is equal to the analytical plus computational error of the asymptotic series. Below this point one should use the ascending series, and above it one should use the asymptotic series. Unhappily the error at the crossover point is often so great that half the number of significant digits of the arithmetic are lost. Sometimes the domain of application of the asymptotic series can be extended significantly by calculating a minimax rational approximation that has the same form as the asymptotic expansion. But some functions, such as the Bessel functions, may require a special third expansion for the transition region between the usable domains of the ascending and descending series.

2.8 Continued Fractions

An alternative to power series expansions, especially for functions of more than one variable, is the continued fraction.[6] To illustrate one of the

[6]A good elementary introduction is C. D. Olds, *Continued Fractions*, Mathematical Association of America, 1963.

main techniques for deriving continued fraction expansions, a case involving Bessel functions will be considered. Bessel functions satisfy the following recurrence relation on their order k:

$$-J_{k-1}(x) + \frac{2k}{x} J_k(x) = J_{k+1}(x) \ .$$

This recurrence will be used to develop a continued fraction for the ratio $J_k(x)/J_{k-1}(x)$.

2.8.1 Continued Fractions from Recurrences

Suppose in general that a function f satisfies a three term recurrence

$$a_k f_{k-1} - b_k f_k = f_{k+1} \ .$$

This is eqivalent to

$$a_k \frac{f_{k-1}}{f_k} = b_k + \frac{f_{k+1}}{f_k}$$

or

$$\frac{f_k}{f_{k-1}} = \frac{a_k}{b_k + \frac{f_{k+1}}{f_k}} \ .$$

Substituting from the same recurrence for f_{k+1}/f_k yields

$$\frac{f_k}{f_{k-1}} = \frac{a_k}{b_k + \frac{a_{k+1}}{b_{k+1} + \frac{f_{k+2}}{f_{k+1}}}} \ .$$

Further substitutions may be continued indefinitely, so long as division by zero is avoided. The notation for a general continued fraction is

$$\frac{a_0}{b_0 + \frac{a_1}{b_1 + \frac{a_2}{b_2 + \cdots + \frac{a_n}{b_n}}}}$$

$$\equiv \frac{a_0}{b_0+} \frac{a_1}{b_1+} \frac{a_2}{b_2+} \frac{a_3}{b_3 + \cdots +} \frac{a_n}{b_n} \ .$$

In this notation, after repeated substitutions of the recurrence relation the continued fraction becomes

$$\frac{f_k}{f_{k-1}} = \frac{a_k}{b_k+} \frac{a_{k+1}}{b_{k+1} + \cdots +} \frac{a_{k+n}}{b_{k+n}+} \frac{f_{k+n+1}}{f_{k+n}} \ .$$

An error will be incurred if the final term f_{k+n+1}/f_{k+n} is omitted. As n approaches ∞ the hope is that this error will become arbitrarily small so that the truncated continued fraction can be used to calculate f_k/f_{k-1} in terms of the a's and b's alone. The conditions for convergence will not be pursued here.

Returning to the Bessel functions, the correspondence of a_k and b_k with the stated recurrence relation is

$$\begin{aligned} a_k &= -1 \\ b_k &= -\frac{2k}{x} . \end{aligned}$$

Hence the required continued fraction is

$$\begin{aligned} \frac{J_k(x)}{J_{k-1}(x)} &= \frac{-1}{-2k/x +} \; \frac{-1}{-2(k+1)/x +} \; \frac{-1}{-2(k+2)/x \; + \cdots} \\ &= \frac{1}{2k/x \; -} \; \frac{1}{2(k+1)/x \; -} \; \frac{1}{2(k+2)/x \; - \cdots} . \end{aligned}$$

2.8.2 Recurrences from Differential Equations

It is sometimes possible to develop a continued fraction directly from the differential equation of a function. Suppose $f(x)$ satisfies the equation

$$f^{(0)} = b_0 f^{(1)} + a_1 f^{(2)}$$

where

$$f^{(n)} \equiv \frac{d^n}{dx^n} f(x)$$

and the coefficients a and b may be functions of x. Differentiating both sides yields

$$f^{(1)} = b_0' f^{(1)} + b_0 f^{(2)} + a_1' f^{(2)} + a_1 f^{(3)}$$

or

$$f^{(1)} = \frac{b_0 + a_1'}{1 - b_0'} f^{(2)} + \frac{a_1}{1 - b_0'} f^{(3)}$$

where the primes denote differentiation with respect to x. Since the coefficients do not involve f, this equation generalizes immediately (by induction) to

$$f^{(k)} = b_k f^{(k+1)} + a_{k+1} f^{(k+2)}$$

where

$$\begin{aligned} b_k &= \frac{b_{k-1} + a_k'}{1 - b_{k-1}'} \\ a_{k+1} &= \frac{a_k}{1 - b_{k-1}'} . \end{aligned}$$

This is a recurrence on the order k of the derivative of f. Dividing through by $f^{(k+1)}$ gives

$$\begin{aligned} \frac{f^{(k)}}{f^{(k+1)}} &= b_k + \frac{a_{k+1}}{f^{(k+1)}/f^{(k+2)}} \\ &= b_k + \frac{a_{k+1}}{b_{k+1}+} \; \frac{a_{k+2}}{b_{k+2}+} \; \frac{a_{k+3}}{b_{k+3}+} \cdots . \end{aligned}$$

2.8.3 Computing Continued Fractions

There are two main methods of evaluating a continued fraction: forward recurrence and iterated division.[7] The forward method finds the kth convergent

$$f_k = \frac{p_k}{q_k}$$

in terms of a numerator p_k and a denominator q_k, from the matrix multiplication

$$\begin{pmatrix} p_k \\ q_k \end{pmatrix} = \begin{pmatrix} p_{k-1} & p_{k-2} \\ q_{k-1} & q_{k-2} \end{pmatrix} \begin{pmatrix} b_k \\ a_k \end{pmatrix}$$

starting with

$$\begin{pmatrix} p_1 \\ q_1 \end{pmatrix} = \begin{pmatrix} a_0 & 0 \\ b_0 & 1 \end{pmatrix} \begin{pmatrix} b_1 \\ a_1 \end{pmatrix} .$$

This first step produces

$$f_1 = \frac{p_1}{q_1} = \frac{a_0 b_1}{b_0 b_1 + a_1} .$$

To prevent arithmetic overflow, it is permissible to multiply *all four* of $p_{k-1}, q_{k-1}, p_{k-2}, q_{k-2}$ by a constant such as 2^{-n} at any step. To decide when to terminate the recurrence, it is often necessary to examine f_k and f_{k-1} after each calculation.

If the number of terms is known in advance, it is often faster to evaluate a continued fraction by iterated division. The algorithm begins with the nth term and proceeds backward:

$$\begin{aligned} s_0 &= b_n \\ s_{k+1} &= b_{n-k-1} + \frac{a_{n-k}}{s_k} \quad \text{for } k = 0, 1, \ldots, n-1 \\ f_n &= \frac{a_0}{s_n} . \end{aligned}$$

2.9 Polynomials

Nearly every function discussed in this book requires evaluation of a polynomial, usually given analytically in the form

$$P(x) = c_0 + c_1 x + c_2 x^2 + \ldots + c_n x^n .$$

The best known general way to compute such a polynomial is by Horner's nested multiplication scheme:

$$P(x) = (\ldots((c_n x + c_{n-1})x + c_{n-2})x + \ldots + c_1)x + c_0 .$$

[7]For more on this subject and a great deal of other information about computing functions, see the classic Hart, John F., E. W. Cheney, Charles L. Lawson, Hans J. Maehly, Charles K. Mesztenyi, John R. Rice, Henry C. Thacher, Jr., and Christoph Witzgall, *Computer Approximations*, Wiley, 1968.

This computational procedure is faster than the first form, and often is numerically more stable as well. It can be used in either real or complex arithmetic. Substantial cancellation error can occur, however, when the polynomial is evaluated near one of its zeros.

Faster expansions can be found in special cases, but these must be analyzed carefully for stability. Special polynomials, such as are employed in Chebyshev expansions, may have alternate preferred methods of computation. These are developed in the relevant sections along with the expansions.

Mathematical function programs can make use of a common subroutine program that evaluates polynomials. This routine may take advantage of whatever special hardware or assembly language features are available. Some computers have the ability to accumulate sums of products with greater than normal precision. For these it is advantageous to supply additional library routines for special polynomial and rational forms such as $xP(x)/Q(x)$.

An important type of computer for numerical work is the vectorized array processor. In a machine of this type arithmetic operations are broken up in to a sequence of pipelined hardware logic stages. A new operation can be started at the beginning of the pipe at each new instruction cycle, but the answer will not appear at the other end until several cycles later. This architecture is very suitable for vector and matrix algebra, but unfortunately it is not very suitable for evaluating polynomials. To "vectorize," as it is called, the computation, the polynomial may be rewritten as a polynomial whose coefficients are polynomials of lower degree. These are evaluated in parallel, keeping the pipeline full, then the results of lower degree are combined.

2.9.1 polevl.c

This program evaluates a real polynomial of degree N. Coefficients are stored in the array *coef* in reverse order, with the Nth degree term first and the constant term last.

```
double polevl( x, coef, N )
double x;
double coef[];
int N;
    {
    double ans;
    int i;
    double *p;

    p = coef;
    ans = *p++;
```

```
    i = N;
    do
        ans = ans * x + *p++;
    while( --i );
    return( ans );
    }
```

The function **p1evl**() assumes that the coefficient of degree N is equal to 1.0 and is omitted from the array. Its calling arguments are otherwise the same as those of **polevl**().

```
double p1evl( x, coef, N )
double x;
double coef[];
int N;
    {
    double ans;
    double *p;
    int i;

    p = coef;
    ans = x + *p++;
    i = N-1;
    do
        ans = ans * x + *p++;
    while( --i );
    return( ans );
    }
```

2.10 Newton-Raphson Iterations

Many algebraic functions can be computed by a rapidly converging iterative procedure for which a formula can be derived by the Newton-Raphson method. The design strategy is to form a function $f(x)$ that has a zero at the desired spot, then expand $f(x)$ in a Taylor series. For example, to derive the Newton iteration for the square root function $x = \sqrt{N}$, you could take $f(x) = x^2 - N$; this has a root $f(x) = 0$ at $x = \sqrt{N}$.

The iteration proceeds by calling the initial or current guess "x" and the improved guess "$x + h$," so that $f(x + h)$ is supposed to be closer to zero than was $f(x)$. The iteration step is found by setting the Taylor series expansion of $f(x + h)$ equal to zero,

$$f(x + h) = f(x) + hf'(x) + ... = 0$$

from which, to first order terms,

$$h = -\frac{f(x)}{f'(x)}$$

so that

$$x + h = x - \frac{f(x)}{f'(x)} .$$

Successful application of this formula requires an initial guess that is sufficiently close to the correct value. Starting with an initial guess for x, the formula yields an improved guess $x + h$. Putting $x + h$ back into the same formula in place of x yields a further improvement. From one iteration to the next the number of correct significant figures tends to double, because the error of the Taylor series approximation is of order h^2.

To find the iteration formula for the square root function, take $f(x) = x^2 - N$, as suggested above, whence $f'(x) = 2x$. This leads to the Heron iteration for the square root,

$$x + h = x - \frac{(x^2 - N)}{2x} = \frac{1}{2}\left(x + \frac{N}{x}\right) .$$

For the function $x = 1/\sqrt{N}$, a corresponding development is

$$\begin{aligned} f(x) &= x^2 - 1/N \\ f'(x) &= 2x \\ x + h &= x - \frac{x^2 - 1/N}{2x} \\ &= \frac{1}{2}\left(x + \frac{1}{Nx}\right) . \end{aligned}$$

A different formula for $1/\sqrt{N}$, one that does not require division, can be derived by starting with a different $f(x)$:

$$\begin{aligned} f(x) &= x^{-2} - N \\ f'(x) &= -2x^{-3} \\ x + h &= x - \frac{x^{-2} - N}{-2x^{-3}} \\ &= \frac{x(3 - Nx^2)}{2} . \end{aligned}$$

2.10.1 Division

A useful way to divide without dividing is the iteration

$$f(x) = x^{-1} - N$$

$$
\begin{aligned}
f'(x) &= -x^{-2} \\
x+h &= x - \frac{x^{-1} - N}{-x^{-2}} \\
&= x(2 - Nx) \ .
\end{aligned}
$$

An application of this iteration to extended precision arithmetic was discussed earlier. This iteration is also useful in array processors and digital signal processing chips that can multiply very quickly. To get a close initial guess, the exponent of the floating point number is first removed by a method such as the one described in the next section. The exponent of the inverse $1/x$ is then the negative of the separated exponent. The significand s of the number x lies between 0.5 and 1.0. A polynomial approximation such as

$$\frac{1}{s} \approx 2.5859s^2 - 5.8182s + 4.2424$$

may be used to find $1/s$ to within about 1%. From this initial guess, the above Newton iteration for s^{-1} will converge to machine accuracy in a few steps.

2.10.2 Exponent Separation

For these iterations to work effectively a fairly close initial guess is required. For the iterations to be detailed below, the first step is to get within a factor of two by integer operations on the exponent of the floating point argument.

Sometimes a standard computer library function is available that will separate a floating point number into its exponent and significand. For example, in C language the operation

$$s = \mathbf{frexp}(x, \&e);$$

finds an integer exponent e and floating point significand s such that $2^e s = x$ and $0.5 \leq s < 1.0$. (See `efrexp.c` in Chapter 3 for an example program.)

Note that for $x = 0$ the value of e is undefined. The routines that use this reduction may therefore have to consider $x = 0$ to be a special case.

If no library routine is available, it will be necessary to determine the exact data structure of the floating point number and use machine instructions or the equivalent to operate on the appropriate bit fields. For an IEEE double precision number the objective is to (1) extract the 12 bit number representing the exponent and subtract 1022 from it, then (2) copy the number to a variable that will contain the significand, setting that variable's exponent bit field equal to 1022.

Having isolated the power of two of the argument, the second step is to employ a simple polynomial approximation to obtain an initial guess for the function that is accurate to about 1%. Then the appropriate iteration can be executed several times to converge to an accurate value.

2.10.3 Square Root

As discussed above, the first step in the square root program is to separate the argument x into an integer exponent e and a floating point significand s such that $2^e s = x$ and $0.5 \leq s < 1.0$. The exponent and the significand of the approximate square root are found separately. For the significand, the following minimax polynomial of degree 1 approximates the square root of $0.5 \leq s \leq 1.0$ with a maximum relative error of $7.47 \cdot 10^{-3}$:

$$y = 0.417308 + 0.590162s \ .$$

The exponent of the square root is half the exponent of the argument. If the argument has an odd exponent, y must be multiplied by

$$\sqrt{2} = 1.4142135623730950488 \ .$$

Then the integer exponent e may be shifted 1 bit to the right to accomplish division by two, with the odd half bit incorporated as just shown into y. The approximate square root is a floating point number w that must now be reconstructed from e and y. In C language this can be accomplished using **ldexp**() to multiply y by 2^e using integer operations on the exponent.

Finally, the Heron iteration can be used with w as the initial guess. Each iteration involves replacing w by $w + x/w$ and dividing the result by 2. The latter operation should be done by an integer operation on the exponent, if possible. To obtain a double precision result, three iterations are necessary. On the final iteration, the form

$$w := \frac{w^2 + x}{2w}$$

may give slightly better rounding accuracy. If this form is used it is necessary to take precautions against overflow when x is within a factor of 2 of the machine's overflow threshold. If $x < 0$ there is no real square root. The routine should regard this case to be a domain error.

The square root of a complex number $z = x + iy$ may be expressed as

$$\begin{aligned} w &= \sqrt{\frac{|z| - x}{2}} \\ \sqrt{z} &= \frac{y}{2w} + iw. \end{aligned}$$

There can be severe cancellation error in $|z| - x$. An alternate procedure that handles the case $|z| = x$ is

$$\begin{aligned} \theta &= \frac{1}{2} \tan^{-1}\left(\frac{y}{x}\right) \\ \sqrt{z} &= \sqrt{|z|}(\cos\theta + i\sin\theta). \end{aligned}$$

When $y = 0$ the arctangent will be quickly and correctly evaluated by a two argument quadrant correct arctangent function whose output ranges from 0 to 2π. The result of either formula may be improved by performing one or two Heron iterations in complex arithmetic. The relative error, measured by the complex absolute value of the complex error, will be about 1 lsb if the Heron iteration is included. The program should treat $y = 0$ as a special case, where the result is $i\sqrt{-x}$ if $x < 0$.

2.10.4 sqrt.c

```
#include "mconf.h"
extern double SQRT2;

double sqrt( x )
double x;
    {
    int e;
    double z,w;
    double frexp(), ldexp();

    if( x <= 0.0 )
        {
        if( x < 0.0 )
            mtherr( "sqrt", DOMAIN );
        return( 0.0 );
        }
    w = x;
```
Separate exponent and significand.
```
    z = frexp( x, &e );
```

Approximate square root of number between 0.5 and 1.
```
    x = 4.173075996388649989089E-1
        + 5.901620670906445829966 3E-1 * z;
```

Adjust for odd powers of 2.
```
    if( (e & 1) != 0 )
        x *= SQRT2;
```

Re-insert the exponent.
```
    x = ldexp( x, e >> 1 );
```

Newton iterations:
```
    x += w/x;
```
divide by 2
```
    x = ldexp( x, -1 );
```

```
x += w/x;
x = ldexp( x, -1 );
x += w/x;
x = ldexp( x, -1 );

return(x);
}
```

2.10.5 Longhand Square Root

By way of comparison, square roots are calculated longhand by a shift and subtract method that is similar to long division. The method depends on the binomial theorem

$$(10a + b)^2 = 100a^2 + 20ab + b^2 \ .$$

In the square root step two digits of the dividend are brought down and adjoined to the remainder from the previous step. The current partial root a is left shifted one place (in the radix of the arithmetic) and multiplied by 2. This value, $20a$, is compared to the remainder (with the two digits adjoined) to determine the next digit b of the partial root. Then $b(20a+b)$ is subtracted from the remainder. By previous steps $(10a)^2$ was already subtracted from the dividend, so after each new step the remainder is the difference between the dividend and the square of the partial root. The method is most suitable for special hardware circuits or for extracting the square root of an integer. The program that follows makes use of the arithmetic routines given in Chapter 1 to produce a strictly rounded floating point result.

2.10.6 esqrt.c

Longhand square root with strict rounding rules.

```
extern unsigned short ezero[], eone[];
unsigned static short rndbit[NQ+1] = {0,0,0,0,0,0,1,0};

int esqrt( x, y )
short *x, *y;
    {
    unsigned short temp[NQ+1], num[NQ+1], sq[NQ+1];
    unsigned short xx[NQ+1];
    int i, j, k, n;
    long m, exp;
```

Check for argument ≤ 0

```
      i = ecmp( x, ezero );
      if( i <= 0 )
            {
            eclear(y);
            if( i < 0 )
                  printf( "esqrt domain error\n" );
            return;
            }
```

Bring in the argument and renormalize if it is denormal.

```
      emovi( x, xx );
      m = xx[1];
      if( m == 0 )
            m -= enormlz( xx );
```

Divide exponent by 2

```
      m -= 0x3ffe;
      exp = (unsigned short )( (m / 2) + 0x3ffe );
```

Adjust if exponent odd

```
      if( (m & 1) != 0 )
            {
            if( m > 0 )
                  exp += 1;
            eshdn1( xx );
            }
      ecleaz( sq );
      ecleaz( num );
      n = 8;
      for( j=0; j<8; j++ )
            {
```

Bring in next word of argument.

```
            if( j < 5 )
                  num[NQ] = xx[j+3];
```

Do additional bit on last outer loop, for roundoff.

```
            if( j == 7 )
                  n = 9;
            for( i=0; i<n; i++ )
                  {
```

Next 2 bits of argument

```
                  eshup1( num );
                  eshup1( num );
```

Shift up answer

```
                  eshup1( sq );
```

Make trial divisor

```
                  for( k=0; k<=NQ; k++ )
                        temp[k] = sq[k];
```

```
            eshup1( temp );
            eaddm( rndbit, temp );
```
Subtract and insert answer bit if it goes in.
```
            if( ecmpm( temp, num ) <= 0 )
                {
                esubm( temp, num );
                sq[NQ-1] |= 1;
                }
            }
        }
```
Adjust for extra, roundoff loop done.
```
    exp -= 1;
```
Sticky bit = 1 if the remainder is nonzero.
```
    k = 0;
    for( i=3; i<=NQ; i++ )
    k |= num[i];
```
Renormalize and round off.
```
    ernorm( sq, k, 0, exp, 64 );
    emovo( sq, y );
    }
```

2.10.7 Cube Root

Range reduction for the cube root function involves determining the power of 2 of the argument. A polynomial of degree 2 applied to the significand, and multiplication by the cube root of 1, 2, or 4 approximates the root to within about 0.1%. Then a Newton iteration is used several times to converge to an accurate result.

If s is the significand of x found as described earlier, the following polynomial aproximates the cube root of $0.5 \leq s < 1$, with peak relative error $= 6.4 \cdot 10^{-4}$:

$$y = -0.191502s^2 + 0.697570s + 0.493296 \ .$$

The exponent of the cube root is the exponent of the argument divided by 3. If the exponent does not divide evenly by 3, then y must be adjusted. There are two cases of adjustment. First, if the argument $x \geq 1$ then multiply y by $\sqrt[3]{2}$ or $\sqrt[3]{4}$ depending on the remainder after integer division by 3. In the second case $x < 1$ and $e < 0$. In this case, *divide* by $\sqrt[3]{2}$ or $\sqrt[3]{4}$ depending on the remainder; the correct value of e for the cube root is then $-(|e|/3)$. After making the appropriate adjustment the floating point number y is multiplied by 2^e to form the intial cube root approximation.

The Newton-Raphson iteration for the cube root may be derived as follows:

$$f(y) \ = \ y^3 - x$$

$$\begin{aligned} f'(y) &= 3y^2 \\ y+h &= y - \frac{y^3 - x}{3y^2} \\ &= y - \frac{1}{3}\left(y - \frac{x}{y^2}\right) . \end{aligned}$$

This iteration is performed three times to achieve a double precision result.

2.10.8 cbrt.c

```
#include "mconf.h"

static double CBRT2 = 1.2599210498948731647 7;
static double CBRT4 = 1.5874010519681994747 5;

double cbrt(x)
double x;
    {
    int e, rem, sign;
    double z;
    double frexp(), ldexp();

    if( x == 0 )
        return( 0.0 );
    if( x > 0 )
        sign = 1;
    else
        {
        sign = -1;
        x = -x;
        }
    z = x;
    x = frexp( x, &e );
    x = (-1.9150215751434832257e-1 * x
        + 6.9757045195246484393e-1) * x
        + 4.9329566506409572486e-1;
```

Argument is 1 or greater.

```
    if( e >= 0 )
        {
        rem = e;
        e /= 3;
        rem -= 3*e;
        if( rem == 1 )
            x *= CBRT2;
        else if( rem == 2 )
```

```
                    x *= CBRT4;
          }
```

Argument is less than 1.

```
     else
          {
          e = -e;
          rem = e;
          e /= 3;
          rem -= 3*e;
          if( rem == 1 )
               x /= CBRT2;
          else if( rem == 2 )
                    x /= CBRT4;
          e = -e;
          }
     x = ldexp( x, e );
     x -= ( x - (z/(x*x)) )/3.0;
     x -= ( x - (z/(x*x)) )/3.0;
     x -= ( x - (z/(x*x)) )/3.0;
     if( sign < 0 )
          x = -x;
     return(x);
     }
```

3

Software Notes

3.1 Design Strategy

Priorities for the mathematical function routines were set in the following order:

1. Numerical accuracy.
2. Portability.
3. Speed of operation.

Unfortunately the language compilers that have been developed to operate in modern desk-top and micro-computers often have serious shortcomings from a numerical viewpoint. Perhaps the most common failing is that *the compiler cannot correctly convert a decimal number into floating point format.* Often, the arithmetic used for decimal to binary conversion has no more than the standard working precision. Inevitably this leads to an accumulation of rounding errors in the floating point result. Even when extended precision arithmetic is used, the binary result is likely to be rounded incorrectly.

There is probably no commercial compiler that cannot read 16 bit integers. Therefore, to ensure that the numbers pass unscathed from source program to object program, all potentially troublesome constants have been supplied as arrays of integers. The book is cluttered with listings of these arrays on the theory that if you are actually going to enter one of the programs into your computer, then you deserve the best possible chance to achieve the results claimed for it.

In developing each routine, some effort was made to examine alternative methods of computing the function. As they stand, the programs should give serviceable results. However, *there is no guarantee that the algorithms used are the best ones possible.* You should consult the very extensive literature on numerical computing. Better algorithms may already be known, or may be found in the future.

Many computers are equipped with floating point arithmetic devices or software equivalents that have *more* than normal working precision internally. However, it is often impossible for the programmer to predict the circumstances under which the compiled object program will actually take advantage of the extra precision capability. The accuracy figures reported in this book were obtained either with an arithmetic that was known to have no such features, or with the compiler and the hardware set in a manner that tended to defeat any such optimizations. If these features are available in your computer, you should be able to enjoy a somewhat improved accuracy by using the extra precision arithmetic. On the other hand, your compiler's optimization features may not work; so it would be wise to try the program with and without the extra capabilities, verifying that the results are correct in all cases.

All the accuracy figures reported for DEC arithmetic were obtained using an LSI-11 computer equipped with a DEC floating point arithmetic chip. Figures for IEEE arithmetic were obtained with a Motorola 68882 floating point chip, set to 53 bit rounding precision. An effort was made to ensure that the extended 64 bit precision mode of the 68882 was never invoked by the test programs.

The goal of portability has been verified by compiling and testing most of the programs on at least half a dozen different computers, with no change allowed in the source code. Pursuit of this goal has led to a sort of least common denominator programming style in which unrealized language standards, complicated expressions, and fancy language constructs are avoided. No apologies are offered for this programming style. An occasional apology may be in order for excessive use of `goto` statements; the purpose of these is to produce more efficient object code.

There was relatively little conflict between the goals of portability and speed of operation. However, you should not expect these programs, written in a high level language, to be as fast as library routines that have been hand crafted in machine language for a specific computer. Short of rewriting the algorithms in assembly language, there is probably not much that could be done to increase the execution speed on any specific machine. If your compiler has a good optimizer, though, you could try recombining some of the algebraic expressions that have been written in a form that breaks them up into several successive source statements.

On the subject of DO loops, it was found that the form

```
i = N;
do{
}while( --i );
```

generated faster object code on some systems than

```
for(i=0; i<N; i++){
}
```

No system was encountered whose for() loops were faster than its do-while() loops. This is the reason for the appearence of do-while() loops in many of the programs.

3.2 Testing

There are several important methods of testing mathematical function programs. All of them should be used when feasible, since each may detect errors that the others miss. The main methods are

1. Manual checking of the output at zeros, singularities, domain boundaries, and other points for which the exact function result is easy to state
2. Manual checking using published tables as the standard
3. Consistency testing against related function programs (for example, $\ln e^x = x$)
4. Comparison with the output of a program that computes the function by a different method
5. Comparison against a program that uses higher precision arithmetic.

Initial tests of a routine involve checking it it in a few cases for which you know the correct answers. From this success you would proceed to verify the output against tables, if any are available. Be warned that published tables are sometimes less than 100% accurate. Nevertheless, tables will seldom be grossly in error, whereas the initial attempt at a program may well contain great blunders. For a high precision comparison, take care that the table or check routine and the program under test receive *exactly* the same inputs. A table of $f(x)$ might give the value of $f(0.3)$, for example. This cannot compare precisely with a value computed in binary floating point arithmetic, since there is no computer number equal to 0.3.

After successful spot checking, the next step is to construct consistency tests using simple relations such as $\sin^2 x + \cos^2 x = 1$ for which you can analyze and predict the results. Measure the peak and rms error over a large number of pseudorandom number arguments. You should satisfy yourself that you understand the usable accuracy of the test itself, as well as the probable accuracy of your function routine, so that the results of the test make sense. Use a variety of consistency tests, as time and energy may permit, to emphasize different parts of the function's domain.

Items 4 and 5 above may be combined, if an extended precision arithmetic is available. In this book, one or more analytical expansions are given for each of the mathematical functions treated.[1] The reason for quoting these expansions, aside from their intrinsic interest, is that each of them was programmed in a high precision arithmetic to become a reference standard for testing. Thus for each program given, there is a corresponding check routine, not given, that produced the reference values for determining the approximation coefficients and also the values used to test the final program. Though the test software has not been included in the book, it is available through other channels.

A suite of test programs for the elementary functions has been published by Cody and Waite (see the reference at the beginning of the chapter on elementary functions). Their suite includes a combination of consistency tests, checking at special points, and testing against extra precision calculations over a limited domain.

An arithmetic test program called PARANOIA, originated by W. M. Kahan, is available[2] in several programming languages. It is very good for detecting errors in the implementation of floating point arithmetic. This program can be difficult to get working on a small computer, but it is well worth the effort if you are concerned about design errors in your computer's arithmetic. Few computers can pass unscathed through the PARANOIA test.

3.3 System Utilities

3.3.1 mconf.h

For most of the programs in the book, the compiler is instructed by the source code statement

```
#include "mconf.h"
```

to read a common **include file** called `mconf.h`. This file contains definitions for error codes that are passed to the common error handling routine **mtherr**() described in the next section.

The include file also has a conditional assembly definition for the type of computer arithmetic to be employed. There are two reasons for this. One is to specify the roundoff error and other parameters of the number system. The other is to invoke expansion coefficients appropriate to the arithmetic. The choices of arithmetic offered are Intel IEEE, DEC, Motorola IEEE, or UNKnown.

For Digital Equipment PDP-11 and VAX computers, certain IBM systems, and others that use numbers with a 56-bit significand, the symbol

[1] Nearly all of these expansions can be found in Abramowitz, Milton, and Irene A. Stegun, editors, *Handbook of Mathematical Functions with Formulas, Graphs, and Mathematical Tables*, National Bureau of Standards Applied Mathematics Series No. 55, U. S. Government Printing Office, 1968. This giant collection is referenced as "AMS55."

[2] PARANOIA is currently obtainable through Richard Karpinski, Computer Center U-76, University of California, San Francisco, CA 94143-0704, USA.

DEC should be defined equal to 1. In this mode, most floating point constants are given as arrays of octal integers to eliminate decimal to binary conversion errors that might be introduced by the compiler.

For many computers, such as the IBM PC, that (supposedly) follow the IEEE Standard for Binary Floating Point Arithmetic (ANSI/IEEE Std 754-1985), the symbol `IIEEE` should be defined equal to 1. The IEEE numbers have 53-bit significands. In this mode, constants are provided as arrays of hexadecimal 16 bit integers.

There are two IEEE number formats in common use. Computers that use Intel processor chips such as the 80287 require the format invoked by the symbol `IIEEE`. A different IEEE format is used by Motorola 68000 processors. It has the same bits, but if the number is interpreted as an array of 16 bit integers, those integers are in reverse order compared to the Intel format. A different symbol, `MIEEE`, should be defined equal to 1 for this mode. To save space, the `MIEEE` style coefficients have not been included in the book. They can be produced by reversing the order of the integers given in each line of `IIEEE` style coefficients.

A great variety of floating point number formats exist in different computers and programming languages. To accommodate other types of double precision computer arithmetic, all constants are also provided in a normal decimal radix that one can hope will be converted correctly to a suitable format by the available C language compiler. To invoke this mode, the symbol UNK is defined equal to 1. To save space, these coefficients are omitted from the computer program listings presented in the book; the same decimal numbers are already given in the text. All four numeric formats are included in the computer readable distribution of the library.

An important difference among these modes is a predefined set of machine arithmetic constants for each. The numbers `MACHEP` (the machine roundoff error), `MAXNUM` (largest number represented), and several other parameters are implied by the configuration symbol. Check the file `const.c` to ensure that these values are correct for your computer.

mconf.h

Constant definitions for math function error conditions

```
#define DOMAIN 1 Argument domain error
#define SING 2 Singularity
#define OVERFLOW 3 Overflow range error
#define UNDERFLOW 4 Underflow range error
#define TLOSS 5 Total loss of precision
#define PLOSS 6 Partial loss of precision
```

Data structure of a complex number

```
typedef struct
    {
    double r;
```

```
    double i;
    }cmplx;
```

Define only one of the following equal to 1
to declare the desired type of computer arithmetic.
All the other arithmetic type definitions
must be commented out.

PDP-11, Pro350, VAX, some IBM mainframes
```
#define DEC 1
```

Intel IEEE, low order words come first
(8086, 8088, 80286, 80386 microprocessors)
```
#define IBMPC 1
```

Motorola IEEE, high order words come first
(68000 family microprocessors)
```
#define MIEEE 1
```

UNKnown arithmetic, invokes coefficients given in
decimal radix. Beware of range boundary
problems (`MACHEP`, `MAXLOG`, etc. in `const.c`) and
roundoff problems in `pow.c` and elsewhere.
```
#define UNK 1
```

3.3.2 mtherr.c

This subroutine may be called to report one of the following error conditions whose code values are defined in the include file `mconf.h`.

Mnemonic	Value	Significance
DOMAIN	1	Argument domain error
SING	2	Function singularity
OVERFLOW	3	Overflow range error
UNDERFLOW	4	Underflow range error
TLOSS	5	Total loss of precision
PLOSS	6	Partial loss of precision
EDOM	33	Unix domain error code
ERANGE	34	Unix range error code

The default version of the file, listed below, prints the function name that was passed to it via the pointer *name*, then prints a description of the error condition. The display is directed to the standard output device. After printing these messages, the routine returns to the calling program. Users may wish to modify the program to abort by calling **exit**() under severe error conditions such as domain errors. Since all error conditions are supposed to pass control to this function, the display may be easily changed, eliminated, or directed to a different error logging device by modifying the one subroutine.

On return from **mtherr**(), the error code bits have been left in a global variable called **merror** for possible inspection and action by higher level programs. In the compiler run-time library there is often a system error code variable as well; you may wish to modify this program so that it uses the error flag word that is already provided by your operating system.

```
#include "mconf.h"

Note, the order of appearance of the following
messages is bound to the error codes defined in mconf.h.

static char *ermsg[7] = {
      "unknown", Error code 0:
      "domain", Error code 1:
      "singularity", et seq.
      "overflow",
      "underflow",
      "total loss of precision",
      "partial loss of precision"
      };

extern int merror = 0; Global error code variable

mtherr( name, code )
char *name;
int code;
      {

Display the string passed by calling program,
which is supposed to be the name of the
function in which the error occurred.
      printf( "\n%s ", name );
Display the error message defined
by the code argument.
      if( (code <= 0) || (code >= 6) )
            code = 0;
      printf( "%s error\n", ermsg[code] );
Leave the error code in a globally accessible place
      merror = code;
      }
```

3.3.3 const.c

This file contains a number of mathematical constants and some needed size parameters of the computer arithmetic. The values are supplied as arrays of hexadecimal integers for IEEE arithmetic, arrays of octal constants for DEC arithmetic, or in a normal decimal notation for other machines. The particular notation used is determined by a symbol (DEC, IIEEE, MIEEE, or UNK) defined in the include file `mconf.h`.

const.c

```
#include "mconf.h"
```

```
#ifdef DEC
```

2^{-56}

```
extern short MACHEP[4] = {
0022200,0000000,0000000,0000000};
```

$\ln 2^{127}$

```
extern short MAXLOG[4] = {
041660,007463,0143742,025733,};
```

$\ln 2^{-128}$

```
extern short MINLOG[4] = {
0141661,071027,0173721,0147572,};
```

2^{127}

```
extern short MAXNUM[4] = {
077777,0177777,0177777,0177777,};
```

$\pi = 3.14159265358979323846$

```
extern short PI[4] = {
040511,007732,0121041,064302,};
```

$\pi/2 = 1.57079632679489661923$

```
extern short PIO2[4] = {
040311,007732,0121041,064302,};
```

$\pi/4 = 0.785398163397448309616$

```
extern short PIO4[4] = {
040111,007732,0121041,064302,};
```

$\sqrt{2} = 1.41421356237309504880$

```
extern short SQRT2[4] = {
040265,002363,031771,0157145,};
```

$\sqrt{1/2} = 0.707106781186547524401$

```
extern short SQRTH[4] = {
040065,002363,031771,0157144,};
```

$1/\ln 2 = 1.44269504088896340735 99$

```
extern short LOG2E[4] = {
040270,0125073,024534,013761,};
```

$\sqrt{2/\pi} = 0.797884560802865355879 89$

```
extern short SQ2OPI[4] = {
040114,041051,0117241,0131204,};
```

$\ln 2 = 0.693147180559945309417$

```
extern short LOGE2[4] = {
040061,071027,0173721,0147572,};
```

```
#ifdef IIEEE
```

2^{-53}

```
extern short MACHEP[4] = {
0x0000,0x0000,0x0000,0x3ca0};
```

$\ln 2^{1024}$

```
extern short MAXLOG[4] = {
0x39ef,0xfefa,0x2e42,0x4086};
```

$\ln 2^{-1022}$

```
extern short MINLOG[4] = {
0xbcd2,0xdd7a,0x232b,0xc086};
```

$2^{1024}(1 - \mathbf{MACHEP})$

```
extern short MAXNUM[4] = {
0xffff,0xffff,0xffff,0x7fef};
extern short PI[4] = {
0x2d18,0x5444,0x21fb,0x4009};
extern short PIO2[4] = {
0x2d18,0x5444,0x21fb,0x3ff9};
extern short PIO4[4] = {
0x2d18,0x5444,0x21fb,0x3fe9};
extern short SQRT2[4] = {
0x3bcd,0x667f,0xa09e,0x3ff6};
extern short SQRTH[4] = {
0x3bcd,0x667f,0xa09e,0x3fe6};
extern short LOG2E[4] = {
0x82fe,0x652b,0x1547,0x3ff7};
extern short SQ2OPI[4] = {
0x3651,0x33d4,0x8845,0x3fe9};
extern short LOGE2[4] = {
0x39ef,0xfefa,0x2e42,0x3fe6};
```

$\frac{1}{2}\ln 2 = 0.34657359027997265470\dot{9}$

```
extern short LOGSQ2[4] = {
037661,071027,0173721,0147572,};
```

$3\pi/4 = 2.35619449019234492885$

```
extern short THPIO4[4] = {
040426,0145743,0174631,007222,};
```

$2/\pi = 0.63661977236758134307553\,5$

```
extern short TWOOPI[4] = {
040042,0174603,067116,042025,};
#endif
```

```
extern short LOGSQ2[4] = {
0x39ef,0xfefa,0x2e42,0x3fd6};

extern short THPIO4[4] = {
0x21d2,0x7f33,0xd97c,0x4002};

extern short TWOOPI[4] = {
0xc883,0x6dc9,0x5f30,0x3fe4};
#endif
```

Decimal values of the typical machine constants are

$$
\begin{aligned}
2^{-56} &= 1.38777878078144567553 \cdot 10^{-17} \\
2^{127} &= 1.70141183460469231730 \cdot 10^{38} \\
\ln 2^{127} &= 88.029691931113054295988 \\
\ln 2^{-128} &= -88.72283911167299960540 \\
2^{-53} &= 1.11022302462515654042 \cdot 10^{-16} \\
2^{1024}(1-2^{-53}) &= 1.79769313486231570815 \cdot 10^{308} \\
\ln 2^{1024} &= 709.782712893383996843 \\
\ln 2^{-1022} &= -708.39641853226410622 4 \\
2^{-1074} &= 4.94065645841246544 \cdot 10^{-324} \\
\ln 2^{-1074} &= -744.440071921381262 3\ .
\end{aligned}
$$

3.4 Arithmetic Utilities

Many of the mathematical routines make use of C functions that convert from one numeric format to another. These may be system or hardware dependent but are usually available in the language's standard software library. The functions used are

Find largest integer $\leq x$, with double precision result.

```
y = floor( x )
```

Separate x into $s\ 2^e$, $0.5 \leq s < 1.0$.

```
s = frexp( x, &e )
```

Mulitply x by 2^e, e an integer.

```
x = ldexp( x, e )
```

Examples of these functions are presented in the following subsections. Another routine of this type is the following, which finds the nearest integer to x, returning it as a double precision number. Note that the program rounds to even if the fractional part is exactly 0.5, in accordance with strict rounding rules.

```
double round(x)
double x;
    {
    double y, r;
    double floor();

    y = floor(x);
    r = x - y;
    if( r > 0.5 )
        goto rndup;
    if( r == 0.5 )
        {
        r = y - 2.0 * floor( 0.5 * y );
        if( r == 1.0 )
            {
rndup:
            y += 1.0;
            }
        }
    return(y);
    }
```

3.4.1 efloor.c

This program finds the largest integer not greater than x, returning it as a floating point number in the format of `ieee.c` (see Section 1.7).

```
static unsigned short bmask[] = {
    0xffff,
    0xfffe,
    0xfffc,
    0xfff8,
    0xfff0,
    0xffe0,
    0xffc0,
    0xff80,
    0xff00,
    0xfe00,
    0xfc00,
    0xf800,
    0xf000,
    0xe000,
    0xc000,
    0x8000,
    0x0000,
```

```
    };

void efloor( x, y )
unsigned short x[], y[];
    {
    register unsigned short *p;
    int e, expon, i;
    unsigned short f[NE];
```

Input remains in external format.

```
    emov( x, f );
    expon = f[NE-1];
    e = (expon & 0x7fff) - (EXONE - 1);
    if( e <= 0 )
        {
        eclear(y);
        goto isitneg;
        }
```

Find number of bits to clear out.

```
    e = NBITS - e;
    emov( f, y );
    if( e <= 0 )
        return;

    p = &y[0];
    while( e >= 16 )
        {
        *p++ = 0;
        e -= 16;
        }
```

Clear the remaining bits using mask table.

```
    *p &= bmask[e];
```

Truncate negative input toward minus infinity.

```
isitneg:
    if( expon & 0x8000 )
        {
        for( i=0; i<NE-1; i++ )
            {
            if( f[i] != y[i] )
                {
                esub( eone, y, y );
                break;
                }
            }
        }
```

```
	}
```

3.4.2 efrexp.c

This program extracts the exponent and significand of a floating point number in the format of `ieee.c` (see Section 1.7). The exponent is returned in a long (32 bit) integer. This makes it possible to handle denormalized numbers properly, since the returned value can become negative without overflow.

```
efrexp( x, exp, s )
unsigned short x[];
long *exp;
unsigned short s[];
	{
	unsigned short xi[NI];
	long li;

	emovi( x, xi );
	li = (long )((short )xi[1]);

	if( li == 0 )
		{
		li -= enormlz( xi );
		}
	xi[1] = 0x3ffe;
	emovo( xi, s );
	*exp = li - 0x3ffe;
	}
```

3.4.3 eldexp.c

This program multiplies by 2^n, using integer operations on the exponent of the floating point number. The arithmetic format is compatible with `ieee.c` (Section 1.7).

```
eldexp( x, pwr2, y )
unsigned short x[];
long pwr2;
unsigned short y[];
	{
	unsigned short xi[NI];
	long li;
	int i;

	emovi( x, xi );
```

```
li = xi[1];
li += pwr2;
i = 0;
emdnorm( xi, i, i, li, 64 );
emovo( xi, y );
}
```

4

Elementary Functions

4.1 e^x

Since e^x grows faster with x than any polynomial, approximations to the exponential function must be restricted to a limited interval. Fortunately the data structure of floating point numbers provides a convenient way to do this. The floating point exponent gives the scale in powers of two (the base of the arithmetic), while the significand lies between 0.5 and 1. Range reduction for the exponential function is accomplished by finding an integer k and a fraction f such that

$$e^x = 2^k e^f \ .$$

Then 2^k is easily calculated in a binary floating representation, and the exponential of $-0.5\ln 2 \le f \le +0.5\ln 2$ can be found by a rational approximation. The relation among x, k, and f is illustrated by[1]

$$2^k e^f = e^{k\ln 2} e^f = e^{f+k\ln 2} = e^x \ .$$

In other words, the decomposition sought is

$$x = f + k\ln 2 \ .$$

Range reduction may use the integer truncation function **floor**(), which returns a floating point number equal to the largest integer that is less than or equal to its argument. In terms of this function the desired integer is calculated by

$$k = \mathbf{floor}(x \ \log_2 e + 0.5) \ .$$

Adding 0.5 has the effect of rounding to the nearest integer. Having found k, the desired fraction may be calculated by

$$\begin{aligned} f = &\ x - k\ln 2 \\ = &\ (x - 0.693359375k) + 0.00021219444005469058277k \ . \end{aligned}$$

[1]For additional discussion see Hart *et al*, *op. cit.*, Chapter 6.

Separating the factor ln 2 into two parts affords a pseudo extended precision calculation of $x - k \ln 2$, as suggested by Cody and Waite.[2] The point of this is that if 0.693359375 times k is exact and has the same exponent as x, then the term in parentheses is computed with no error. Adding the term $0.002 \ldots k$ commits at worst two rounding errors. By contrast, the direct computation $x - k \ln 2$ has a potentially disastrous cancellation error. The constants are

$$\begin{aligned} \ln 2 &= 0.693147180559945309417232121458176568075500 \\ \log_2 e &= 1.442695040888963407359924681001892137426 = 1/\ln 2 \, . \end{aligned}$$

The exponential of the fractional part is found by the following approximation in Padé form:

$$e^f \approx \frac{Q(f^2) + f\ P(f^2)}{Q(f^2) - f\ P(f^2)} \, .$$

This is better computed in the equivalent form

$$p(x) = \frac{2fP(f^2)}{Q(f^2) - fP(f^2)} \approx e^f - 1$$

where

$$\begin{aligned} P(t) = {} & \\ & 1.26183092834458542160 \cdot 10^{-4} t^2 \\ & +3.02996887658430129200 \cdot 10^{-2} t \\ & +1.00000000000000000000 \, , \end{aligned}$$

$$\begin{aligned} Q(t) = {} & \\ & 3.00227947279887615146 \cdot 10^{-6} t^3 \\ & +2.52453653553222894311 \cdot 10^{-3} t^2 \\ & +2.27266044198352679519 \cdot 10^{-1} t \\ & +2.00000000000000000005 \, . \end{aligned}$$

This approximation has a theoretical peak relative error of $2.4 \cdot 10^{-20}$. Finally,

$$e^x \approx 2^k (p(x) + 1) \, .$$

A C language library function **ldexp**(x, k) (see `ieee.c` in Chapter 1) can be used to find 2^k.

Underflow occurs for large negative x, and overflow occurs for large positive x. These two error thresholds may not be equally balanced about 1.0, depending on how the computer arithmetic is implemented. To avoid confusion among users of the program, it may be preferable to adopt a single value `MAXLOG` for the largest permissible logarithm, and restrict the allowable domain for the exponential function to $-$`MAXLOG` $\leq x \leq$ +`MAXLOG`. If the arithmetic supports denormalized tiny numbers, then the largest negative logarithm `MINLOG` will be significantly different from $-$`MAXLOG`. In this case the proper policy is to ensure that the routine will compute exp(`MINLOG`) and include an explanation of the concept of a denormal number in the user documentation.

[2]For most of the functions in this chapter the reader should consult also: Cody, William J., Jr. and William Waite, *Software Manual for the Elementary Functions*, Prentice-Hall, 1980.

Error amplification in the exponential function can be a serious matter. The error propagation from argument to result involves

$$\exp(x(1+\delta)) = \exp(x)(1 + x\delta + \ldots)$$

for a relative error δ in the argument x. Thus the relative error in x is multiplied by x in the function output. While the program can give an accurate result for arguments that are exactly represented by a computer number, the result contains amplified roundoff error if the argument is not exactly represented.

For *a priori* computation, the identity

$$e^x = \frac{1 + \tanh \frac{1}{2}x}{1 - \tanh \frac{1}{2}x}$$

is useful in conjunction with the following continued fraction:

$$\tanh x = \frac{x}{1+} \frac{x^2}{3+} \frac{x^2}{5+} \frac{x^2}{7+} \cdots .$$

This continued fraction has good stability and converges rapidly if $|x| < 1$.

In the complex plane, if $z = x + iy$, then by Euler's formula

$$e^z = e^x \cos y + i e^x \sin y.$$

This expression may be used directly in calculation of the complex exponential.

4.1.1 exp.c

```
#include "mconf.h"

#ifdef DEC

static short P[] = {

0035004,0050004,0016315,0134545,
0036770,0033415,0105201,0034462,
0040200,0000000,0000000,0000000
};
static short Q[] = {
0033511,0075304,0141275,0061006,
0036045,0071261,0155620,0021143,
0037550,0134156,0006512,0174363,
0040400,0000000,0000000,0000000
};
static short sc1[] = {
0040061,0100000,0000000,0000000};

#ifdef IIEEE

static short P[] = {

0xb72d,0x8399,0x8a00,0x3f20,
0x2726,0xb150,0x06e1,0x3f9f,
0x0000,0x0000,0x0000,0x3ff0
};
static short Q[] = {
0xac41,0x9857,0x2f58,0x3ec9,
0x044c,0x3b72,0xae56,0x3f64,
0x5f1e,0xc1a9,0x170d,0x3fcd,
0x0000,0x0000,0x0000,0x4000
};
static short sc1[] = {
0x0000,0x0000,0x3000,0x3fe6};
```

```
#define C1 (*(double *)sc1)
static short sc2[] = {
0035136,0100202,0161410,0062503};
#define C2 (*(double *)sc2)
#endif
```

```
#define C1 (*(double *)sc1)
static short sc2[] = {
0x0ca8,0x5c61,0xd010,0x3f2b};
#define C2 (*(double *)sc2)
#endif
```

```
static char fname[] = { "exp" };
extern double LOG2E, MAXLOG, MINLOG, MAXNUM;

double exp(x)
double x;
    {
    double px, qx, xx;
    int k;
    double polevl();
    double floor(), frexp(), ldexp();
```

Check for overflow. The routine **mtherr()** displays an error message comprising the function name and the type of error. It may also take other actions such as aborting the program depending on the severity of the error.

```
    if( x > MAXLOG)
        {
        mtherr( fname, OVERFLOW );
        return( MAXNUM );
        }
    if( x < MINLOG )
        {
        mtherr( fname, UNDERFLOW );
        return(0.0);
        }
```

Express $e^x = e^g 2^k = \exp(g + k \ln 2)$

```
    px = x * LOG2E;
    qx = floor( px + 0.5 );
    k = qx;
    x -= qx * C1;
    x += qx * C2;
```

Calculate $e^x - 1 = 2xP(x^2)/(Q(x^2) - xP(x^2))$

```
    xx = x * x;
    px = x * polevl( xx, P, 2 );
    x = px/( polevl( xx, Q, 3 ) - px );
    x = ldexp( x, 1 );
    x = x + 1.0;
    x = ldexp( x, k );
    return(x);
```

```
}
```

4.2 $\ln x$

To compute the natural logarithm function, the argument x is first separated into its exponent k and significand f. The logarithm may then be calculated from

$$\ln x = \ln(2^k f) = \ln f + \ln 2^k = \ln f + k \ln 2$$

where $\ln f$ is found by a rational approximation. The library function

$$f = \mathbf{frexp(x, \&k)}$$

may be used to find the integer k and floating point fraction f such that $x = 2^k f$ and $0.5 \le f < 1$.

Less accuracy is required of the expansion for $\ln f$ when the exponent k is large. On the other hand when $k = 0$, $\ln f$ should have the maximum possible accuracy. If $|k| > 2$ the form of the following expansion for $\ln x$ is very suitable.

$$\ln x = 2\left(\frac{x-1}{x+1}\right) + \frac{2}{3}\left(\frac{x-1}{x+1}\right)^3 + \frac{2}{5}\left(\frac{x-1}{x+1}\right)^5 + \dots .$$

This expansion is best for x near 1, so f should be transformed from the interval $[0.5, 1.0)$ to the interval $[1/\sqrt{2}, \sqrt{2})$ which is symmetrical about 1. This can be done by testing whether $f < 1/\sqrt{2}$. If it is, then multiply f by 2, setting

$$z = 2\,\frac{2f-1}{2f+1} = \frac{f-0.5}{0.5+0.5(f-0.5)} .$$

At the same time, subtract 1 from the exponent k so that the overall scale does not change. Otherwise, $1/\sqrt{2} \le f < 1$. In this case set

$$z = 2\,\frac{f-1}{f+1} = \frac{f-1}{0.5f+0.5} .$$

The constant needed is

$$\frac{\sqrt{2}}{2} = 0.70710678118654752440 .$$

Now approximate the logarithm of f by

$$\ln f \approx z + z^3 \frac{R(z^2)}{S(z^2)}$$

where

$R(t) =$
$-7.8958027888479915412 4 \cdot 10^{-1}t^2$
$+1.6386664569955807976 7 \cdot 10^{1}t$
$-6.4140995295871562295 1 \cdot 10^{1}$,

$S(t) =$
$1.0t^3$
$-3.5672279825632431254 9 \cdot 10^{1}t^2$
$+3.1209376637224418030 3 \cdot 10^{2}t$
$-7.6969194355046000860 4 \cdot 10^{2}$.

Finally, recombine $\ln f$ with the exponent by

$$\ln x \approx (\ln f - 0.000212194440054690582767 9k) + 0.693359375k \ .$$

This is a pseudo extended arithmetic calculation of $\ln f + k \ln 2$, as recommended by Cody and Waite (see the discussion under the section on the exponential function).

If x is close to 1, there is an alternative expansion that can be computed to a higher relative accuracy, though it requires a longer calculation. It is based on the series

$$\ln(1 + x) = x - \frac{1}{2}x^2 + \frac{1}{3}x^3 - \dots \ .$$

If the argument x is exact, there is no error in the first term, and only the rounding error of x^2 in the second term. The procedure is as follows. As before, test whether $f < \sqrt{2}/2$. If it is, then set

$$z = 2f - 1$$

and decrement k to preserve the scale. Otherwise, set

$$z = f - 1 \ .$$

Then use the rational approximation

$$\ln f \approx z - 0.5z^2 + z^3 \frac{P(z)}{Q(z)}$$

where

$P(t) =$
$+4.5848294845814344351 4 \cdot 10^{-5}t^6$
$+0.4985310672540507242 70t^5$
$+6.5631209376999287593 0t^4$
$+29.787742509798692589 1t^3$
$+60.612713446776725803 0t^2$
$+56.734928739175428548 7t$
$+19.889244657287407215 9$,

$Q(t) =$
$+1.0t^6$
$+15.031418263425000324 9t^5$
$+82.741044922243521702 1t^4$
$+220.66438498212192921 8t^3$
$+307.25418997953005826 3t^2$
$+214.95558669642294776 5t$
$+59.667733971862221630 0$.

The theoretical relative error of the rational form is $1.8 \cdot 10^{-18}$, when regarded as an approximation to $(\ln x - x + 0.5x^2)/(x^2)$. The last step is to recombine with the exponent as in the first method.

The power series in $(x - 1)/(x + 1)$, given above, is a fast expansion that is suitable for *a priori* computation of the logarithm. However the accuracy suffers slightly from rounding errors in computing $(x-1)/(x+1)$. An approach that seems slightly more accurate is to compute $\ln(1+x)$ using the continued fraction

$$\ln\left(\frac{1+w}{1-w}\right) = \frac{2w}{1-}\,\frac{w^2}{3-}\,\frac{4w^2}{5-}\,\frac{9w^2}{7-}\cdots$$

with the substitution

$$w = \frac{x}{x+2}\,.$$

This method was used in finding the approximation coefficients.

There are two procedural error conditions for the real logarithm function: a singularity at $x = 0$ and a domain error if $x < 0$.

If $z = x + iy$, $|z| = \sqrt{x^2 + y^2}$, then the complex logarithm is

$$\ln z = \ln|z| + i\tan^{-1}\frac{y}{x}\,,$$

where the arctangent ranges from $-\pi$ to $+\pi$.

4.2.1 log.c

```
#include "mconf.h"
```

$\ln(1+x) = x - 0.5x^2 + x^3 P(x)/Q(x)$, $1/\sqrt{2} <= x < \sqrt{2}$

```
#ifdef DEC
static short P[] = {
0034500,0046473,0051374,0135174,
0037777,0037566,0145712,0150321,
0040722,0002426,0031543,0123107,
0041356,0046513,0170752,0004346,
0041562,0071553,0023536,0163343,
0041542,0170221,0024316,0114216,
0041237,0016454,0046611,0104602
};
```

```
#ifdef IIEEE
static short P[] = {
0x974f,0x6a5f,0x09a7,0x3f08,
0x5a1a,0xd979,0xe7ee,0x3fdf,
0x74c9,0xc66c,0x40a2,0x401a,
0x411d,0x7e3d,0xc9a9,0x403d,
0xdcdc,0x64eb,0x4e6d,0x404e,
0xd312,0x2519,0x5e12,0x404c,
0x3130,0x89b1,0xe3a5,0x4033
};
```

The leading term 1.0 is omitted from the array Q.

```
static short Q[] = {
0041160,0100260,0067736,0102424,
0041645,0075552,0036563,0147072,
0042134,0125025,0021132,0025320,
0042231,0120211,0046030,0103271,
0042126,0172241,0052151,0120426,
0041556,0125702,0072116,0047103
};
#endif
```

```
static short Q[] = {
0xd0a2,0x0dfb,0x1016,0x402e,
0x79c7,0x47ae,0xaf6d,0x4054,
0x455a,0xa44b,0x9542,0x406b,
0x10d7,0x2983,0x3411,0x4073,
0x3423,0x2a8d,0xde94,0x406a,
0xc9c8,0x4e89,0xd578,0x404d
};
#endif
```

$\ln x = z + z^3 P(z)/Q(z)$, $z = 2(x-1)/(x+1)$, $1/\sqrt{2} <= x < \sqrt{2}$

```
#if DEC
static short R[12] = {
0140112,0020756,0161540,0072035,
0041203,0013743,0114023,0155527,
0141600,0044060,0104421,0050400,
};
```

```
#if IIEEE
static short R[12] = {
0x0e84,0xdc6c,0x443d,0xbfe9,
0x7b6b,0x7302,0x62fc,0x4030,
0x2a20,0x1122,0x0906,0xc050,
};
```

The leading term 1.0 is omitted from the array S.

```
static short S[12] = {
0141416,0130152,0017543,0064122,
0042234,0006000,0104527,0020155,
0142500,0066110,0146631,0174731,
};
#endif
```

```
static short S[12] = {
0x6d0a,0x43ec,0xd60d,0xc041,
0xe40e,0x112a,0x8180,0x4073,
0x3f3b,0x19b3,0x0d89,0xc088,
};
#endif
```

```
#define SQRTH 0.70710678118654752440
static char fname[] = {"log"};
extern double LOGE2, SQRT2, MAXLOG, MINLOG;

double log(x)
double x;
    {
    int e;
    short *q;
    double y, z;
    double frexp(), ldexp(), polevl(), plevl();
```

Test for domain errors.

```
    if( x <= 0.0 )
        {
        if( x == 0.0 )
            mtherr( fname, SING );
        else
            mtherr( fname, DOMAIN );
        return( MINLOG );
        }
```

Separate significand from exponent

```
    x = frexp( x, &e );
```

Logarithm using $\ln x \approx z + z^3 P(z)/Q(z)$,
where $z = 2(x-1)/x+1)$

```
    if( (e > 2) || (e < -2) )
        {
        if( x < SQRTH )
            {
            e -= 1;
            z = x - 0.5;
```

```
                y = 0.5 * z + 0.5;
                }
        else
                {
                z = x - 0.5;
                z -= 0.5;
                y = 0.5 * x + 0.5;
                }
        x = z / y;
        z = x*x;
        z = x + x * ( z * polevl( z, R, 2 )
                / p1evl( z, S, 3 ) );
        goto ldone;
        }
```

Logarithm using $\ln(1+x) = x - .5x^2 + x^3P(x)/Q(x)$

```
if( x < SQRTH )
        {
```

Compute $2x - 1$

```
        e -= 1;
        x = ldexp( x, 1 ) - 1.0;
        }
else
        {
        x = x - 1.0;
        }
z = x*x;
y = x * ( z * polevl( x, P, 6 ) / p1evl( x, Q, 6 ) );
y = y - ldexp( z, -1 ); y - 0.5 * z
z = x + y;
```

Recombine with exponent term

```
ldone:
        if( e != 0 )
                {
                y = e;
                z = z - y * 2.121944400546905827679e-4;
                z = z + y * 0.693359375;
                }
        return( z );
        }
```

4.3 Argument Transformation for Circular Functions

Since practical expansions for the trigonometric functions apply only to a very limited span of the argument, a transformation is required that reduces the argument to the interval of approximation. For the method described here, a reduction to the interval from 0 to $\pi/4$ is required. It is also necessary to know the octant between 0 and 2π in which the argument lies.

In the following, $[x]$ denotes the largest integer that is less than or equal to x. This should be computed with the largest possible dynamic range, preferably by operating directly on the floating point number to produce a floating point integer valued result. Supplementary routines, variously named `AINT` (in Fortran and Basic) or **floor** (in C), are available for this purpose. Therefore it is not necessary to limit the argument to the size of the computer's integer variable.

The computer routine for the odd functions $\sin x$, $\tan x$, $\cot x$ may begin by saving the sign and taking the absolute value of the argument x. This does not improve the approximation accuracy, but it does simplify the range reduction. To reduce the argument to the interval between 0 and $\pi/4$ first find the integer part of $4x/\pi$ by

$$y = \left[\frac{x}{\pi/4}\right]$$

using the **floor()** function to get an integer valued floating point result y. Then y can be converted without overflow to an integer variable j that is equal to the correct octant by

$$j = y - 2^3[2^{-3}y].$$

Consideration of the odd octants 1, 3, 5, 7 may be eliminated by testing if j is odd. If it is, then add 1 both to j (mod 8) and to y. This operation has the benefit of mapping all function zeros to the origin.

In the case of the sine and cosine functions the next step is to reflect angles greater than π about the origin. If j is greater than 3 then subtract 4 from j and invert the sign of the final function value. For the tangent and cotangent this step is omitted, but a further adjustment will be made after computing the function approximation (see below).

For the sine and cosine routines, j is now either 0 or 2. The sine expansion will be needed if x is nearest to 0 or π (i.e., if $j = 0$), whereas the cosine expansion will be used if x is nearest $\pi/2$ or $3\pi/2$ (i.e., $j = 2$).

From the integer valued y, the argument z of the trigonometric approximation may be computed accurately by

$$\begin{aligned} z &= x \bmod \pi/4 \\ &= x - \frac{\pi}{4}y \\ &= ((x - p_1 y) - p_2 y) - p_3 y\,. \end{aligned}$$

An extended precision arithmetic is implemented, in which

$$\pi/4 = p_1 + p_2 + p_3$$

where $p_1 >> p_2 >> p_3$. The specific values of p_i depend on the particular computer arithmetic; p_1 and p_2 may be chosen so that a little more than half of the lower digits or bits of the significand are zero. Then the products $p_1 y$ and $p_2 y$ are exact for y up to about $\sqrt{1/\mathtt{MACHEP}}$. In IEEE double precision, this will yield a result accurate to nearly full machine precision for x up to $1.073741824 \cdot 10^9$. Suitable choices for p_i, in decimal and hexadecimal array eqivalents, are indicated in the program listings.

The exactness of the range reduction can be confusing to the user, since one might expect $\tan \mathtt{PI}/2$, where `PI` is some computer number, to be infinite, or $\sin \mathtt{PI}$ to be zero. However, the program correctly computes the tangent or sine of the number given it.

The extended precision arithmetic fails if the partial results are not exact. In the IEEE example partial loss of significance occurs for $x > 2^{30}$. The loss is not gradual, but jumps rather suddenly to about 1 part in 10^7. The computed result may be meaningless for $x > 2^{49} = 5.6 \cdot 10^{14}$, beyond which the absolute error of dividing by $\pi/4$ approaches or exceeds 1. The computer routine should generate warnings or error signals for these conditions.

4.4 Sine and cosine

Fundamental power series expansions for the circular sine and cosine are

$$\begin{aligned} \sin x &= x - \frac{x^3}{3!} + \frac{x^5}{5!} - \frac{x^7}{7!} + \dots \\ \cos x &= 1 - \frac{x^2}{2!} + \frac{x^4}{4!} - \frac{x^6}{6!} + \dots . \end{aligned}$$

These are suitable for *a priori* computation, as there is no cancellation when $|x| < \pi/4$ even though the terms alternate in sign. Convergence is a bit slow, however, and can be speeded up only by reducing the range to an extremely small interval about zero.

The polynomial approximations for the sine and cosine in the range $[-\pi/4, +\pi/4]$ are

$$\begin{aligned} \sin x &\approx x + x^3 S(x^2) \\ \cos x &\approx 1 - x^2 C(x^2) \end{aligned}$$

where

$$S(t) = 1.58962301576546568060 \cdot 10^{-10}t^5 - 2.50507477628578072866 \cdot 10^{-8}t^4 + 2.75573136213857245213 \cdot 10^{-6}t^3 - 1.98412698295895385996 \cdot 10^{-4}t^2 + 8.33333333332211858878 \cdot 10^{-3}t - 1.66666666666666307295 \cdot 10^{-1},$$

$$C(t) = 1.13678171380010505367 \cdot 10^{-11}t^6 - 2.08758833757646780967 \cdot 10^{-9}t^5 + 2.75573155429816368675 \cdot 10^{-7}t^4 - 2.48015872936186303093 \cdot 10^{-5}t^3 + 1.38888888888806666760 \cdot 10^{-3}t^2 - 4.16666666666666348141 \cdot 10^{-2}t + 4.99999999999999999798 \cdot 10^{-1}.$$

The approximation for $\sin x$ has a relative error of $3.6 \cdot 10^{-18}$. The expression $x^2C(x^2)$ has relative error $4.0 \cdot 10^{-19}$, considered as an approximation to $1 - \cos x$. The first peak in the error curve of $tC(t)$ is essentially at $t = 0$, but the apparent error goes to zero in finite precision arithmetic. This makes it difficult to level the peaks in the error curve correctly. A 100-decimal arithmetic was used to calculate the coefficients of $C(t)$, with error peaks leveled to a part in 10^{23}. The sine approximation was also computed in 100-decimal arithmetic. However, the extra precision has no practical effect on the accuracy of the computer program since the relative contribution of the higher degree terms of the polynomial is small.

4.4.1 sin.c

```
#include "mconf.h"

#ifdef UNK
static double DP1 = 7.85398125648498535156E-1;
static double DP2 = 3.77489470793079817668E-8;
static double DP3 = 2.69515142907905952645E-15;
static double LOSSTH = 1.073741824e9;
#endif

#ifdef DEC
static short sincof[] = {
0030056,0143750,0177214,0163153,
0131727,0027455,0044510,0175352,
0033470,0167432,0131752,0042414,
0135120,0006400,0146776,0174027,
0036410,0104210,0104207,0137202,
0137452,0125252,0125252,0125103
};
```

```
#ifdef IIEEE
static short sincof[] = {
0x9ccd,0x1fd1,0xd8fd,0x3de5,
0x1f5d,0xa929,0xe5e5,0xbe5a,
0x48a1,0x567d,0x1de3,0x3ec7,
0xdf03,0x19bf,0x01a0,0xbf2a,
0xf7d0,0x1110,0x1111,0x3f81,
0x5548,0x5555,0x5555,0xbfc5
};
```

Values for extended precision reduction modulo $\pi/4$

$7.85398162901401519775 39 \cdot 10^{-1}$

```
static short P1[] =
0040111,0007732,0120000,0000000};
```

$4.96046786979675857764 9 \cdot 10^{-10}$

```
static short P2[] =
0030410,0055060,0100000,0000000};
```

$7.85398125648498535156 \cdot 10^{-1}$

```
static short P1[] = {
0x0000,0x4000,0x21fb,0x3fe9};
```

$3.77489470793079817668 \cdot 10^{-8}$

```
static short P2[] = {
0x0000,0x0000,0x442d,0x3e64};
```

```
2.86059436305491589838 · 10^-18
static short P3[] =
0021523,0011431,0105056,0001560};
#endif
```

```
2.69515142907905952645 · 10^-15
static short P3[] = {
0x5170,0x98cc,0x4698,0x3ce8};
#endif
```

```
#define DP1 *(double *)P1
#define DP2 *(double *)P2
#define DP3 *(double *)P3
static double LOSSTH = 1.073741824e9;
```

```
#if DEC
static short coscof[28] = {
0027107,0176030,0153276,0031137,
0131017,0072476,0007450,0105310,
0032623,0171174,0070066,0146400,
0134320,0006400,0147355,0163313,
0035666,0005540,0133012,0165067,
0137052,0125252,0125252,0125206,
0040000,0000000,0000000,0000000,
};
#endif
```

```
#if IIEEE
static short coscof[28] = {
0xc64c,0x1ad7,0xff83,0x3da8,
0x1159,0xc1e5,0xeea7,0xbe21,
0xd9a0,0x8e06,0x7e4f,0x3e92,
0xbcd9,0x19dd,0x01a0,0xbefa,
0x5d47,0x16c1,0xc16c,0x3f56,
0x5551,0x5555,0x5555,0xbfa5,
0x0000,0x0000,0x0000,0x3fe0,
};
#endif
```

```
extern double PIO4;

double sin(x)
double x;
	{
	double y, z, zz;
	int rflg, j, sign;
	double polevl(), floor(), ldexp();

	sign = 1;
	if( x < 0 )
		{
		x = -x;
		sign = -1;
		}
	if( x > lossth )
		{
		mtherr( "sin", TLOSS );
		return(0.0);
		}
	y = floor( x/PIO4 );
```

Strip high bits of integer part to prevent integer overflow

```
	z = ldexp( y, -4 );
	z = floor(z);
	z = y - ldexp( z, 4 );
```

```
    j = z;
    if( j & 1 )
        {
        j += 1;
        y += 1.0;
        }
    j = j & 07;
    if( j > 3)
        {
        sign = -sign;
        j -= 4;
        }
    z = ((x - y * DP1) - y * DP2) - y * DP3;
    zz = z * z;
    if( (j==2) || (j==1) )
        {
        y = 1.0 - zz * polevl( zz, coscof, 6 );
        }
    else
        {
        y = z + z * (zz * polevl( zz, sincof, 5 ));
        }
    if(sign < 0)
        y = -y;
    return(y);
    }
```

4.4.2 cos.c

The cosine program is the same as the sine program, except that the sine and cosine approximations are interchanged.

```
double cos(x)
double x;
    {
    double y, z, zz;
    long i;
    int j, sign, refl;
    double polevl(), floor(), ldexp();

    sign = 1;
    if( x < 0 )
        x = -x;
    if( x > lossth )
```

```
        {
        mtherr( "cos", TLOSS );
        return(0.0);
        }
    y = floor( x/PIO4 );
    z = ldexp( y, -4 );
    z = floor(z);
    z = y - ldexp( z, 4 );
    i = z;
    if( i & 1 )
        {
        i += 1;
        y += 1.0;
        }
    j = i & 07;
    if( j > 3)
        {
        j -=4;
        sign = -sign;
        }
    if( j > 1 )
        sign = -sign;
    z = ((x - y * DP1) - y * DP2) - y * DP3;
    zz = z * z;
    if( (j==1) || (j==2) )
        {
        y = z + z * (zz * polevl( zz, sincof, 5 ));
        }
    else
        {
        y = 1.0 - zz * polevl( zz, coscof, 6 );
        }
    if(sign < 0)
        y = -y;
    return(y);
    }
```

4.5 Tangent and Cotangent

For the tangent and cotangent on the interval $|x| < \pi/4$ the following rational approximation may be employed:

$$\tan x \approx x + x^3 \frac{P(x^2)}{Q(x^2)}$$

where

$P(t) =$
$-1.30936939181383777646 \cdot 10^4 t^2$
$+1.15351664838587416140 \cdot 10^6 t$
$-1.79565251976484877988 \cdot 10^7$,

$Q(t) =$
$1.0t^4$
$+1.36812963470692954678 \cdot 10^4 t^3$
$-1.32089234440210967447 \cdot 10^6 t^2$
$+2.50083801823357915839 \cdot 10^7 t$
$-5.38695755929454629881 \cdot 10^7$.

This has a theoretical relative error of $2.0 \cdot 10^{-20}$.

After computing the approximation the integer octant variable j must be examined (see the earlier section on range reduction).

If j is either 2 or 6, then the function value should be negated in the case of the cotangent or in the case of the tangent it should be inverted and negated.

Otherwise j is 0 or 4, and the value should be inverted for the cotangent or left alone for the tangent.

For all the odd functions the final result must be multiplied by the previously saved sign of the argument. The cases can be made clear by sketching a diagram of the functions.

In all usual computer arithmetics there is no exact representation of the number π. Since the reduction mod $\pi/4$ is more exact than the machine precision, the tangent program with argument in radians *cannot produce an overflow condition.* An error escape is required only for the cotangent of $x = 0$.

For *a priori* calculation, a suitable continued fraction is

$$\tan x = \frac{x}{1-} \frac{x^2}{3-} \frac{x^2}{5-} \frac{x^2}{7-} \cdots .$$

Formulas for the complex tangent and cotangent are given in the next section.

4.5.1 tan.c

```
#include "mconf.h"

#ifdef DEC
static short P[] = {
0143514,0113306,0111171,0174674,
0045214,0147545,0027744,0167346,
0146210,0177526,0114514,0105660
};
```

```
#ifdef IIEEE
static short P[] = {
0x3f38,0xd24f,0x92d8,0xc0c9,
0x9ddd,0xa5fc,0x99ec,0x4131,
0x9176,0xd329,0x1fea,0xc171
};
```

The leading coefficient 1.0 is omitted from the array Q.

```
static short Q[] = {
0043525,0142457,0072633,0025617,
0145241,0036742,0140525,0162256,
0046276,0146176,0013526,0143573,
0146515,0077401,0162762,0150607
```

```
static short Q[] = {
0x6572,0xeeb3,0xb8a5,0x40ca,
0xbc96,0x582a,0x27bc,0xc134,
0xd8ef,0xc2ea,0xd98f,0x4177,
0x5a31,0x3cbe,0xafe0,0xc189
```

```
};                                      };
Coefficients for reduction mod π/4
7.853981629014015197753 9·10^-1        7.85398125648498535156·10^-1
static short P1[] =                     static short P1[] = {
0040111,0007732,0120000,0000000        0x0000,0x4000,0x21fb,0x3fe9
};                                      };
4.960467869796758577649·10^-10         3.77489470793079817668·10^-8
static short P2[] =                     static short P2[] = {
0030410,0055060,0100000,0000000        0x0000,0x0000,0x442d,0x3e64
};                                      };
2.860594363054915898381 0^-18          2.69515142907905952645·10^-15
static short P3[] =                     static short P3[] = {
0021523,0011431,0105056,0001560        0x5170,0x98cc,0x4698,0x3ce8
};                                      };
#define DP1 *(double *)P1
#define DP2 *(double *)P2
#define DP3 *(double *)P3
static double LOSSTH = 1.073741824e9;
#endif                                  #endif
#ifdef UNK
static double DP1 = 7.853981554508209228515625E-1;
static double DP2 = 7.94662735614792836714E-9;
static double DP3 = 3.06161699786838294307E-17;
static double LOSSTH = 1.073741824e9;
#endif


extern double MAXNUM;
extern double PIO4;

double tan(x)
double x;
    {
    double tancot();

    return( tancot(x,0) );
    }


double cot(x)
double x;
    {
    double tancot();

    if( x == 0.0 )
        {
        mtherr( "cot", SING );
```

```
        return( MAXNUM );
        }
    return( tancot(x,1) );
    }


static double tancot( xx, cotflg )
double xx;
int cotflg;
    {
    double x, y, z, zz;
    int j, sign;
    double polevl(), p1evl(), floor(), ldexp();

    if( xx < 0 )
        {
        x = -xx;
        sign = -1;
        }
    else
        {
        x = xx;
        sign = 1;
        }
    if( x > lossth )
        {
        if( cotflg )
            mtherr( "cot", TLOSS );
        else
            mtherr( "tan", TLOSS );
        return(0.0);
        }
```

Compute x mod $\pi/4$

```
    y = floor( x/PIO4 );
    z = ldexp( y, -3 );
    z = floor(z);
    z = y - ldexp( z, 3 );
    j = z;
```

Map zeros and singularities to origin.

```
    if( j & 1 )
        {
        j += 1;
        y += 1.0;
        }
    z = ((x - y * DP1) - y * DP2) - y * DP3;
```

```
zz = z * z;
if( zz > 1.0e-14 )
     y = z+z * (zz * polevl( zz, P, 2 )/p1evl(zz, Q, 4));
else
     y = z;
if( j & 2 )
     {
     if( cotflg )
          y = -y;
     else
          y = -1.0/y;
     }
else
     {
     if( cotflg )
          y = 1.0/y;
     }
if( sign < 0 )
     y = -y;
return( y );
}
```

4.6 Complex Circular Functions

In the complex plane, formulas for the sine and cosine functions of $z = x+iy$ are

$$\begin{aligned}\sin z &= \sin x \cosh y + i \cos x \sinh y \\ \cos z &= \cos x \cosh y - i \sin x \sinh y \ .\end{aligned}$$

These may be used directly for computation, though it may be desirable to check for potential overflow conditions. An improvement in computational speed can be achieved by calculating both sinh and cosh with only one call to the exponential function, provided that $|x| > 0.5$, as follows:

$$\begin{aligned}f &= e^x \\ g &= 0.5/f \\ h &= 0.5f \\ \sinh x &= h - g \\ \cosh x &= h + g \ .\end{aligned}$$

When $|x| \leq 0.5$ the rational approximations for sinh and cosh must be used.

Analytically, the tangent and cotangent are given by

$$\begin{aligned}\tan z &= \frac{\sin 2x + i \sinh 2y}{\cos 2x + \cosh 2y} \\ \cot z &= \frac{\sin 2x - i \sinh 2y}{\cosh 2y - \cos 2x}\end{aligned} .$$

On the real axis the denominator is zero at odd or even multiples of $\pi/2$. Special care is required to evaluate the denominator series near these points. If the absolute value of the denominator is less than 0.25, the following procedure may be used to improve the relative accuracy. The first step is to subtract the nearest multiple of π from $2x$. This can be done with extended precision by

$$\begin{aligned}w &= \begin{cases} x/\pi + 0.5, & x \geq 0 \\ x/\pi - 0.5, & x < 0 \end{cases} \\ k &= [w] \\ t &= ((x - 3.14159265160560607910k) \\ &\quad -1.98418714791870343106 \cdot 10^{-9}k) \\ &\quad -1.14423774522196636802 \cdot 10^{-17}k.\end{aligned}$$

Setting $u = 2y$, and with t as just computed, an expansion for the denominator is

$$\cosh 2y - \cos 2x = \sum_{k=1}^{\infty} \frac{u^{2k} + (-1)^{k-1} t^{2k}}{(2k)!} .$$

4.7 $\sin^{-1} x$

Like $\sin x$, the principal value of $\arcsin x = \sin^{-1} x$ is an odd function. A basic power series expansion is

$$\sin^{-1} z = z + \frac{z^3}{2 \cdot 3} + \frac{1 \cdot 3 z^5}{2 \cdot 4 \cdot 5} + \frac{1 \cdot 3 \cdot 5 z^7}{2 \cdot 4 \cdot 6 \cdot 7} + \dots .$$

In double precision arithmetic $\sin^{-1} x \approx x$ if $|x| < 10^{-7}$. For $|x|$ up to 0.5, use the rational approximation

$$\sin^{-1} x \approx x + x^3 \frac{P(x^2)}{Q(x^2)}$$

where

$$P(t) = -0.69682259994868617421 7t^4 +10.152982860898720997 22t^3 -39.734077139157875729 4t^2 +57.291214470984649613 4t -27.414820046592570802 0 ,$$

$$Q(t) = 1.0t^5 -23.836824500517748824 2t^4 +151.095072703128995631t^3 -382.340216045978957023t^2 +417.767300951716199422t -164.488920279555473283 .$$

This has a theoretical relative error of $9.1 \cdot 10^{-20}$. As x approaches 1, $\sin^{-1} x$ gradually approaches $\pi/2$ according to

$$\sin^{-1} x = \frac{\pi}{2} - 2 \sin^{-1} \sqrt{\frac{1-x}{2}} .$$

This relation is derived from the identity

$$\sin^{-1} A \pm \sin^{-1} B = \sin^{-1} \left(A\sqrt{1-B^2} \pm B\sqrt{1-A^2} \right) .$$

The transformation should be used for range reduction if $|x| > 0.5$. For real x the function is undefined if $|x| > 1$; this case requires an error escape.

4.7.1 asin.c

```
#include "mconf.h"

#ifdef DEC
static short P[] = {
0140062,0061367,0042744,0127464,
0041042,0107250,0070611,0005470,
0141436,0167661,0165345,0131200,
0041545,0025064,0020124,0000411,
0141333,0050615,0026056,0134351
};
```

Leading coefficient 1.0 is omitted from Q.

```
static short Q[] = {
0141276,0130721,0005461,0134336,
0042027,0014126,0127506,0127155,
0142277,0025614,0031413,0103353,
0042320,0161066,0165346,0163707,
0142044,0076451,0160443,0005274
};
#endif

#ifdef IIEEE
static short P[] = {
0x95e7,0xe8bc,0x4c5e,0xbfe6,
0x2167,0x0e31,0x51d5,0x4024,
0xb650,0x3d5c,0xddf6,0xc043,
0x8021,0x840a,0xa546,0x404c,
0xd71d,0xa585,0x6a31,0xc03b
};
static short Q[] = {
0x371c,0x2166,0xd63a,0xc037,
0xd5ce,0xd5e8,0xe30a,0x4062,
0x70dd,0x8661,0xe571,0xc077,
0xdcf9,0xdd5c,0x1c46,0x407a,
0x6158,0x3c24,0x8fa5,0xc064
};
#endif


double asin(x)
double x;
    {
    double a, p, z, zz;
```

```
double sqrt(), polevl(), p1evl();
short sign, flag;
extern double PIO2;

if( x > 0 )
     {
     sign = 1;
     a = x;
     }
else
     {
     sign = -1;
     a = -x;
     }
if( a > 1.0 )
     {
     mtherr( "asin", DOMAIN );
     return( 0.0 );
     }
if( a < 1.0e-7 )
     {
     z = a;
     goto done;
     }
```

$\sin^{-1} x = \pi/2 - 2\sin^{-1}\sqrt{\frac{1}{2}(1-x)}$

```
if( a > 0.5 )
     {
     zz = 0.5 -a;
     zz = (zz + 0.5)/2.0;
     z = sqrt( zz );
     flag = 1;
     }
else
     {
     z = a;
     zz = z * z;
     flag = 0;
     }
p = zz * polevl( zz, P, 4)/p1evl( zz, Q, 5);
z = z * p + z;
if( flag != 0 )
     {
     z = z + z;
     z = PIO2 - z;
```

```
          }
done:
     if( sign < 0 )
          z = -z;
     return(z);
     }
```

4.8 $\cos^{-1} x$

Analytically, the arccosine is given by

$$\cos^{-1} x = \frac{\pi}{2} - \sin^{-1} x.$$

This expression may be used for computation when $|x| < 0.5$. But if $|x|$ is near 1 there is cancellation error in subtracting $\sin^{-1} x$ from $\pi/2$. In this region the identities

$$\cos^{-1} x = \pi - 2\sin^{-1}\sqrt{\frac{1+x}{2}}$$

for $x < -0.5$, and

$$\cos^{-1} x = 2\sin^{-1}\sqrt{\frac{1-x}{2}},$$

for $x > +0.5$, should be employed. The arcsine routine described in the previous section may be used with the argument as transformed by these expressions.

4.8.1 acos.c

```
extern double PIO2, PI;

double acos(x)
double x;
     {
     double asin(), sqrt();

     if( x < -1.0 )
          goto domerr;
cos^-1 x = π - 2 sin^-1 √(½(1 + x))
     if( x < -0.5)
          return( PI - 2.0 * asin( sqrt(0.5*(1.0+x)) ) );
     if( x > 1.0 )
          {
domerr:   mtherr( "acos", DOMAIN );
```

```
        return( 0.0 );
        }
```

$\cos^{-1} x = 2\sin^{-1}\sqrt{\tfrac{1}{2}(1-x)}$

```
    if( x > 0.5 )
        return( 2.0 * asin( sqrt(0.5*(1.0-x) ) ) );
    return( PIO2 - asin(x) );
    }
```

4.9 $\tan^{-1} x$

Computation of $\tan^{-1} x$ may use the following transformations to reduce the argument to the interval from 0 to $\tan \pi/8$.

$$\tan^{-1} x = \frac{\pi}{2} + \tan^{-1}\left(-\frac{1}{x}\right)$$

$$\tan^{-1} x = \frac{\pi}{4} + \tan^{-1}\left(\frac{x-1}{x+1}\right)$$

These are based on the identity

$$\tan^{-1} a \pm \tan^{-1} b = \tan^{-1}\left(\frac{a \pm b}{1 \mp ab}\right).$$

Since $\tan^{-1} x$ is an odd function, the first step of the computer program may be to save the sign of x and to replace x by its absolute value. One of the above transformations may then be applied. In doing so, an auxiliary offset value c is established; this will be added to the arctangent of the reduced argument. The possibilities are

1. If $x > \tan(3\pi/8)$, then replace x by $-1/x$ and set $c = \pi/2$.
2. If $\tan(3\pi/8) \geq x > \tan(\pi/8)$, then replace x by $(x-1)/(x+1)$ and set $c = \pi/4$.
3. Otherwise, set $c = 0$ and leave x alone.

Then insert c and the transformed value of x into the following rational approximation:

$$\tan^{-1} x \approx c + x + x^3 \frac{P(x^2)}{Q(x^2)}$$

where

$P(t) =$
$-0.8409808780644997160 01t^3$
$-8.838608370237723942 79t^2$
$-21.847621308131670572 4t$
$-14.830705034043894699 3$,

$Q(t) =$
$1.0t^4$
$+15.497412467530726755 2t^3$
$+62.790655576265301726 3t^2$
$+92.238132985621440648 5t$
$+44.492115102131943846 5$.

x	y	Action
≥ 0	0	return 0
0	> 0	return $\pi/2$
< 0	0	return π
0	< 0	return $3\pi/2$
> 0	> 0	set $w = 0$
< 0	$\neq 0$	set $w = \pi$
> 0	< 0	set $w = 2\pi$

Table 4.1: **atan2()** Cases

The theoretical relative error is $3.0 \cdot 10^{-18}$. Finally, the saved sign of the input argument must be restored by negating the output if the original x was negative. The constants of interest are

$$\begin{aligned}\tan(3\pi/8) &= \sqrt{2}+1 = 2.41421356237309504880\\ \tan(\pi/4) &= 1\\ \tan(\pi/8) &= \sqrt{2}-1 = 0.41421356237309504880\ .\end{aligned}$$

Note that the range reduction algorithm transforms $x = 1$ to $x = 0$ with $c = \pi/4$, so $\tan^{-1} 1 = \pi/4$ to full machine accuracy. For a priori computation use the continued fraction

$$\tan^{-1} z = \frac{z}{1+}\,\frac{z^2}{3+}\,\frac{4z^2}{5+}\,\frac{9z^2}{7+}\,\frac{16z^2}{9+\cdots}$$

together with reduction of the range at least to the interval $[0, \tan \pi/4]$. Convergence of the continued fraction is much faster with range reduction to $\tan(\pi/8)$, but this means that full precision values of the constants must be available.

A quadrant correct arctangent, usually bearing a name such as **atan2()**, is a function of the two perpendicular sides x and y of a right triangle. These may have positive, zero, or negative length. Therefore several cases must be considered, according to Table 4.1. In the last three cases of the table the program returns $w + \tan^{-1}(y/x)$. This yields a result, in radians, between 0 and 2π. Conventions seem to vary regarding the order of the subroutine arguments and the range of the result. Sometimes it is from $-\pi$ to π, as in the ANSI C language standard.

4.9.1 atan.c

```
#include "mconf.h"

#ifdef DEC
static short P[] = {
0140127,0045205,0153731,0030027,
0141015,0065360,0116105,0025210,
0141256,0143755,0127056,0105716,
0141155,0045221,0056235,0072437
};
```

```
#ifdef IIEEE
static short P[] = {
0x2603,0xbafb,0xe950,0xbfea,
0xa551,0x1388,0xad5e,0xc021,
0xd17a,0xb5c5,0xd8fd,0xc035,
0xaea4,0x2b93,0xa952,0xc02d
};
```

The leading coefficient 1.0 of Q is omitted from the array.

```
static short Q[] = {
0041167,0172546,0143212,0126224,
0041573,0024641,0116611,0153210,
0041670,0074754,0110422,0127624,
0041461,0173755,0002566,0014374
};
```

```
static short Q[] = {
0x5592,0xd8d1,0xfeac,0x402e,
0x3ad1,0x33b1,0x6534,0x404f,
0x55f2,0x9222,0x0f3d,0x4057,
0xc31f,0xa0ae,0x3efd,0x4046
};
```

$\tan 3\pi/8 = \sqrt{2}+1$

```
static short T3P8A[] = {
040432,0101171,0114774,0167462,};
#define T3P8 *(double *)T3P8A
```

```
static short T3P8A[] = {
0x9de6,0x333f,0x504f,0x4003};
#define T3P8 *(double *)T3P8A
```

$\tan \pi/8 = \sqrt{2}-1$

```
static short TP8A[] =
037724,011714,0147747,074622,};
#define TP8 *(double *)TP8A
#endif
```

```
static short TP8A[] = {
0xef32,0x99fc,0x8279,0x3fda};
#define TP8 *(double *)TP8A
#endif
```

```
double atan(x)
double x;
    {
    extern double PIO2, PIO4;
    double y,z,p,q;
    double polevl(), p1evl();
    short sign;

    sign = 1;
    if( x < 0.0 )
        {
        sign = -1;
        x = -x;
        }
    if( x > T3P8 )
        {
        y = PIO2;
```

```
        x = -( 1.0/x );
        }
else if( x > TP8 )
                {
                y = PIO4;
                x = (x-1.0)/(x+1.0);
                }
        else
                y = 0.0;
z = x * x;
y = y + ( polevl( z, P, 3 ) / p1evl( z, Q, 4 ) )
        * z * x + x;
if( sign < 0 )
        y = -y;
return(y);
}
```

4.9.2 atan2.c

Quadrant correct arctangent. This version has a result that ranges from 0 to 2π. In some systems, including ANSI standard C language, the arguments x and y occur in the opposite order and the result ranges from $-\pi$ to π.

```
extern double PI, PIO2;

double atan2( x, y )
double x, y;
        {
        double z, w;
        short code;
        double atan();

        code = 0;
        if( x < 0.0 )
                code = 2;
        if( y < 0.0 )
                code |= 1;
        if( x == 0.0 )
                {
                if( code & 1 )
                        return( 3.0*PIO2 );
                if( y == 0.0 )
                        return( 0.0 );
                return( PIO2 );
                }
```

```
if( y == 0.0 )
     {
     if( code & 2 )
          return( PI );
     return( 0.0 );
     }
switch( code )
     {
     case 0:  w = 0.0; break;
     case 1:  w = 2.0 * PI; break;
     case 2:
     case 3:  w = PI; break;
     }
z = atan( y/x );
return( w + z );
}
```

4.10 Complex Inverse Circular Functions

Versions of the inverse functions for complex arguments can be constructed from the following identities.

$$\begin{aligned}
\sin^{-1} z &= -i\ln(iz+\sqrt{1-z^2}) \\
\cos^{-1} z &= \pi/2 - \sin^{-1} z \\
\tan^{-1} z &= \frac{1}{2}\tan^{-1}\left(\frac{2x}{1-x^2-y^2}\right) + k\pi + i\frac{1}{4}\ln\left(\frac{x^2+(y+1)^2}{x^2+(y-1)^2}\right) .
\end{aligned}$$

The arctangent requires error escapes in case either of the denominators in the formula is equal to zero, i.e., if $z^2 = -1$. The real part of the arctangent may be reduced by subtracting the nearest multiple of π. The arcsine of $z = x + iy$ is undefined if $y = 0$ and $x > 1$. For z near the origin, the power series given earlier for $\sin^{-1} z$ should be used, substituting the identity

$$z^2 = x^2 - y^2 - 2ixy$$

into the Taylor series.

4.11 $\sinh x$

Analytically, the hyperbolic sine is defined by

$$\sinh x = \frac{e^x - e^{-x}}{2} .$$

The program for this computes

$$
\begin{aligned}
w &= e^{|x|} \\
y &= 0.5w - 0.5/w \ .
\end{aligned}
$$

Then $\sinh x = y$ if $x \geq 0$, or $-y$ if x is negative. The exponential term whose argument is positive dominates the result. If $x < 0$, this woud be e^{-x}, corresponding to $0.5/w$. The effect of the rounding error from dividing is made smaller by ensuring that it is always the other term, $0.5w$, that dominates. This is the reason for using $|x|$ instead of x.

When x is near zero, both the terms of the numerator are nearly equal, so there will be severe cancellation error. Therefore if $|x| \leq 1$ use the following rational approximation:

$$
\sinh x \approx x + x^3 \frac{P(x^2)}{Q(x^2)}
$$

where

$P(t) =$	$Q(t) =$
$-7.89474443963537015605 \cdot 10^{-1} t^3$	$+1.0 t^3$
$-1.63725857525983828727 \cdot 10^{2} t^2$	$-2.77711081420602794433 \cdot 10^{2} t^2$
$-1.15614435765005216044 \cdot 10^{4} t$	$+3.61578279834431989373 \cdot 10^{4} t$
$-3.51754964808151394800 \cdot 10^{5}$,	$-2.11052978884890840399 \cdot 10^{6}$.

The theoretical relative error is $3.4 \cdot 10^{-20}$. There will be overflow for $|x|$ near the maximum size logarithm possible in the arithmetic. An overflow error escape is required for this.

The power series

$$
\sinh x = x + \frac{x^3}{3!} + \frac{x^5}{5!} + \frac{x^7}{7!} + \ldots
$$

may be used for *a priori* computation when x is small.

4.11.1 sinh.c

```
#include "mconf.h"

#ifdef DEC                                  #ifdef IIEEE
static short P[] = {                        static short P[] = {
0140112,0015377,0042731,0163255,            0x3cd6,0xe8bb,0x435f,0xbfe9,
0142043,0134721,0146177,0123761,            0xf4fe,0x398f,0x773a,0xc064,
0143464,0122706,0034353,0006017,            0x6182,0xc71d,0x94b8,0xc0c6,
0144653,0140536,0157665,0054045             0xab05,0xdbf6,0x782b,0xc115
};                                          };
```

Leading coefficient 1.0 is omitted from Q.

```
static short Q[] = {                        static short Q[] = {
0142212,0155404,0133513,0022040,            0x6484,0x96e9,0x5b60,0xc071,
```

```
0044015,0036723,0173271,0011053,
0145400,0150407,0023710,0001034
};
#endif
```

```
0x2245,0x7ed7,0xa7ba,0x40e1,
0x0044,0xe4f9,0x1a20,0xc140
};
#endif
```

```
extern double MAXNUM, MAXLOG;

double sinh(x)
double x;
    {
    double a;
    double fabs(), exp(), polevl(), p1evl();

    a = fabs(x);
    if( a > MAXLOG )
        {
        mtherr( "sinh", DOMAIN );
        if( x > 0 )
            return( MAXNUM );
        else
            return( -MAXNUM );
        }
    if( a > 1.0 )
        {
        a = exp(a);
        a = 0.5*a - (0.5/a);
        if( x < 0 )
            a = -a;
        return(a);
        }
    a *= a;
    return( x + x * a * (polevl(a,P,3)/p1evl(a,Q,3)) );
    }
```

4.12 $\cosh x$

This function requires no special approximations. It can be computed directly from

$$\cosh x = \frac{e^x + e^{-x}}{2}.$$

As with $\sinh x$, an overflow escape must be provided for $|x|$ too large. The limitations for IEEE and DEC arithmetic are, respectively,

$$\ln 2^{1024} = 709.782712893383996843$$

$$\ln 2^{127} = 88.02969193111305429598 8 .$$

Technically, while e^x does overflow at this point, neither $\sinh x$ nor $\cosh x$ will overflow until a slightly larger $|x|$. A refinement (see Cody and Waite) can extend the acceptable domain slightly.

4.12.1 cosh.c

```
#include "mconf.h"
extern double MAXLOG, MAXNUM;

double cosh(x)
double x;
	{
	double y;
	double exp(), ldexp();

	if( x < 0 )
		x = -x;
	if( x > MAXLOG )
		{
		mtherr( "cosh", OVERFLOW );
		return( MAXNUM );
		}
	y = exp(x);
	y = y + 1.0/y;
	y = ldexp( y, -1 );
	return( y );
	}
```

4.13 tanh x

The hyperbolic tangent is defined by

$$\begin{aligned} \tanh x &= \frac{\sinh x}{\cosh x} \\ &= \frac{e^x - e^{-x}}{e^x + e^{-x}} \\ &= 1 - \frac{2}{e^{2x} + 1} . \end{aligned}$$

This expression may be used for computation if $|x| \geq 5/8$. $\tanh x$ is an odd function; when x is negative, the arithmetic is more accurate if the formula is used with $|x|$ and the result negated. There will be cancellation error for small $|x|$, so a rational approximation is needed:

$$\tanh x \approx x + x^3 \frac{P(x^2)}{Q(x^2)}$$

where

$P(t) =$
$-0.96439917942505223862 8t^2$
$-99.28772310019185865 64t$
-1614.68768441708447952 ,

$Q(t) =$
$+1.0t^3$
$+112.81167849163293140 2t^2$
$+2235.48839060100448583t$
$+4844.06305325125486048$.

Theoretical relative error is $5.6 \cdot 10^{-19}$. Possible overflow in e^x must be handled. If x is larger than half the maximum possible logarithm, the program may simply approximate $\tanh x = 1$ or -1 according as x is large and positive or large and negative.

A continued fraction for $\tanh x$ was given in the section on the exponential function.

4.13.1 tanh.c

```
#include "mconf.h"

#ifdef DEC
static short P[] = {
0140166,0161335,0053753,0075126,
0141706,0111520,0070463,0040552,
0142711,0153001,0101300,0025430
};
Leading coefficient 1.0 omitted from Q
static short Q[] = {
0041741,0117624,0051300,0156060,
0043013,0133720,0071251,0127717,
0043227,0060201,0021020,0020136
};
#endif

#ifdef IIEEE
static short P[] = {
0x6f4b,0xaafd,0xdc5b,0xbfee,
0x682d,0x0e26,0xd26a,0xc058,
0x0563,0x3058,0x3ac0,0xc099
};

static short Q[] = {
0x1b86,0x8a58,0x33f2,0x405c,
0x35fa,0x0e55,0x76fa,0x40a1,
0x040c,0x2442,0xec10,0x40b2
};
#endif


extern double MAXLOG;

double tanh(x)
double x;
    {
    double s, z;
    double fabs(), exp(), polevl(), p1evl();

    z = fabs(x);
    if( z > 0.5 * MAXLOG )
```

```
        {
        if( x > 0 )
            return( 1.0 );
        else
            return( -1.0 );
        }
    if( z >= 0.625 )
        {
tanh x = 1 - 2/(e^{2x} + 1)
        s = exp(2.0*z);
        z = 1.0 - 2.0/(s + 1.0);
        if( x < 0 )
            z = -z;
        }
    else
        {
tanh x ≈ x + x^3 P(x^2)/Q(x^2)
        s = x * x;
        z = polevl( s, P, 2 )/p1evl(s, Q, 3);
        z = x * s * z;
        z = x + z;
        }
    return( z );
    }
```

4.14 $\sinh^{-1} x$

Since

$$\sinh x = \frac{e^x - e^{-x}}{2} ,$$

the inverse function can be computed by

$$\sinh^{-1} x = \ln \left(x + \sqrt{x^2 + 1} \right) .$$

This identity can be verified by exponentiation. Note $\sinh^{-1} x$ is an odd function of x, so the computer program may save the sign of x and operate with $|x|$ thereafter, restoring the sign at the end. For large x ($> 10^8$ in double precision), the function may be approximated with sufficient accuracy by

$$|\sinh^{-1} x| \approx \ln |x| + \ln 2.$$

For $|x| < 0.5$ use the rational approximation

$$\sinh^{-1} x \approx x + x^3 \frac{P(x^2)}{Q(x^2)}$$

where

$P(t) =$
$-4.33231683752342103572 \cdot 10^{-3} t^4$
$-5.91750212056387121207 \cdot 10^{-1} t^3$
$-4.37390226194356683570 t^2$
$-9.09030533308377316566 t$
-5.56682227230859640450 ,

$Q(t) =$
$+1.0 t^4$
$+12.8757002067426453537 t^3$
$+48.6042483805291788324 t^2$
$+69.5722521337257608734 t$
$+33.4009336338516356383$.

Theoretical relative error is $5.1 \cdot 10^{-19}$. The continued fraction

$$\sinh^{-1} x = \sqrt{1+x^2}\left(\frac{x}{1+}\;\frac{1\cdot 2x^2}{3+}\;\frac{1\cdot 2x^2}{5+}\;\frac{3\cdot 4x^2}{7+}\;\frac{3\cdot 4x^2}{9+}\cdots\right)$$

may be used in a high precision routine for checking at small x.

4.14.1 asinh.c

```
#include "mconf.h"

#ifdef DEC
static short P[] = {
0136215,0173033,0110410,0105475,
0140027,0076361,0020056,0164520,
0140613,0173401,0160136,0053142,
0141021,0070744,0000503,0176261,
0140662,0021550,0073106,0133351
};
```

The leading coefficient 1.0 is omitted from the array Q.

```
static short Q[] = {
0041116,0001336,0034120,0173054,
0041502,0065300,0013144,0021231,
0041613,0022376,0035516,0153063,
0041405,0115216,0054265,0004557
};
#endif

#ifdef IIEEE
static short P[] = {
0x1168,0x7221,0xbec3,0xbf71,
0xdd2a,0x2405,0xef9e,0xbfe2,
0xcacc,0x3c0b,0x7ee0,0xc011,
0x7f96,0x8028,0x2e3c,0xc022,
0xd6dd,0x0ec8,0x446d,0xc016
};
static short Q[] = {
0x1ec5,0xc70a,0xc05b,0x4029,
0x8453,0x02cc,0x4d58,0x4048,
0xdac6,0xc769,0x649f,0x4051,
0xa12e,0xcb16,0xb351,0x4040
};
#endif


extern double LOGE2;

double asinh(x)
double x;
    {
    double a, z;
    int sign;
    double log(), sqrt(), polevl(), p1evl();

    if( x < 0.0 )
```

```
        {
        sign = -1;
        x = -x;
        }
    else
        sign = 1;
    if( x > 1.0e8 )
        return( sign * (log(x) + LOGE2) );
    z = x * x;
    if( x < 0.5 )
```
$\sinh^{-1} x \approx x + x^3 P(x^2)/Q(x^2)$
```
        {
        a = ( polevl(z, P, 4)/p1evl(z, Q, 4) ) * z;
        a = a * x + x;
        if( sign < 0 )
            a = -a;
        return(a);
        }
```
$\sinh^{-1} x = \ln(x + \sqrt{x^2+1})$
```
    a = sqrt( z + 1.0 );
    return( sign * log(x + a) );
    }
```

4.15 $\cosh^{-1} x$

Since

$$\cosh x = \frac{e^x + e^{-x}}{2} ,$$

the inverse hyperbolic cosine can be written

$$\cosh^{-1} x = \ln\left(x + \sqrt{x^2 - 1}\right)$$

as can be verified by exponentiating. This formula is suitable for computation if $x > 1.5$, though $x^2 - 1$ should be computed in the form $(x-1)(x+1)$. As x grows large, $\cosh^{-1} x$ approaches $\ln 2x$ as

$$\cosh^{-1} x = \ln 2x - \frac{1}{4x^2} - \dots .$$

In double precision arithmetic this approximation is accurate for $x > 10^8$, if computed in the form

$$\cosh^{-1} x \approx \ln x + \ln 2 .$$

If $x < 1$ then $\cosh^{-1} x$ is undefined; an error escape must be supplied.

When x is near 1 the logarithmic form is unsuitable because there is cancellation in calculating the argument of the logarithm. For $1 \le x \le 1.5$ it is best to use a rational approximation

$$\cosh^{-1} x \approx \frac{P(x-1)}{Q(x-1)} \sqrt{x-1}$$

where

$P(t) =$
$+1.18801130533544501356 \cdot 10^2 t^4$
$+3.94726656571334401102 \cdot 10^3 t^3$
$+3.43989375926195455866 \cdot 10^4 t^2$
$+1.08102874834699867335 \cdot 10^5 t$
$+1.10855947270161294369 \cdot 10^5$,

$Q(t) =$
$+1.0 t^5$
$+1.86145380837903397292 \cdot 10^2 t^4$
$+4.15352677227719831579 \cdot 10^3 t^3$
$+2.97683430363289370382 \cdot 10^4 t^2$
$+8.29725251988426222434 \cdot 10^4 t$
$+7.83869920495893927727 \cdot 10^4$.

Theoretical relative error, when regarded as an approximation to

$$\cosh^{-1}(1+x)/\sqrt{x} ,$$

is $4.2 \cdot 10^{-19}$.

4.15.1 acosh.c

```
#include "mconf.h"

#ifdef DEC
static short P[] = {
0041755,0115055,0144002,0146444,
0043166,0132103,0155150,0150302,
0044006,0057360,0003021,0162753,
0044323,0021557,0175225,0056253,
0044330,0101771,0040046,0006636
};
Leading 1.0 is omitted from Q.
static short Q[] = {
0042072,0022467,0126670,0041232,
0043201,0146066,0152142,0034015,
0043750,0110257,0121165,0026100,
0044242,0007103,0034667,0033173,
0044231,0014576,0175573,0017472
};
#endif

#ifdef IIEEE
static short P[] = {
0x59a4,0xb900,0xb345,0x405d,
0x1a18,0x7b4d,0xd688,0x40ae,
0x3cbd,0x00c2,0xcbde,0x40e0,
0xab95,0xff52,0x646d,0x40fa,
0xc1b4,0x2804,0x107f,0x40fb
};

static short Q[] = {
0x0853,0xf5b7,0x44a6,0x4067,
0x4702,0xda8c,0x3986,0x40b0,
0xa588,0xf44e,0x1215,0x40dd,
0xe6cf,0x6736,0x41c8,0x40f4,
0x63e7,0xdf6f,0x232f,0x40f3
};
#endif


extern double LOGE2;

double acosh(x)
double x;
```

```
{
double a, z;
double log(), sqrt(), polevl(), p1evl();

if( x < 1.0 )
	{
	mtherr( "acosh", DOMAIN );
	return(0.0);
	}
if( x > 1.0e8 )
	return( log(x) + LOGE2 );
z = x - 1.0;
if( z < 0.5 )
```

$\cosh^{-1} x \approx \sqrt{(x-1)} P(x-1)/Q(x-1)$

```
	{
	a = sqrt(z)
		* (polevl(z, P, 4) / p1evl(z, Q, 5) );
	return( a );
	}
```

$\cosh^{-1} x = \ln(x + \sqrt{x^2 - 1})$

```
a = sqrt( z*(x+1.0) );
return( log(x + a) );
}
```

4.16 $\tanh^{-1} x$

Analytically,

$$\tanh^{-1} x = \frac{1}{2} \ln \left(\frac{1+x}{1-x} \right)$$

which is an odd function of x. This expression may be used for computation if $x \geq 0.5$. The function is undefined for $x \geq 1$. An error escape must be supplied for this condition. When $|x| < 0.5$, use the rational approximation

$$\tanh^{-1} x \approx x + x^3 \frac{P(x^2)}{Q(x^2)}$$

where

$P(t) =$
$-0.85407433192966930519\,6t^4$
$+12.0426861384072379242t^3$
$-46.1252884198732692637t^2$
$+65.4566728676544377376t$
-30.9092539379866942570 ,

$Q(t) =$
$+1.0t^5$
$-19.5638849376911654834t^4$
$+108.938092147140262656t^3$
$-249.839401325893582852t^2$
$+252.006675691344555838t$
-92.7277618139601130017 .

Theoretical relative error is $2.0 \cdot 10^{-19}$. In double precision, if $|x| < 10^{-7}$, then $\tanh^{-1} x \approx x$.

4.16.1 atanh.c

```
#include "mconf.h"

#ifdef DEC
static short P[] = {
0140132,0122235,0105775,0130300,
0041100,0127327,0124407,0034722,
0141470,0100113,0115607,0130535,
0041602,0164721,0003257,0013673,
0141367,0043046,0166673,0045750
};
```

Leading coefficient 1.0 is omitted from Q.

```
static short Q[] = {
0141234,0101326,0015460,0134564,
0041731,0160115,0116451,0032045,
0142171,0153343,0000532,0167226,
0042174,0000665,0077604,0000310,
0141671,0072235,0031114,0074377
};
#endif

#ifdef IIEEE
static short P[] = {
0xb618,0xb17f,0x5493,0xbfeb,
0xe73a,0xf520,0x15da,0x4028,
0xf62c,0x7370,0x1009,0xc047,
0xe2f7,0x20d5,0x5d3a,0x4050,
0x697d,0xddb7,0xe8c4,0xc03e
};
static short Q[] = {
0x172f,0xc366,0x905a,0xc033,
0x2685,0xb3a5,0x3c09,0x405b,
0x5dd3,0x602b,0x3adc,0xc06f,
0x8019,0xaff0,0x8036,0x406f,
0x8f20,0xa649,0x2e93,0xc057
};
#endif


extern double MAXNUM;

double atanh(x)
double x;
    {
    double s, z;
    int i, n;
    double fabs(), log(), polevl(), p1evl();

    z = fabs(x);
    if( z >= 1.0 )
        {
```

$\tanh^{-1} 1 = \infty$

```
        if( x == 1.0 )
            return( MAXNUM );
```

$\tanh^{-1} -1 = -\infty$

```
        if( x == -1.0 )
            return( -MAXNUM );
        mtherr( "atanh", DOMAIN );
        return( MAXNUM );
        }
```

```
        if( z < 1.0e-7 )
                return(x);
        if( z < 0.5 )
                {
tanh^-1 x ≈ x + x^3 P(x^2)/Q(x^2)
                z = x * x;
                s = x + x * z
                        * (polevl(z, P, 4) / p1evl(z, Q, 5));
                return(s);
                }
tanh^-1 x = 1/2 ln(1 + x)/(1 - x)
        return( 0.5 * log((1.0+x)/(1.0-x)) );
        }
```

4.17 Power Function

An ability to raise x to the yth power is included as an intrinsic operation in many computer languages. Expressed in terms of the logarithm and exponential functions, this operation can be implemented as

$$\begin{aligned} w &= x^y \\ &= (e^{\ln x})^y \\ &= e^{y \ln x} . \end{aligned}$$

Because of error amplification by the exponential function this formulation can have rather large relative error when implemented with the standard library routines. If a somewhat extended precision arithmetic, such as the IEEE extended double format, is available, it should by all means be used to evaluate the logarithm and exponential functions for this application.

An extended precision algorithm developed by Cody may be regarded as a preferred method for general real x and y when no other means of extending the arithmetic precision is available.

The following special cases must be checked for:

$$\begin{aligned} x^0 &= 1, \ x \neq 0 \\ 0^0 &= 1 \\ 0^y &= \infty, \ y < 0 \end{aligned}$$

A domain error exists if x is negative and y is an even root. Detection of cases such as the cube root of a negative number is difficult, owing to the inexactness of computer numbers like $y = 1/3$. Therefore the program rejects negative x unless y is an integer. An overflow signal is required if the exponential function would overflow, for sufficiently large $y \ln x$.

4.17.1 Real Exponent

Cody and Waite's exponentiation method uses a lookup table and extended precision arithmetic to achieve an extra three bits of accuracy in both the logarithm and the exponential.

If $x = s \cdot 2^e$ then

$$\begin{aligned}\log_2 x &= \log_2(s \cdot 2^e)\\ &= e + \log_2(as/a)\\ &= e + \log_2 a + \log_2 s/a\\ &= e + \log_2 a + \log_2(1 + (s - a)/a) \ .\end{aligned}$$

The term a is equal to $2^{-k/16}$ for some value of k between 1 and 16. Values of a with extended precision accuracy are stored in a lookup table. A rational approximation is used to find the last term using an expansion for $\ln(1 + x)$. It has several extra bits of absolute accuracy since $(s - a)/a$ is small.

Now the product $y \log_2 x$ is computed in extended precision arithmetic. This is done by separating y into a large part which is an integer multiple of 1/16 plus a small part less than 1/16. The product is eventually expressed as an integer multiple of 1/16 plus a residual term.

The answer is the base 2 exponential of this number. It is computed as the product of three parts: an integer power of two; 2 raised to the $n/16$ power, which can be found in the lookup table; and the exponential of the small residual, which is computed by a rational approximation.

A complete program is presented below. It uses the more accurate of the two forms presented earlier for computing logarithms. For additional details, see the Cody and Waite reference given at the beginning of this chapter.

4.17.2 pow.c

```
#include "mconf.h"
static char fname[] = {"pow"};

#define SQRTH 0.70710678118654752440
```

$(\ln x - x + \frac{1}{2}x^2)/x^2$

```
#ifdef UNK
static double P[] = {
4.97778295871696322025E-1,
3.73336776063286838734E0,
7.69994162726912503298E0,
4.66651806774358464979E0
};
```

Leading term 1.0 of Q omitted.

```
static double Q[] = {
9.33340916416696166113E0,
2.79999886606328401649E1,
3.35994905342304405431E1,
1.39995542032307539578E1
};
```

$2^{-i/16}$, IEEE 53 bit precision

```
static double A[] = {
1.00000000000000000000E0,
9.57603280698573700036E-1,
9.17004043204671215328E-1,
8.78126080186649726755E-1,
8.40896415253714502036E-1,
8.05245165974627141736E-1,
7.71105412703970372057E-1,
7.38413072969749673113E-1,
7.07106781186547572737E-1,
6.77127773468446325644E-1,
6.48419777325504820276E-1,
6.20928906036742001007E-1,
5.94603557501360513449E-1,
5.69394317378345782288E-1,
5.45253866332628844837E-1,
5.22136891213706877402E-1,
5.00000000000000000000E-1
};
```

Less significant word of $2^{-i/16}$, IEEE 53 bit precision

```
static double B[] = {
0.00000000000000000000E0,
1.64155361212281360176E-17,
4.09950501029074826006E-17,
3.97491740484881042808E-17,
-4.83364665672645672553E-17,
1.26912513974441574796E-17,
1.99100761573282305549E-17,
-1.52339103990623557348E-17,
0.00000000000000000000E0
};
static double R[] = {
1.49664108433729301083E-5,
1.54010762792771901396E-4,
1.33335476964097721140E-3,
9.61812908476554225149E-3,
5.55041086645832347466E-2,
2.40226506959099779976E-1,
6.93147180559945308821E-1
};
```

```
#define douba(k) A[k]
#define doubb(k) B[k]
#define MEXP 2031.0
#endif

#ifdef DEC
static short P[] = {
0037776,0156313,0175332,0163602,
0040556,0167577,0052366,0174245,
0040766,0062753,0175707,0055564,
0040625,0052035,0131344,0155636,
};
Leading term 1.0 omitted in Q.
static short Q[] = {
0041025,0052644,0154404,0105155,
0041337,0177772,0007016,0047646,
0041406,0062740,0154273,0020020,
0041137,0177054,0106127,0044555,
};
static short A[] = {
0040200,0000000,0000000,0000000,
0040165,0022575,0012444,0103314,
0040152,0140306,0163735,0022071,
0040140,0146336,0166052,0112341,
0040127,0042374,0145326,0116553,
0040116,0022214,0012437,0102201,
0040105,0063452,0010525,0003333,
0040075,0004243,0117530,0006067,
0040065,0002363,0031771,0157145,
0040055,0054076,0165102,0120513,
0040045,0177326,0124661,0050471,
0040036,0172462,0060221,0120422,
0040030,0033760,0050615,0134251,
0040021,0141723,0071653,0010703,
0040013,0112701,0161752,0105727,
0040005,0125303,0063714,0044173,
0040000,0000000,0000000,0000000
};
static short B[] = {
0000000,0000000,0000000,0000000,
0021473,0040265,0153315,0140671,
0121074,0062627,0042146,0176454,
0121413,0003524,0136332,0066212,
0121767,0046404,0166231,0012553,
0121257,0015024,0002357,0043574,
0021736,0106532,0043060,0056206,
0121310,0020334,0165705,0035326,
0000000,0000000,0000000,0000000
};
```

```
#ifdef IIEEE
static short P[] = {
0x5cf0,0x7f5b,0xdb99,0x3fdf,
0xdf15,0xea9e,0xddef,0x400d,
0xeb6f,0x7f78,0xccbd,0x401e,
0x9b74,0xb65c,0xaa83,0x4012,
};

static short Q[] = {
0x914e,0x9b20,0xaab4,0x4022,
0xc9f5,0x41c1,0xffff,0x403b,
0x6402,0x1b17,0xccbc,0x4040,
0xe92e,0x918a,0xffc5,0x402b,
};
static short A[] = {
0x0000,0x0000,0x0000,0x3ff0,
0x90da,0xa2a4,0xa4af,0x3fee,
0xa487,0xdcfb,0x5818,0x3fed,
0x529c,0xdd85,0x199b,0x3fec,
0xd3ad,0x995a,0xe89f,0x3fea,
0xf090,0x82a3,0xc491,0x3fe9,
0xa0db,0x422a,0xace5,0x3fe8,
0x0187,0x73eb,0xa114,0x3fe7,
0x3bcd,0x667f,0xa09e,0x3fe6,
0x5429,0xdd48,0xab07,0x3fe5,
0x2a27,0xd536,0xbfda,0x3fe4,
0x3422,0x4c12,0xdea6,0x3fe3,
0xb715,0x0a31,0x06fe,0x3fe3,
0x6238,0x6e75,0x387a,0x3fe2,
0x517b,0x3c7d,0x72b8,0x3fe1,
0x890f,0x6cf9,0xb558,0x3fe0,
0x0000,0x0000,0x0000,0x3fe0
};
static short B[] = {
0x0000,0x0000,0x0000,0x0000,
0x3707,0xd75b,0xed02,0x3c72,
0xcc81,0x345d,0xa1cd,0x3c87,
0x4b27,0x5686,0xe9f1,0x3c86,
0x6456,0x13b2,0xdd34,0xbc8b,
0x42e2,0xafec,0x4397,0x3c6d,
0x82e4,0xd231,0xf46a,0x3c76,
0x8a76,0xb9d7,0x9041,0xbc71,
0x0000,0x0000,0x0000,0x0000
};
```

```
static short R[] = {
0034173,0014076,0137624,0115771,
0035041,0076763,0003744,0111311,
0035656,0141766,0041127,0074351,
0036435,0112533,0073611,0116664,
0037143,0054106,0134040,0152223,
0037565,0176757,0176026,0025551,
0040061,0071027,0173721,0147572
};
#define douba(k) (*(double *)&A[(k)<<2])
#define doubb(k) (*(double *)&B[(k)<<2])
#define MEXP 2031.0
#endif
```

```
static short R[] = {
0x937f,0xd7f2,0x6307,0x3eef,
0x9259,0x60fc,0x2fbe,0x3f24,
0xef1d,0xc84a,0xd87e,0x3f55,
0x33b7,0x6ef1,0xb2ab,0x3f83,
0x1a92,0xd704,0x6b08,0x3fac,
0xc56d,0xff82,0xbfbd,0x3fce,
0x39ef,0xfefa,0x2e42,0x3fe6
};
#define MEXP 16383.0
#endif
```

$\log 2(e) - 1$

```
#define LOG2EA 0.44269504088896340736
```

log2(e) - 1

```
#define LOG2EA 0.44269504088896340736

#define F W
#define Fa Wa
#define Fb Wb
#define G W
#define Ga Wa
#define Gb u
#define H W
#define Ha Wb
#define Hb Wb

extern double MAXNUM;

double pow( x, y )
double x, y;
      {
      double w, z, W, Wa, Wb, ya, yb;
      double u, v;
```

F, Fa, Fb, G, Ga, Gb, H, Ha, Hb reuse space already declared

```
      int e, i, nflg;
      double floor(), fabs(), frexp(), ldexp();
      double reduc(), polevl(), p1evl(), powi();
```

flag = 1 if $x < 0$ raised to integer power

```
      nflg = 0;
      w = floor(y);
```

```
if( (w == y) && (fabs(w) < 32768.0) )
        {
        i = w;
        w = powi( x, i );
        return( w );
        }
```

Special cases for $x = 0$

```
if( x <= 0.0 )
        {
        if( x == 0.0 )
                {
                if( y == 0.0 )
                        return( 1.0 );
                else
                        return( 0.0 );
                }
        else
                {
                if( w != y )
                        {
```

Reject noninteger power of negative number

```
                        mtherr( fname, DOMAIN );
                        return(0.0);
                        }
                nflg = 1;
                x = fabs(x);
                }
        }
```

Find significand in antilog table A[]

```
x = frexp( x, &e );
i = 1;
if( x <= douba(9) )
        i = 9;
if( x <= douba(i+4) )
        i += 4;
if( x <= douba(i+2) )
        i += 2;
if( x >= douba(1) )
        i = -1;
i += 1;
```

Find $(x - A[i])/A[i]$ in order to compute $\ln(x/A[i])$.

```
x -= douba(i);
x -= doubb(i/2);
x /= douba(i);
```

Rational approximation for log(1+v)

$\ln(1+v) = v - v^2/2 + v^3 P(v)/Q(v)$

```
    z = x*x;
    w = x * ( z * polevl( x, P, 3 )
            / p1evl( x, Q, 4 ) );
    w = w - ldexp( z, -1 ); w - 0.5 * z
```

Convert to base 2 logarithm.

```
    w = w + LOG2EA * w;
```

Add x, while multiplying by $\log_2 e$.

```
    z = w + LOG2EA * x;
    z = z + x;
```

Compute exponent term of the base 2 logarithm.

```
    w = -i;
    w = ldexp( w, -4 );
    w += e;
```

Now base 2 log of x is $w + z$.

Multiply base 2 log by y, in extended precision.

Separate y into large part ya and small part yb less than $1/16$.

```
    ya = reduc(y);
    yb = y - ya;
    F = z * y + w * yb;
    Fa = reduc(F);
    Fb = F - Fa;
    G = Fa + w * ya;
    Ga = reduc(G);
    Gb = G - Ga;
    H = Fb + Gb;
    Ha = reduc(H);
    w = ldexp( Ga+Ha, 4 );
    Test the power of 2 for overflow.
    if( w > MEXP )
        {
        mtherr( fname, OVERFLOW );
        return( MAXNUM );
        }
    if( w < -MEXP )
        {
        mtherr( fname, UNDERFLOW );
        return( 0.0 );
        }
    e = w;
    Hb = H - Ha;
    if( Hb > 0.0 )
        {
        e += 1;
```

```
        Hb -= 0.0625;
        }
```

Now the product $y \log_2 x = Hb + e/16$.

Compute base 2 exponential of Hb, where $-0.0625 <= Hb <= 0$.

```
    z = Hb * polevl( Hb, R, 6 );   2^Hb - 1
```

Express e/16 as an integer plus a negative number of 16ths.

Find lookup table entry for the fractional power of 2.

```
    if( e < 0 )
        i = 0;
    else
        i = 1;
    i = e/16 + i;
    e = 16*i - e;
    w = douba( e );
```

$2^{-e}(1 + (2^{Hb} - 1))$

```
    z = w + w * z;
```

Multiply by integer power of 2.

```
    z = ldexp( z, i );
    if( nflg )
        {
```

If $x < 0$, find out if the integer exponent is odd or even.

```
        w = ldexp( y, -1 );
        w = floor(w);
        w = ldexp( w, 1 );
        if( w != y )
            z = -z;   Odd exponent
        }
    return( z );
    }
```

Subroutine finds a multiple of 1/16 that is within 1/16 of x.

```
static double reduc(x)
double x;
    {
    double t;
    double ldexp(), floor();

    t = ldexp( x, 4 );
    t = floor( t );
    t = ldexp( t, -4 );
    return(t);
    }
```

4.17.3 Integer Exponent

Even when calculated in somewhat extended precision arithmetic, the exponentiated logarithm cannot approach the accuracy of simple arithmetic when calculating small integer powers. Therefore an alternative algorithm for integer exponents is desirable. If y is an integer, then

$$x^y = x \cdot x \cdots x \ ,$$

a y-fold multiplication. By regrouping the product terms, the total number of multiplications can be made proportional to $\log_2 y$. The procedure depends on expressing y in terms of its binary integer representation

$$y = \sum_{k=0}^{n} b_k \ 2^k$$

where $b_k = 1$ if the corresponding bit of the integer is 1, or 0 if the bit is 0. Then

$$\begin{aligned} w &= x^y \\ &= x^{b_0 \ 2^0} \ x^{b_1 \ 2^1} \cdots x^{b_n \ 2^n} \\ &= \prod_{k=0}^{n} x^{b_k \ 2^k} \end{aligned}$$

which is a product of 2^kth powers of x. This product has a maximum number of terms equal to the number of binary bits in an integer.

For computation a variable u is used to accumulate the 2^kth power of x. On the zeroth step $u = x$. On this step $w = 1$ if the least significant bit of y is a zero or $w = x$ if the bit is a 1. Then the integer y is shifted down 1 bit to access its next more significant bit. On the kth step u is squared. This forms the 2^kth power of x. If the kth bit of y is a 1, then w is multiplied by the new value of u; otherwise w is left alone. Then y is shifted down 1 bit to access its $k + 1$st bit. After each step, the shifted-down value of y is tested to see if it has any nonzero bits remaining; if there are none, the algorithm terminates immediately.

The computer algorithm presumes that x and y are both positive. Therefore, if $y < 0$, then replace y by $|y|$ and invert the result. If the result extends into the range of denormal tiny numbers, then there would be overflow while calculating the positive power. In that event x may be replaced by $1/x$; but the accuracy will be diminished by amplification of the roundoff error in $1/x$. If $x < 0$ then replace x by $|x|$ and negate the answer for odd integer powers of the negative base. Overflow detection may use the library routine **frexp**(x, &p) to find the nearest power of two larger than x. Then if $(p - 1)y$ exceeds the largest legal power of two (1023 in IEEE arithmetic), the routine can take an error exit and return the appropriate underflow or overflow value. This test is highly granular and may fail. A better test approximation for the logarithm of the result is shown in the program.

4.17.4 powi.c

```
#include "mconf.h"
extern double MAXNUM, MAXLOG, MINLOG, LOGE2;

double powi( x, nn )
double x;
int nn;
    {
    int n, e, sign, asign, lx;
    short *p;
    double w, y, s;
    double log(), frexp();

    if( x == 0.0 )
        {
        if( nn == 0 )
            return( 1.0 );
        else if( nn < 0 )
                return( MAXNUM );
        else
            return( 0.0 );
        }
    if( nn == 0 )
        return( 1.0 );
    if( x < 0.0 )
        {
        asign = -1;
        x = -x;
        }
    else
        asign = 0;
    if( nn < 0 )
        {
        sign = -1;
        n = -nn;
        }
    else
        {
        sign = 0;
        n = nn;
```

```
        }
Overflow detection
Calculate approximate logarithm of answer.
    s = frexp( x, &lx );
    e = (lx - 1)*n;
    if( (e == 0) || (abs(e) > 64) )
        {
        s = (s - 7.0710678118654752e-1)
            / (s + 7.0710678118654752e-1);
        s = (2.9142135623730950 * s - 0.5 + lx)
            * nn * LOGE2;
        }
    else
        {
        s = LOGE2 * e;
        }
    if( s > MAXLOG )
        {
        mtherr( "powi", OVERFLOW );
        y = MAXNUM;
        goto done;
        }
    if( s < MINLOG )
        return(0.0);
Handle tiny denormal answer, but with less accuracy.
The demarcation should be the gradual underflow threshold.
    if( s < (-MAXLOG+2.0) )
        {
        x = 1.0/x;
        sign = 0;
        }
First bit of the power
    if( n & 1 )
        y = x;
    else
        {
        y = 1.0;
        asign = 0;
        }
    w = x;
    n >>= 1;
    while( n )
        {
        w = w * w; arg to the 2-to-the-kth power
        if( n & 1 ) if that bit is set, then include in product
```

```
                y *= w;
            n >>= 1;
            }
done:
        if( asign )
            y = -y; odd power of negative number
        if( sign )
            y = 1.0/y;
        return(y);
        }
```

4.18 Testing

The tables give experimental peak and root mean square (rms) relative errors of the computer programs presented in this chapter. Each function was tested by generating reference function values with a high precision check routine that used extended precision arithmetic. The reference values were compared to the output of the program under test, which used arithmetic of normal working precision. Thus small fractions of a bit of roundoff error could be measured accurately. In the tables, $f(z)$ denotes a complex function, $f(x)$ a real function. Table 4.2 gives the accuracy for IEEE arithmetic. Table 4.3 is for DEC arithmetic. The following consistency tests among the double precision functions were also carried out.

$$
\begin{aligned}
\ln(e^x) &= x \\
\sin(\sin^{-1}(z)) &= z \\
\cos(\cos^{-1}(z)) &= z \\
\tan^{-1}(\tan(z)) &= z \\
\tan z \cot z &= 1 + i0 \\
\sqrt{z^2} &= z
\end{aligned}
$$

The number of pseudorandom arguments tried in each test is given under the heading, "Trials." In the tests over the interval $[1, \mathtt{MAXNUM}]$, the logarithms of the random arguments were uniformly distributed over the interval $[0, \mathtt{MAXLOG}]$. Because of the relatively large gaps in the range of the exponential function, it is possible that some programs, particularly the logarithm, will seem more accurate than they really are. Therefore, extra steps were taken to improve the distribution of exponentiated random numbers. The arguments were produced by generating two random numbers r_1 and r_2, each between 1 and 2, then forming the test argument x by

$$
\begin{aligned}
u &= \exp[(\mathtt{MAXLOG} - \mathtt{MINLOG})(r_1 - 1) + \mathtt{MINLOG}] \\
x &= u - 10^{-13} u (r_2 - 1) \ .
\end{aligned}
$$

The function **pow**(x, y) was tested in two ways: first, $1/26 < x, y < 26$ with $\ln x$ uniformly distributed and y uniformly distributed; second, $0.99 < x < 1.01$, $0 < y < 8700$, with x and y uniformly distributed. **powi**(x, y) was tested in a similar way, except in the second case $1 \le x \le 2$ and $-1022 \le y \le 1023$. Larger relative error than indicated in the complex function $\sin^{-1} z$ can be observed for z near zero. The error criterion for the complex logarithm was absolute if $|\ln z| < 1$ and relative otherwise. Large relative error in the complex logarithm can be observed for z near $1 + i0$.

4.19 Single Precision Polynomial Approximations

A growing number of microprocessor and digital signal processing devices have combinatorial logic circuits that perform very fast floating point multiplication and addition. Division in these devices is very much slower than multiplication, so the balance comes out in favor of omitting division altogether in most numerical algorithms.

This section gives polynomial approximations suitable for the construction of single precision routines for the elementary functions. Except for minor changes, the range reductions and other procedures are the same as in the double precision routines discussed earlier in this chapter. Therefore the computer programs are omitted.

Tests were run over the basic intervals of the approximations using a Motorola 68882 arithmetic coprocessor set to IEEE 24 bit rounding precision. In all but one case, the peak relative error was less than 1 lsb.

4.19.1 $\cos x$

Single precision circular radian cosine
Test interval: $[-\pi/4, +\pi/4]$
Random trials: 10000
Peak relative error: $8.3 \cdot 10^{-8}$
RMS relative error: $2.2 \cdot 10^{-8}$

$$
\begin{aligned}
z &= x^2 \\
\cos x &\approx \\
&\quad (2.4462803166 \cdot 10^{-5} z^3 - 1.3887580023 \cdot 10^{-3} z^2 \\
&\quad +4.1666650433 \cdot 10^{-2} z - 4.9999999968 \cdot 10^{-1}) z + 1.0 \,.
\end{aligned}
$$

4.19.2 $\cosh^{-1} x$

Single precision inverse hyperbolic cosine
Test interval: $[1.0, 1.5]$
Random trials: 10000
Peak relative error: $1.7 \cdot 10^{-7}$

Table 4.2: Accuracy of Elementary Functions in IEEE arithmetic.

Function	Domain	Trials	Peak	RMS
$\sqrt{x}$	0,`MAXNUM`	30000	$1.7 \cdot 10^{-16}$	$6.3 \cdot 10^{-17}$
$\sqrt{z}$	$-10, +10$	100000	$3.2 \cdot 10^{-16}$	$7.6 \cdot 10^{-17}$
$\sqrt[3]{x}$	0, `MAXNUM`	30000	$1.5 \cdot 10^{-16}$	$5.0 \cdot 10^{-17}$
$\exp x$	$-, +$`MAXLOG`	35000	$2.3 \cdot 10^{-16}$	$5.7 \cdot 10^{-17}$
$\exp z$	$-10, +10$	30000	$3.0 \cdot 10^{-16}$	$8.7 \cdot 10^{-17}$
$\ln x$	$0.5, 2.0$	35000	$1.8 \cdot 10^{-16}$	$5.6 \cdot 10^{-17}$
$\ln x$	1, `MAXNUM`	10000	$1.1 \cdot 10^{-16}$	$4.8 \cdot 10^{-17}$
$\ln z$	$-10, +10$	30000	$5.2 \cdot 10^{-16}$	$1.0 \cdot 10^{-16}$
$\sin x$	$-, +1.07 \cdot 10^9$	40000	$2.2 \cdot 10^{-16}$	$5.6 \cdot 10^{-17}$
$\cos x$	$-, +1.07 \cdot 10^9$	30000	$2.0 \cdot 10^{-16}$	$5.6 \cdot 10^{-17}$
$\tan x$	$-, +1.07 \cdot 10^9$	30000	$2.9 \cdot 10^{-16}$	$8.1 \cdot 10^{-17}$
$\cot x$	$-, +1.07 \cdot 10^9$	30000	$2.9 \cdot 10^{-16}$	$8.2 \cdot 10^{-17}$
$\tan^{-1} z$	$-10, +10$	30000	$2.3 \cdot 10^{-15}$	$8.5 \cdot 10^{-17}$
$\cos^{-1} z$	$-10, +10$	30000	$1.8 \cdot 10^{-14}$	$2.2 \cdot 10^{-15}$
$\sin^{-1} z$	$-10, +10$	30000	$2.2 \cdot 10^{-14}$	$2.7 \cdot 10^{-15}$
$\tan^{-1} x$	$-10, 10$	30000	$3.1 \cdot 10^{-16}$	$7.9 \cdot 10^{-17}$
`atan2`(x)	$-10, 10$	30000	$3.6 \cdot 10^{-16}$	$7.1 \cdot 10^{-17}$
$\cos^{-1} x$	$-1, 1$	30000	$2.7 \cdot 10^{-16}$	$7.1 \cdot 10^{-17}$
$\sin^{-1} x$	$-1, 1$	30000	$4.8 \cdot 10^{-16}$	$8.4 \cdot 10^{-17}$
$\cot z$	$-10, +10$	30000	$9.2 \cdot 10^{-16}$	$1.2 \cdot 10^{-16}$
$\tan z$	$-10, +10$	30000	$7.2 \cdot 10^{-16}$	$1.2 \cdot 10^{-16}$
$\sin z$	$-10, +10$	30000	$3.8 \cdot 10^{-16}$	$1.0 \cdot 10^{-16}$
$\cos z$	$-10, +10$	30000	$3.8 \cdot 10^{-16}$	$1.0 \cdot 10^{-16}$
$\sinh x$	$-, +$`MAXLOG`	30000	$2.6 \cdot 10^{-16}$	$5.7 \cdot 10^{-17}$
$\cosh x$	$-, +$`MAXLOG`	30000	$2.6 \cdot 10^{-16}$	$5.7 \cdot 10^{-17}$
$\tanh x$	$-2, 2$	30000	$2.5 \cdot 10^{-16}$	$5.8 \cdot 10^{-17}$
$\sinh^{-1} x$	$-1, 1$	30000	$3.7 \cdot 10^{-16}$	$7.8 \cdot 10^{-17}$
$\sinh^{-1} x$	$1, 3$	30000	$2.5 \cdot 10^{-16}$	$6.7 \cdot 10^{-17}$
$\cosh^{-1} x$	$1, 3$	30000	$4.6 \cdot 10^{-16}$	$8.7 \cdot 10^{-17}$
$\tanh^{-1} x$	$0, 1$	30000	$1.9 \cdot 10^{-16}$	$5.2 \cdot 10^{-17}$
`pow`(x, y)	$-26, 26$	30000	$4.2 \cdot 10^{-16}$	$7.7 \cdot 10^{-17}$
`pow`(x, y)	$0, 8700$	30000	$1.5 \cdot 10^{-14}$	$2.1 \cdot 10^{-15}$
`powi`(x, y)	$-26, 26$	30000	$2.0 \cdot 10^{-15}$	$3.8 \cdot 10^{-16}$
`powi`(x, y)	$-1022, 1023$	30000	$8.6 \cdot 10^{-14}$	$1.6 \cdot 10^{-14}$

Table 4.3: Accuracy of Elementary Functions in DEC arithmetic.

Function	Domain	Trials	Peak	RMS
$\sqrt{x}$	$0, 30$	2000	$2.1 \cdot 10^{-17}$	$5.2 \cdot 10^{-18}$
$\sqrt{z}$	$-10, +10$	25000	$3.2 \cdot 10^{-17}$	$9.6 \cdot 10^{-18}$
$\sqrt[3]{x}$	$0, 8$	50000	$1.8 \cdot 10^{-17}$	$5.6 \cdot 10^{-18}$
$\exp x$	0,`MAXLOG`	38000	$3.0 \cdot 10^{-17}$	$6.2 \cdot 10^{-18}$
$\exp z$	$-10, +10$	8700	$3.7 \cdot 10^{-17}$	$1.1 \cdot 10^{-17}$
$\ln x$	$0.5, 2.0$	20000	$2.0 \cdot 10^{-17}$	$7.0 \cdot 10^{-18}$
$\ln x$	0, `MAXNUM`	27700	$2.1 \cdot 10^{-17}$	$5.7 \cdot 10^{-18}$
$\ln z$	$-10, +10$	7000	$8.5 \cdot 10^{-17}$	$1.9 \cdot 10^{-17}$
$\sin x$	0,10	20000	$2.5 \cdot 10^{-17}$	$7.1 \cdot 10^{-18}$
$\sin x$	$0, 1.07 \cdot 10^9$	28000	$2.8 \cdot 10^{-17}$	$7.1 \cdot 10^{-18}$
$\cos x$	$0, +1.07 \cdot 10^9$	17000	$3.0 \cdot 10^{-17}$	$7.2 \cdot 10^{-18}$
$\tan x$	$0, +1.07 \cdot 10^9$	44000	$4.0 \cdot 10^{-17}$	$1.0 \cdot 10^{-17}$
$\cot x$	$0, +1.07 \cdot 10^9$	44000	$4.0 \cdot 10^{-17}$	$1.0 \cdot 10^{-17}$
$\cos z$	$-10, +10$	8400	$4.5 \cdot 10^{-17}$	$1.3 \cdot 10^{-17}$
$\sin z$	$-10, +10$	8400	$5.3 \cdot 10^{-17}$	$1.3 \cdot 10^{-17}$
$\tan z$	$-10, +10$	5200	$7.1 \cdot 10^{-17}$	$1.6 \cdot 10^{-17}$
$\cot z$	$-10, +10$	3000	$6.5 \cdot 10^{-17}$	$1.6 \cdot 10^{-17}$
$\sin^{-1} x$	$-1, 1$	20000	$5.7 \cdot 10^{-17}$	$1.2 \cdot 10^{-17}$
$\cos^{-1} x$	$-1, 1$	13000	$3.2 \cdot 10^{-17}$	$7.8 \cdot 10^{-18}$
$\tan^{-1} x$	$0, 10$	9500	$4.2 \cdot 10^{-17}$	$8.6 \cdot 10^{-18}$
$\sin^{-1} z$	$-10, +10$	10100	$2.1 \cdot 10^{-15}$	$3.4 \cdot 10^{-16}$
$\cos^{-1} z$	$-10, +10$	5200	$1.6 \cdot 10^{-15}$	$2.8 \cdot 10^{-16}$
$\tan^{-1} z$	$-10, +10$	5900	$1.3 \cdot 10^{-16}$	$7.8 \cdot 10^{-18}$
$\sinh x$	$0, 2$	10000	$4.2 \cdot 10^{-17}$	$8.3 \cdot 10^{-18}$
$\cosh x$	$0, 2$	20000	$2.9 \cdot 10^{-17}$	$8.6 \cdot 10^{-18}$
$\tanh x$	$0, 2$	5000	$2.9 \cdot 10^{-17}$	$6.4 \cdot 10^{-18}$
$\sinh^{-1}$	$0, 2$	15000	$4.7 \cdot 10^{-17}$	$1.3 \cdot 10^{-17}$
$\cosh^{-1} x$	$1, 3$	10000	$4.1 \cdot 10^{-17}$	$1.1 \cdot 10^{-17}$
$\tanh^{-1} x$	$0, 1$	10000	$3.0 \cdot 10^{-17}$	$6.5 \cdot 10^{-18}$
`pow`(x, y)	$-26, 26$	18000	$4.4 \cdot 10^{-17}$	$8.9 \cdot 10^{-18}$
`pow`(x, y)	$0, 8700$	1000	$1.3 \cdot 10^{-15}$	$2.1 \cdot 10^{-16}$

RMS relative error: $5.0 \cdot 10^{-8}$

$$
\begin{aligned}
w &= x-1 \\
\cosh^{-1} x &\approx \\
& (1.7596881071 \cdot 10^{-3} w^4 - 7.5272886713 \cdot 10^{-3} w^3 \\
& +2.6454905019 \cdot 10^{-2} w^2 - 1.1784741703 \cdot 10^{-1} w \\
& +1.4142135263)\sqrt{w} \; .
\end{aligned}
$$

4.19.3 $\exp x$

Single precision exponential function
Test interval: $[-0.5, +0.5]$
Random trials: 80000
Peak relative error: $7.6 \cdot 10^{-8}$
RMS relative error: $2.8 \cdot 10^{-8}$

$$
\begin{aligned}
e^x &\approx \\
& (1.9875691500 \cdot 10^{-4} x^5 + 1.3981999507 \cdot 10^{-3} x^4 \\
& +8.3334519073 \cdot 10^{-3} x^3 + 4.1665795894 \cdot 10^{-2} x^2 \\
& +1.6666665459 \cdot 10^{-1} x + 5.0000001201 \cdot 10^{-1}) x^2 \\
& +x + 1.0 \; .
\end{aligned}
$$

4.19.4 $\ln x$

Single precision natural logarithm
Test interval: $[\sqrt{2}/2, \sqrt{2}]$
Random trials: 10000
Peak relative error: $7.1 \cdot 10^{-8}$
RMS relative error: $2.7 \cdot 10^{-8}$

$$
\begin{aligned}
z &= x-1 \\
\ln x &\approx \\
& (7.0376836292 \cdot 10^{-2} z^8 - 1.1514610310 \cdot 10^{-1} z^7 \\
& +1.1676998740 \cdot 10^{-1} z^6 - 1.2420140846 \cdot 10^{-1} z^5 \\
& +1.4249322787 \cdot 10^{-1} z^4 - 1.6668057665 \cdot 10^{-1} z^3 \\
& +2.0000714765 \cdot 10^{-1} z^2 - 2.4999993993 \cdot 10^{-1} z \\
& +3.3333331174 \cdot 10^{-1}) z^3 - 0.5 z^2 + z \; .
\end{aligned}
$$

4.19.5 $\sin x$

Single precision circular radian sine
Test interval: $[-\pi/4, +\pi/4]$
Random trials: 10000
Peak relative error: $6.8 \cdot 10^{-8}$
RMS relative error: $2.6 \cdot 10^{-8}$

$$\begin{aligned} z &= x^2 \\ \sin x &\approx \\ &(-1.9515295891 \cdot 10^{-4} z^2 + 8.3321608736 \cdot 10^{-3} z \\ &-1.6666654611 \cdot 10^{-1}) x^3 + x\ . \end{aligned}$$

4.19.6 $\sin^{-1} x$

Single precision circular radian arcsine
Test interval: $[-0.5, +0.5]$
Random trials: 10000
Peak relative error: $6.7 \cdot 10^{-8}$
RMS relative error: $2.5 \cdot 10^{-8}$

$$\begin{aligned} z &= x^2 \\ \sin^{-1} x &\approx \\ &(4.2163199048 \cdot 10^{-2} z^4 + 2.4181311049 \cdot 10^{-2} z^3 \\ &+4.5470025998 \cdot 10^{-2} z^2 + 7.4953002686 \cdot 10^{-2} z \\ &+1.6666752422 \cdot 10^{-1}) x^3 + x\ . \end{aligned}$$

4.19.7 Square Root

Single precision square root
Test interval: $[0.5, 2.0]$
Random trials: 50000
Peak relative error: $8.8 \cdot 10^{-8}$
RMS relative error: $3.3 \cdot 10^{-8}$

By integer operations on the exponent, the range is reduced to $0.5 \leq x < 2.0$. This interval is covered by three polynomial approximations. The subintervals given here for the three approximations are probably not the best possible choices.

$$\begin{aligned} \sqrt{2} &< x \leq 2 \\ w &= x - 2 \\ \sqrt{x} &\approx \end{aligned}$$

$$9.8843065718 \cdot 10^{-4} w^6 + 7.9479950957 \cdot 10^{-4} w^5$$
$$-3.5890535377 \cdot 10^{-3} w^4 + 1.1028809744 \cdot 10^{-2} w^3$$
$$-4.4195203560 \cdot 10^{-2} w^2 + 3.5355338194 \cdot 10^{-1} w$$
$$+1.41421356237 \, .$$

$$\begin{aligned} \sqrt{2}/2 &\le x \le \sqrt{2} \\ w &= x - 1 \\ \sqrt{x} &\approx \end{aligned}$$

$$(1.35199291026 \cdot 10^{-2} w^5 - 2.26657767832 \cdot 10^{-2} w^4$$
$$+2.78720776889 \cdot 10^{-2} w^3 - 3.89582788321 \cdot 10^{-2} w^2$$
$$+6.24811144548 \cdot 10^{-2} w - 1.25001503933 \cdot 10^{-1}) w^2$$
$$+0.5w + 1.0 \, .$$

$$\begin{aligned} \tfrac{1}{2} &\le x < \sqrt{2}/2 \\ w &= x - \tfrac{1}{2} \\ \sqrt{x} &\approx \end{aligned}$$

$$-3.9495006054 \cdot 10^{-1} w^6 + 5.1743034569 \cdot 10^{-1} w^5$$
$$-4.3214437330 \cdot 10^{-1} w^4 + 3.5310730460 \cdot 10^{-1} w^3$$
$$-3.5354581892 \cdot 10^{-1} w^2 + 7.0710676017 \cdot 10^{-1} w$$
$$+7.07106781187 \cdot 10^{-1} \, .$$

4.19.8 $\tan x$

Single precision circular radian tangent
Test interval: $[-\pi/4, +\pi/4]$
Random trials: 10000
Peak relative error: $8.7 \cdot 10^{-8}$
RMS relative error: $2.8 \cdot 10^{-8}$

$$\begin{aligned} z &= x^2 \\ \tan x &\approx \end{aligned}$$

$$(9.38540185543 \cdot 10^{-3} z^5 + 3.11992232697 \cdot 10^{-3} z^4$$
$$+2.44301354525 \cdot 10^{-2} z^3 + 5.34112807005 \cdot 10^{-2} z^2$$
$$+1.33387994085 \cdot 10^{-1} z + 3.33331568548 \cdot 10^{-1}) x^3$$
$$+x \, .$$

4.19.9 $\tan^{-1} x$

Single precision circular radian arctangent

Test interval: $[-\tan(\pi/8), +\tan(\pi/8)]$
Random trials: 10000
Peak relative error: $7.7 \cdot 10^{-8}$
RMS relative error: $2.9 \cdot 10^{-8}$

$$\begin{aligned} z &= x^2 \\ \tan^{-1} x &\approx \\ & (8.05374449538 \cdot 10^{-2} z^3 - 1.38776856032 \cdot 10^{-1} z^2 \\ & +1.99777106478 \cdot 10^{-1} z - 3.33329491539 \cdot 10^{-1}) x^3 \\ & +x \,. \end{aligned}$$

4.19.10 $\tanh x$

Single precision hyperbolic tangent
Test interval: $[-0.625, +0.625]$
Random trials: 10000
Peak relative error: $7.2 \cdot 10^{-8}$
RMS relative error: $2.6 \cdot 10^{-8}$

$$\begin{aligned} z &= x^2 \\ \tanh x &\approx \\ & (-5.70498872745 \cdot 10^{-3} z^4 + 2.06390887954 \cdot 10^{-2} z^3 \\ & -5.37397155531 \cdot 10^{-2} z^2 + 1.33314422036 \cdot 10^{-1} z \\ & -3.33332819422 \cdot 10^{-1}) x^3 + x \,. \end{aligned}$$

4.19.11 $\tanh^{-1} x$

Single precision inverse hyperbolic tangent
Test interval: $[-0.5, +0.5]$
Random trials: 10000
Peak relative error: $8.2 \cdot 10^{-8}$
RMS relative error: $3.0 \cdot 10^{-8}$

$$\begin{aligned} z &= x^2 \\ \tanh^{-1} x &\approx \\ & (1.81740078349 \cdot 10^{-1} z^4 + 8.24370301058 \cdot 10^{-2} z^3 \\ & +1.46691431730 \cdot 10^{-1} z^2 + 1.99782164500 \cdot 10^{-1} z \\ & +3.33337300303 \cdot 10^{-1}) x^3 + x \,. \end{aligned}$$

5

Probability Distributions and Related Functions

Statistics and probability rank among the more important areas of applied mathematics. This chapter has the theme, or objective, to compute some commonly encountered probability distribution functions (that is, integrals of probability densities). To accomplish this goal several related functions are required. These include the gamma and beta functions and their associated integrals. They are important in their own right and have applications in areas other than statistics. Their relation to statistics is that a number of frequently used probability distributions can be obtained from the incomplete beta and gamma integrals by simple change of variables. This connection was a considerable preoccupation of Bayes and Pierce, who prepared tables of the functions; but it is little discussed in present day texts on statistics.

Owing to the many threads of independent development, the nomenclature of statistical functions varies. Check the definitions carefully. The term "inverse" of a distribution will sometimes be used here to denote the functional inverse $x = P^{-1}(y)$ of a distribution function $y = P(x)$. This is the value of the random variable x, whose density is $p_x(x)$, for which the probability

$$P(x) = \int_{-\infty}^{x} p_x(t)\, dt$$

is equal to y.

5.1 n!

The factorial

$$
\begin{aligned}
n! &= \prod_{k=1}^{n} k \\
&= 1 \cdot 2 \cdot 3 \cdots n
\end{aligned}
$$

for integer n appears ubiquitously in combinatorial mathematics. 0! is defined to be 1. If n is negative, $n!$ may be regarded as undefined, or else as infinitely large. The function can be calculated by the recursion

$$n! = n\ (n-1)!$$

though the subroutine linkages for such a computer program would make it unnecessarily slow. A better approach is to use a simple do loop of the form

```
y = 1.0;
for( i=1; i<=n; i++ )
      y = y * i;
```

On some computers it may be faster to include a separate double precision variable, to which 1.0 is added each time, than to perform the implicit conversion of i to double precision.

For small n the values of $n!$ may simply be tabulated. Table 5.1 gives the values for n up to 33. In DEC arithmetic this fills the whole range of possible double precision numbers, so no computation is required.

In IEEE double precision arithmetic, the factorial does not overflow until after 170!. For large k the gamma function (see the next section), which satisfies

$$\Gamma(x+1) = x\Gamma(x)$$

for arbitrary x, can be used to find $n!$ through

$$n! = \Gamma(n+1)\ .$$

The program shown calls the gamma function for values of n greater than 55. In the range $56 \le n \le 170$ the peak relative error is $1.4 \cdot 10^{-15}$.

For values that are only somewhat beyond the range of the table, computing $n!$ by

$$n! = m!(m+1)(m+2)\cdots(n-1)n$$

where $m!$ is the largest tabulated value, is faster than computing the gamma function, and no less accurate.

Table 5.1: Table of Factorials

n	$n!$
0	$1.0 \cdot 10^0$
1	$1.0 \cdot 10^0$
2	$2.0 \cdot 10^0$
3	$6.0 \cdot 10^0$
4	$2.4 \cdot 10^1$
5	$1.2 \cdot 10^2$
6	$7.2 \cdot 10^2$
7	$5.04 \cdot 10^3$
8	$4.032 \cdot 10^4$
9	$3.6288 \cdot 10^5$
10	$3.6288 \cdot 10^6$
11	$3.99168 \cdot 10^7$
12	$4.790016 \cdot 10^8$
13	$6.2270208 \cdot 10^9$
14	$8.71782912 \cdot 10^{10}$
15	$1.307674368 \cdot 10^{12}$
16	$2.0922789888 \cdot 10^{13}$
17	$3.55687428096 \cdot 10^{14}$
18	$6.402373705728 \cdot 10^{15}$
19	$1.21645100408832 \cdot 10^{17}$
20	$2.43290200817664 \cdot 10^{18}$
21	$5.109094217170944 \cdot 10^{19}$
22	$1.12400072777760768 \cdot 10^{21}$
23	$2.585201673888497664 \cdot 10^{22}$
24	$6.2044840173323943936 \cdot 10^{23}$
25	$1.5511210043330985984 \cdot 10^{25}$
26	$4.03291461126605635584 \cdot 10^{26}$
27	$1.088886945041835216076 8 \cdot 10^{28}$
28	$3.0488834461171386050150 4 \cdot 10^{29}$
29	$8.841761993739701954543616 \cdot 10^{30}$
30	$2.6525285981219105863630848 \cdot 10^{32}$
31	$8.22283865417792281772556288 \cdot 10^{33}$
32	$2.631308369336935301672180121 6 \cdot 10^{35}$
33	$8.683317618811886495518194401 28 \cdot 10^{36}$

5.1.1 fac.c

Factorial function.

```
#include "mconf.h"
extern double MAXNUM;
```

```
#ifdef DEC
static short factbl[] = {
0040200,0000000,0000000,0000000,
0040200,0000000,0000000,0000000,
0040400,0000000,0000000,0000000,
0040700,0000000,0000000,0000000,
0041300,0000000,0000000,0000000,
0041760,0000000,0000000,0000000,
0042464,0000000,0000000,0000000,
0043235,0100000,0000000,0000000,
0044035,0100000,0000000,0000000,
0044661,0030000,0000000,0000000,
0045535,0076000,0000000,0000000,
0046430,0042500,0000000,0000000,
0047344,0063740,0000000,0000000,
0050271,0112146,0000000,0000000,
0051242,0060731,0040000,0000000,
0052230,0035673,0126000,0000000,
0053230,0035673,0126000,0000000,
0054241,0137567,0063300,0000000,
0055265,0173546,0051630,0000000,
0056330,0012711,0101504,0100000,
0057407,0006635,0171012,0150000,
0060461,0040737,0046656,0030400,
0061563,0135223,0005317,0101540,
0062657,0027031,0127705,0023155,
0064003,0061223,0041723,0156322,
0065115,0045006,0014773,0004410,
0066246,0146044,0172433,0173526,
0067414,0136077,0027317,0114261,
0070566,0044556,0110753,0045465,
0071737,0031214,0032075,0036050,
0073121,0037543,0070371,0064146,
0074312,0132550,0052561,0116443,
0075512,0132550,0052561,0116443,
0076721,0005423,0114035,0025014
};
#define MAXFAC 33
#endif
```

```
#ifdef IBMPC
static short factbl[] = {
0x0000,0x0000,0x0000,0x3ff0,
0x0000,0x0000,0x0000,0x3ff0,
0x0000,0x0000,0x0000,0x4000,
0x0000,0x0000,0x0000,0x4018,
0x0000,0x0000,0x0000,0x4038,
0x0000,0x0000,0x0000,0x405e,
0x0000,0x0000,0x8000,0x4086,
0x0000,0x0000,0xb000,0x40b3,
0x0000,0x0000,0xb000,0x40e3,
0x0000,0x0000,0x2600,0x4116,
0x0000,0x0000,0xaf80,0x414b,
0x0000,0x0000,0x08a8,0x4183,
0x0000,0x0000,0x8cfc,0x41bc,
0x0000,0xc000,0x328c,0x41f7,
0x0000,0x2800,0x4c3b,0x4234,
0x0000,0x7580,0x0777,0x4273,
0x0000,0x7580,0x0777,0x42b3,
0x0000,0xecd8,0x37ee,0x42f4,
0x0000,0xca73,0xbeec,0x4336,
0x9000,0x3068,0x02b9,0x437b,
0x5a00,0xbe41,0xe1b3,0x43c0,
0xc620,0xe9b5,0x283b,0x4406,
0xf06c,0x6159,0x7752,0x444e,
0xa4ce,0x35f8,0xe5c3,0x4495,
0x7b9a,0x687a,0x6c52,0x44e0,
0x6121,0xc33f,0xa940,0x4529,
0x7eeb,0x9ea3,0xd984,0x4574,
0xf316,0xe5d9,0x9787,0x45c1,
0x6967,0xd23d,0xc92d,0x460e,
0xa785,0x8687,0xe651,0x465b,
0x2d0d,0x6e1f,0x27ec,0x46aa,
0x33a4,0x0aae,0x56ad,0x46f9,
0x33a4,0x0aae,0x56ad,0x4749,
0xa541,0x7303,0x2162,0x479a
};
#define MAXFAC 170
#endif
```

```
extern double MAXNUM;
```

```
double fac(i)
int i;
    {
    double x, f, n;
    int j;
    double gamma();

    if( i < 0 )
        {
        mtherr( "fac", SING );
        return( MAXNUM );
        }
    if( i > MAXFAC )
        {
        mtherr( "fac", OVERFLOW );
        return( MAXNUM );
        }
```

Get answer from table for small i.

```
    if( i < 34 )
        {
#ifdef UNK
        return( factbl[i] );
#else
            return( *(double *)(&factbl[4*i]) );
#endif
        }
```

Use gamma function for large i.

```
    if( i > 55 )
        {
        x = i + 1;
        return( gamma(x) );
        }
```

Compute directly for intermediate i.

```
    n = 34.0;
    f = 34.0;
    for( j=35; j<=i; j++ )
        {
        n += 1.0;
        f *= n;
        }
#ifdef UNK
    f *= factbl[33];
#else
        f *= *(double *)(&factbl[4*33]);
#endif
```

```
return( f );
}
```

5.2 $\Gamma(x)$

The gamma function, which generalizes the factorial to noninteger arguments, can be defined by the integral

$$\Gamma(x) = \int_0^\infty e^{-t} t^{x-1} dt \; .$$

For sufficiently large x, the following asymptotic expansion may be used to compute the logarithm of the gamma function.

$$\ln\Gamma(x) \approx \tfrac{1}{2}\ln 2\pi - x + (x - \tfrac{1}{2})\ln x + \sum_{k=1}^{\infty} \frac{B_{2k}}{2k(2k-1)x^{2k-1}}$$

where the B_{2k} are Bernoulli numbers. These numbers may be found by solving the equations

$$\begin{aligned}
B_0 &= 1 \\
B_0 + 2B_1 &= 0 \\
B_0 + 3B_1 + 3B_2 &= 0 \\
B_0 + 4B_1 + 6B_2 + 4B_3 &= 0 \\
\sum_{k=0}^{n-1} \frac{n!}{k!\,(n-k)!} B_k &= 0 \; .
\end{aligned}$$

The equations become quite ill-conditioned as k becomes large, though all the B_k are rational numbers.

The error of the asymptotic expansion is less than the first omitted term. For *a priori* calculation when x is small, the recurrence

$$\Gamma(x+1) = x\,\Gamma(x)$$

is used to transform x to a sufficiently large value. Written out with exact expressions for the coefficients, the formula is

$$\begin{aligned}
\ln\Gamma(x) \approx\; & \tfrac{1}{2}\ln 2\pi + (x - \tfrac{1}{2})\ln x - x \\
& + \tfrac{1}{12}x^{-1} - \tfrac{1}{360}x^{-3} + \tfrac{1}{1260}x^{-5} - \tfrac{1}{1680}x^{-7} + \tfrac{1}{1188}x^{-9} \\
& - \tfrac{691}{360360}x^{-11} + \tfrac{1}{156}x^{-13} - \tfrac{3617}{122400}x^{-15} + \tfrac{43867}{244188}x^{-17} \\
& - \tfrac{174611}{125400}x^{-19} + \tfrac{77683}{5796}x^{-21} - \tfrac{236364091}{1506960}x^{-23} + \tfrac{657931}{300}x^{-25} \\
& - \tfrac{3392780147}{9396}x^{-27} + \tfrac{1723168255201}{2492028}x^{-29} - \tfrac{7709321041217}{505920}x^{-31} \\
& + \tfrac{151628697551}{396}x^{-33} - \tfrac{26315271553053477373}{2418179400}x^{-35}
\end{aligned}$$

$$+\frac{154210205991661}{444}x^{-37} - \frac{261082718496449122051}{21106800}x^{-39}$$
$$+\frac{152009764391807085252691}{3109932}x^{-41} - \frac{2530297234481911294093}{118680}x^{-43} + \ldots$$

Exponentiating the formula for $\ln\Gamma(x)$ yields Stirling's formula for the gamma function, which is

$$\begin{aligned}
\Gamma(x) \quad \approx \quad & \sqrt{2\pi}\; x^{x-\frac{1}{2}} e^{-x} \left(1 + \tfrac{1}{12}x^{-1} + \tfrac{1}{288}x^{-2} - \tfrac{139}{51840}x^{-3} - \tfrac{571}{2488320}x^{-4}\right. \\
& + \tfrac{163879}{209018880}x^{-5} + \tfrac{5246819}{75246796800}x^{-6} - \tfrac{534703531}{902961561600}x^{-7} \\
& - \tfrac{4483131259}{86684309913600}x^{-8} + \tfrac{432261921612371}{514904800886784000}x^{-9} \\
& + \tfrac{6232523202521089}{86504006548979712000}x^{-10} - \tfrac{25834629665134204969}{13494625021640835072000}x^{-11} \\
& - \tfrac{1579029138854919086429}{9716130015581401251840000}x^{-12} + \tfrac{746590869962651602203151}{116593560186976815022080000}x^{-13} \\
& + \tfrac{1511513601028097903631961}{2798245444487443560529920000}x^{-14} \\
& - \tfrac{8849272268392873147705987190261}{299692087104605205332754432000000}x^{-15} \\
& - \tfrac{142801712490607530608130701097701}{57540880724084199423888850944000000}x^{-16} \\
& + \tfrac{2355444393109967510921431436000087153}{13119320805091197468646658015232000000}x^{-17} \\
& \left. + \tfrac{2346608607351903737647919577082115121863}{1558575311644834259275222972209561600000000}x^{-18} - \ldots\right)
\end{aligned}$$

This expansion converges about twice as slowly as the one for $\ln\Gamma(x)$.

The range of $\Gamma(x)$ is greater than that of the exponential function. Thus overflow can occur even for $\ln\Gamma(x)$. Having subroutines for both $\Gamma(x)$ and $\ln\Gamma(x)$ permits expressions involving a ratio of gamma functions or expressions such as $e^{-x}\Gamma(x)$ to be calculated over a wider domain than if only $\Gamma(x)$ were available. If x is negative, overflow of $\Gamma(x)$ may occur for nearly integer values of x.

For large x, $\Gamma(x)$ could be computed by finding $w = \ln\Gamma(x)$, then $\Gamma(x) = e^{w}$. However, Stirling's formula gives a more accurate result. It can be computed in the form

$$\Gamma(x) \approx \frac{\sqrt{2\pi}\; x^{x-\frac{1}{2}}}{e^{x}} \{1 + (1/x)S(1/x)\}$$

where $S(1/x)$ is a polynomial approximation to the asymptotic series. Note that, in the exponentiated term, subtracting $\frac{1}{2}$ from x is an exact arithmetic operation, as no nonzero bit is shifted out from the significand. Despite this exactness of the argument, the error of the approximation is dominated by the accuracy to which x^{x} can be calculated. Using the extended precision Cody and Waite **pow**() routine from the previous chapter, the rms error in $\Gamma(x)$ for $33 < x < 143$ is $2.8 \cdot 10^{-16}$. When x^{x} is computed by **exp**(x **log**(x)), the rms error of $\Gamma(x)$ in the same interval is $3.2 \cdot 10^{-14}$ — an increase of two orders of magnitude. When $\Gamma(x)$ is computed by **exp**(**lgam**(x)) the error is $4.4 \cdot 10^{-14}$.

Though x^x overflows before $\Gamma(x)$, the power function can still be used at large x by computing $w = x^{x/2}$ whence $x^x = w^2$. See the program listings for details. The coefficients for Stirling's formula are

$S(t) =$
$7.87311395793093628397 \cdot 10^{-4}t^4$
$-2.29549961613378126380 \cdot 10^{-4}t^3$
$-2.68132617805781232825 \cdot 10^{-3}t^2$
$+3.47222221605458667310 \cdot 10^{-3}t$
$+8.33333333333482257126 \cdot 10^{-2}$

and

$$\sqrt{2\pi} = 2.50662827463100050242 .$$

The form $1/x\ S(1/x)$ has an absolute error of $1.4 \cdot 10^{-18}$ for $33 \leq x \leq 172$.

For $|x| \leq 34$, the argument may be transformed to the interval $2 \leq x \leq 3$ by repeated application of one of the equivalent recurrence formulas

$$\begin{aligned} \Gamma(x) &= (x-1)\Gamma(x-1) \\ \Gamma(x) &= \Gamma(x+1)/x . \end{aligned}$$

The first recurrence is used while $x > 3$, the second while $x < 2$. In the computer these are implemented by the following assignments starting with $w = 1$. While $x > 3$,

$$\begin{aligned} x &:= x - 1 \\ w &:= wx . \end{aligned}$$

While $x < 2$,

$$\begin{aligned} w &:= w/x \\ x &:= x + 1 . \end{aligned}$$

In the latter case there is a possiblity of passing through $x = 0$. If during the recurrence one finds that $|x| < 10^{-7}$ then with sufficient accuracy

$$\Gamma(x) \approx \frac{w}{x + \gamma x^2}$$

where γ is Euler's constant. Having reduced x to the interval [2,3], if x is exactly 2 or exactly 3, then $\Gamma(x) = w$. Intermediate values may be handled by the rational approximation

$$\Gamma(x) \approx (x-2)\,\frac{P(x-2)}{Q(x-2)}$$

where

$$
\begin{aligned}
P(t) = {} & \\
& +1.60119522476751861407 \cdot 10^{-4} t^6 \\
& +1.19135147006586384913 \cdot 10^{-3} t^5 \\
& +1.04213797561761569935 \cdot 10^{-2} t^4 \\
& +4.76367800457137231464 \cdot 10^{-2} t^3 \\
& +2.07448227648435975150 \cdot 10^{-1} t^2 \\
& +4.94214826801497100753 \cdot 10^{-1} t \\
& +9.99999999999999996796 \cdot 10^{-1} \; ,
\end{aligned}
$$

$$
\begin{aligned}
Q(t) = {} & \\
& -2.31581873324120129819 \cdot 10^{-5} t^7 \\
& +5.39605580493303397842 \cdot 10^{-4} t^6 \\
& -4.45641913851797240494 \cdot 10^{-3} t^5 \\
& +1.18139785222060435552 \cdot 10^{-2} t^4 \\
& +3.58236398605498653373 \cdot 10^{-2} t^3 \\
& -2.34591795718243348568 \cdot 10^{-1} t^2 \\
& +7.14304917030273074085 \cdot 10^{-2} t \\
& +1.00000000000000000320 \; .
\end{aligned}
$$

This has a theoretical relative error of $6.4 \cdot 10^{-18}$.

The following reflection formula should be used in both the gamma and log gamma routines when x is large and negative:

$$\Gamma(x)\,\Gamma(1-x) = \frac{\pi}{\sin \pi x}\;.$$

For the log gamma function, this is implemented by

$$
\begin{aligned}
s &= \sin(-\pi x) \\
u &= -x|s| \\
\ln|\Gamma(x)| &= \ln \pi - \ln u - \ln \Gamma(-x)\;.
\end{aligned}
$$

If x is a nonpositive integer, then $u = 0$ and an error escape must be provided. The sine of πx can be calculated accurately by performing some of the range reduction before calling the sine subroutine. Do this by subtracting x from its nearest integer neighbor. *Then* multiply by π and call the sine. By this means x is mapped to the interval from $-\pi/2$ to $+\pi/2$, and integer x becomes zero exactly. If the input argument $x < 0$ is regarded as exact, this scheme yields low *relative* error in $\Gamma(x)$ at all points in the domain.

When computing either $\Gamma(x)$ or $\ln \Gamma(x)$, the sign of $\Gamma(x)$ should be returned in a separate integer variable. It is the opposite of the sign of s. The computer subroutine will return $\ln|\Gamma(x)|$ as a double precision result and also the sign of $\Gamma(x)$ as an integer. This makes it possible for other programs to avoid overflow in intermediate calculations by using $\ln \Gamma(x)$ with known sign, instead of having to use $\Gamma(x)$.

To compute $\ln \Gamma(x)$ the domain must be partitioned into several segments. For $x \geq 13$, the logarithmic version of Stirling's formula may be used in one of the following forms.

$$q = (x - 0.5)\ln x - x + \ln \sqrt{2\pi}$$

where the constant $\ln \sqrt{2\pi} = 0.91893853320467274178$. If $x > 10^8$ then $\ln \Gamma(x) \approx q$. For $10^8 \geq x \geq 10^3$, set $v = x^{-2}$. Then

$$\ln \Gamma(x) \approx q + \frac{v^2/1260 - v/360 + 1/12}{x}$$

where the constants may be expressed as multiplicative equivalents to reduce the number of division operations. For $10^3 > x \geq 13$ use

$$\ln \Gamma(x) \approx q + \frac{A(v)}{x}$$

where
$A(t) =$
$8.11614167470508450300 \cdot 10^{-4} t^4$
$-5.95061904284301438324 \cdot 10^{-4} t^3$
$+7.93650340457716943945 \cdot 10^{-4} t^2$
$-2.77777777730099687205 \cdot 10^{-3} t$
$+8.33333333333331927722 \cdot 10^{-2}$.

This form has an absolute error of $9.8 \cdot 10^{-19}$ for x between 13 and 1000.

If x is in the range $(-34, +13)$ then reduce it by recurrence to the half open interval [2,3) using the method described above for $\Gamma(x)$. Test w as found by the recurrence. It may be either positive or negative, corresponding to the sign of $\Gamma(x)$. If it is negative, then negate it and also negate the sign variable that will be returned as a subroutine result. At this point, if x is exactly 2, then $\ln \Gamma(x) = \ln w$. Otherwise,

$$\ln \Gamma(x) \approx \ln w + (x-2) \frac{B(x-2)}{C(x-2)}$$

where

$B(t) =$
$-1.37825152569120859100 \cdot 10^3 t^5$
$-3.88016315134637840924 \cdot 10^4 t^4$
$-3.31612992738871184744 \cdot 10^5 t^3$
$-1.16237097492762307383 \cdot 10^6 t^2$
$-1.72173700820839662146 \cdot 10^6 t$
$-8.53555664245765465627 \cdot 10^5$,

$C(t) =$
$+1.0 t^6$
$-3.51815701436523470549 \cdot 10^2 t^5$
$-1.70642106651881159223 \cdot 10^4 t^4$
$-2.20528590553854454839 \cdot 10^5 t^3$
$-1.13933444367982507207 \cdot 10^6 t^2$
$-2.53252307177582951285 \cdot 10^6 t$
$-2.01889141433532773231 \cdot 10^6$.

Table 5.2 gives test results for the gamma and log gamma routines. The error criterion for log gamma is absolute if its value is less than 1; otherwise the relative error criterion applies.

5.2.1 gamma.c

```
#include "mconf.h"

#ifdef DEC
static short P[] = {
0035047,0162701,0146301,0005234,
0035634,0023437,0032065,0176530,
0036452,0137157,0047330,0122574,
```

```
#ifdef IIEEE
static short P[] = {
0x2153,0x3998,0xfcb8,0x3f24,
0xbfab,0xe686,0x84e3,0x3f53,
0x14b0,0xe9db,0x57cd,0x3f85,
```

Function	Arith.	Domain	Trials	Peak	RMS
gamma	DEC	$-34, 34$	10000	$1.3 \cdot 10^{-16}$	$2.5 \cdot 10^{-17}$
gamma	IEEE	$-170, -33$	20000	$2.3 \cdot 10^{-15}$	$3.3 \cdot 10^{-16}$
gamma	IEEE	$-33, 33$	20000	$9.4 \cdot 10^{-16}$	$2.2 \cdot 10^{-16}$
gamma	IEEE	$33, 171.6$	20000	$2.3 \cdot 10^{-15}$	$3.2 \cdot 10^{-16}$
lgam	DEC	$0, 3$	7000	$5.2 \cdot 10^{-17}$	$1.3 \cdot 10^{-17}$
lgam	DEC	$2.718, 2.0 \cdot 10^{36}$	5000	$3.9 \cdot 10^{-17}$	$9.9 \cdot 10^{-18}$
lgam	IEEE	$0, 3$	28000	$5.4 \cdot 10^{-16}$	$1.1 \cdot 10^{-16}$
lgam	IEEE	$2.718, 2.6 \cdot 10^{305}$	40000	$3.5 \cdot 10^{-16}$	$8.3 \cdot 10^{-17}$
lgam	IEEE	$-200, -4$	10000	$4.8 \cdot 10^{-16}$	$1.3 \cdot 10^{-16}$

Table 5.2: Accuracy of gamma functions

```
0037103,0017310,0143041,0017232,
0037524,0066516,0162563,0164605,
0037775,0004671,0146237,0014222,
0040200,0000000,0000000,0000000
};
static short Q[] = {
0134302,0041724,0020006,0116565,
0035415,0072121,0044251,0025634,
0136222,0003447,0035205,0121114,
0036501,0107552,0154335,0104271,
0037022,0135717,0014776,0171471,
0137560,0034324,0165024,0037021,
0037222,0045046,0047151,0161213,
0040200,0000000,0000000,0000000
};
static short STIR[20] = {
0035516,0061622,0144553,0112224,
0135160,0131531,0037460,0165740,
0136057,0134460,0037242,0077270,
0036143,0107070,0156306,0027751,
0037252,0125252,0125252,0146064
};
#define MAXSTIR 26.77
static short SQT[4] = {
0040440,0066230,0177661,0034055
};
#define SQTPI *(double *)SQT
#endif
```

```
0x23d3,0x18c4,0x63d9,0x3fa8,
0x7d31,0xdcae,0x8da9,0x3fca,
0xe312,0x3993,0xa137,0x3fdf,
0x0000,0x0000,0x0000,0x3ff0
};
static short Q[] = {
0xd3af,0x8400,0x487a,0xbef8,
0x2573,0x2915,0xae8a,0x3f41,
0xb44a,0xe750,0x40e4,0xbf72,
0xb117,0x5b1b,0x31ed,0x3f88,
0xde67,0xe33f,0x5779,0x3fa2,
0x87c2,0x9d42,0x071a,0xbfce,
0x3c51,0xc9cd,0x4944,0x3fb2,
0x0000,0x0000,0x0000,0x3ff0
};
static short STIR[20] = {
0x7293,0x592d,0xcc72,0x3f49,
0x1d7c,0x27e6,0x166b,0xbf2e,
0x4fd7,0x07d4,0xf726,0xbf65,
0xc5fd,0x1b98,0x71c7,0x3f6c,
0x5986,0x5555,0x5555,0x3fb5
};
#define MAXSTIR 143.01608
static short SQT[4] = {
0x2706,0x1ff6,0x0d93,0x4004
};
#define SQTPI *(double *)SQT
#endif
```

```
#ifdef DEC
#define MAXGAM 34.84425627277176174
#endif
#ifdef IIEEE
#define MAXGAM 171.624376956302725
#endif
```

Globally declared symbol is used to return sign of $\Gamma(x)$.

```
int sgngam = 0;
extern int sgngam;
extern double MAXLOG, MAXNUM, PI;
```

Gamma function computed by Stirling's formula.
The polynomial STIR is valid for $33 \leq x \leq 172$.

```
static double stirf(x)
double x;
     {
     double y, w, v;
     double pow(), exp(), polevl();
     double log();

     w = 1.0/x;
     w = 1.0 + w * polevl( w, STIR, 4 );
     y = exp(x);
     if( x > MAXSTIR )
          {
```

Avoid overflow in pow()

```
          v = pow( x, 0.5 * x - 0.25 );
          y = v * (v / y);
          }
     else
          {
          y = pow( x, x - 0.5 ) / y;
          }
     y = SQTPI * y * w;
     return( y );
     }
```

Main routine for $\Gamma(x)$

```
double gamma(x)
double x;
     {
     double p, q, z;
     int i;
```

```
    double fabs(), lgam(), log(), exp(), gamma(), sin();
    double floor(), polevl(), p1evl(), stirf();

    sgngam = 1;
    q = fabs(x);
    if( q > 33.0 )
         {
         if( x < 0.0 )
              {
              p = floor(q);
              if( p == q )
                   goto goverf;
              i = p;
              if( (i & 1) == 0 )
                   sgngam = -1;
              z = q - p;
              if( z > 0.5 )
                   {
                   p += 1.0;
                   z = q - p;
                   }
              z = q * sin( PI * z );
              if( z == 0.0 )
                   {
goverf:
                   mtherr( "gamma", OVERFLOW );
                   return( sgngam * MAXNUM);
                   }
              z = fabs(z);
              z = PI/(z * stirf(q) );
              }
         else
              {
              z = stirf(x);
              }
         return( sgngam * z );
         }
    z = 1.0;
    while( x >= 3.0 )
         {
         x -= 1.0;
         z *= x;
         }
    while( x < 0.0 )
         {
```

```
		if( x > -1.E-9 )
			goto small;
		z /=x;
		x += 1.0;
		}
	while( x < 2.0 )
		{
		if( x < 1.e-9 )
			goto small;
		z /=x;
		x += 1.0;
		}
	if( (x == 2.0) || (x == 3.0) )
		return(z);
	x -= 2.0;
	p = polevl( x, P, 6 );
	q = polevl( x, Q, 7 );
	return( z * p / q );
small:
	if( x == 0.0 )
		{
		mtherr( "gamma", SING );
		return( MAXNUM );
		}
	else
		return( z/((1.0 + 0.5772156649015329 * x) * x) );
	}
```

5.2.2 lgam.c

Logarithm of gamma function.

```
#include "mconf.h"

#ifdef DEC
static short A[] = {
0035524,0141201,0034633,0031405,
0135433,0176755,0126007,0045030,
0035520,0006371,0003342,0172730,
0136066,0005540,0132605,0026407,
0037252,0125252,0125252,0125132
};
static short B[] = {
0142654,0044014,0077633,0035410,
0144027,0110641,0125335,0144760,
0144641,0165637,0142204,0047447,
0145215,0162027,0146246,0155211,
```

```
#ifdef IIEEE
static short A[] = {
0x6661,0x2733,0x9850,0x3f4a,
0xe943,0xb580,0x7fbd,0xbf43,
0x5ebb,0x20dc,0x019f,0x3f4a,
0xa5a1,0x16b0,0xc16c,0xbf66,
0x554b,0x5555,0x5555,0x3fb5
};
static short B[] = {
0x6761,0x8ff3,0x8901,0xc095,
0xb93e,0x355b,0xf234,0xc0e2,
0x89e5,0xf890,0x3d73,0xc114,
0xdb51,0xf994,0xbc82,0xc131,
```

```
0145322,0026110,0010317,0110130,  0xf20b,0x0219,0x4589,0xc13a,
0145120,0061472,0120300,0025363,  0x055e,0x5418,0x0c67,0xc12a
};                                };
```

Leading 1.0 is omitted from array C.

```
static short C[] = {               static short C[] = {
0142257,0164150,0163630,0112622,  0x12b2,0x1cf3,0xfd0d,0xc075,
0143605,0050153,0156116,0135272,  0xd757,0x7b89,0xaa0d,0xc0d0,
0144527,0056045,0145642,0062332,  0x4c9b,0xb974,0xeb84,0xc10a,
0145213,0012063,0106250,0001025,  0x0043,0x7195,0x6286,0xc131,
0145432,0111254,0044577,0115142,  0xf34c,0x892f,0x5255,0xc143,
0145366,0071133,0050217,0005122   0xe14a,0x6a11,0xce4b,0xc13e
};                                };
```

$\ln \sqrt{2\pi}$

```
static short LS2P[] = {            static short LS2P[] = {
040153,037616,041445,0172645      0xbeb5,0xc864,0x67f1,0x3fed
};                                };
#endif                            #endif
#define LS2PI *(double *)LS2P
#ifdef DEC
#define MAXLGM 2.035093e36
#endif
#ifdef IIEEE
#define MAXLGM 2.556348e305
#endif

double lgam(x)
double x;
    {
    double p, q, z;
    int i;
    double fabs(), sin(), log(), gamma(), polevl();
    double p1evl(), floor();

    sgngam = 1;
    if( x < -34.0 )
        {
        q = -x;
        p = floor(q);
        if( p == q )
            goto loverf;
        i = p;
        if( (i & 1) == 0 )
            sgngam = -sgngam;
        z = q - p;
        if( z > 0.5 )
            {
            p += 1.0;
```

```
                z = p - q;
                sgngam = -sgngam;
                }
            z = q * sin( PI * z );
            if( z == 0.0 )
                goto loverf;
            z = log(PI) - log( z ) - lgam(q);
            return( z );
            }
        if( x < 13.0 )
            {
            z = 1.0;
            while( x >= 3.0 )
                {
                x -= 1.0;
                z *= x;
                }
            while( x < 2.0 )
                {
                if( x == 0.0 )
                    goto loverf;
                z /=x;
                x += 1.0;
                }
            if( z < 0.0 )
                {
                sgngam = -1;
                z = -z;
                }
            else
                sgngam = 1;
            if( x == 2.0 )
                return( log(z) );
            x -= 2.0;
            p = x * polevl( x, B, 5 ) / p1evl( x, C, 6);
            return( log(z) + p );
            }
        if( x > MAXLGM )
            {
loverf:
            mtherr( "lgam", OVERFLOW );
            return( sgngam * MAXNUM );
            }
        q = ( x - 0.5 ) * log(x) - x + LS2PI;
        if( x > 1.0e8 )
```

```
        return( q );
    p = 1.0/(x*x);
    if( x >= 1000.0 )
        q += (( 7.9365079365079365079365e-4 * p
        - 2.7777777777777777777778e-3) *p
        + 0.0833333333333333333333) / x;
    else
        q += polevl( p, A, 4 ) / x;
    return( q );
    }
```

5.3 Incomplete Gamma Integral

The incomplete gamma integral is defined by

$$\begin{aligned} P(a,x) &= \frac{1}{\Gamma(a)} \int_0^x e^{-t} t^{a-1} dt \\ &= \frac{x^a e^{-x}}{\Gamma(a+1)} {}_1F_1(1, a+1, x) \end{aligned}$$

where ${}_1F_1$ is the confluent hypergeometric function (see Section 7.1). Its complement is

$$P^*(a,x) = 1 - P(a,x)$$

which is the area under the right hand tail. The integral is related by change of variables to the gamma, Poisson, and χ^2 distributions. The connection with the Poisson distribution becomes evident on expanding by partial integration (see the power series below). Programs using the integral to compute these distribution functions are given in the following sections.

For $x \le 0$ or $a \le 0$, the program may return $P^*(a,x) = 1$. If $x > 1$ and $x > a$ the following continued fraction may be used.

$$\begin{aligned} P^*(x) &= \\ & c \, \frac{1}{x+} \, \frac{1-a}{1+} \, \frac{1}{x+} \, \frac{2-a}{1+} \, \frac{2}{x+\cdots} = \\ & c \, \frac{x-a}{x+1+} \, \frac{a-2}{x-a+4+} \, \frac{2a-6}{x-a+6+} \, \frac{3a-12}{x-a+8+} \, \frac{4a-20}{x-a+10+\cdots} \end{aligned}$$

where the normalizing factor c is

$$\begin{aligned} c &= \frac{x^a e^{-x}}{\Gamma(a)} \\ &= \exp\left(a \ln x \ - x - \ln\Gamma(a)\right) . \end{aligned}$$

An underflow escape should be provided if the exponential function would underflow, and the program may return 0 in this case. The area under

Function	Arithmetic	Domain	Trials	Peak	RMS
igam	DEC	0, 30	4000	$4.4 \cdot 10^{-15}$	$6.3 \cdot 10^{-16}$
igam	IEEE	0, 30	10000	$3.6 \cdot 10^{-14}$	$5.1 \cdot 10^{-15}$
igamc	DEC	0, 30	2000	$2.7 \cdot 10^{-15}$	$4.0 \cdot 10^{-16}$
igamc	IEEE	0, 30	60000	$1.4 \cdot 10^{-12}$	$6.3 \cdot 10^{-15}$
igami	DEC	0, 0.5	3400	$8.8 \cdot 10^{-16}$	$1.3 \cdot 10^{-16}$
igami	IEEE	0, 0.5	10000	$1.1 \cdot 10^{-14}$	$1.0 \cdot 10^{-15}$

Table 5.3: Gamma integrals, relative error

the left tail, $P(a,x)$, is 0 when $x \leq 0$ or $a \leq 0$. If $x > 1$ and $x > a$ then the continued fraction above may be used to form $P(a,x) = 1 - P^*(a,x)$. The following power series expansion is applicable when $0 < x < 1$ or $x < a$:

$$\begin{aligned} P(a,x) &= x^a e^{-x} \sum_{k=0}^{\infty} \frac{x^k}{\Gamma(a+k+1)} \\ &= \frac{x^a e^{-x}}{\Gamma(a+1)} \left\{ 1 + \sum_{k=1}^{\infty} \frac{x^k}{(a+1)(a+2)\cdots(a+k)} \right\} . \end{aligned}$$

Table 5.3 gives experimental results for the gamma integral programs. The functional inverse igami.c was tested for a ranging from 0 to 30 and x ranging from 0 to 0.5.

5.3.1 igamc.c

Right tail of incomplete gamma function

```
#include "mconf.h"

#define BIG 1.44115188075855872E+17
extern double MACHEP, MAXLOG;

double igamc( a, x )
double a, x;
    {
    double ans, c, yc, ax, y, z;
    double pk, pkm1, pkm2, qk, qkm1, qkm2;
    double r, t;
```

```
    double lgam(), exp(), log(), fabs();
    double igam();
    static double big = BIG;

    if( (x <= 0) || ( a <= 0) )
         return( 1.0 );
    if( (x < 1.0) || (x < a) )
         return( 1.0 - igam(a,x) );
    ax = a * log(x) - x - lgam(a);
    if( ax < -MAXLOG )
         {
         mtherr( "igamc", UNDERFLOW );
         return( 0.0 );
         }
    ax = exp(ax);
continued fraction
    y = 1.0 - a;
    z = x + y + 1.0;
    c = 0.0;
    pkm2 = 1.0;
    qkm2 = x;
    pkm1 = x + 1.0;
    qkm1 = z * x;
    ans = pkm1/qkm1;
    do
         {
         c += 1.0;
         y += 1.0;
         z += 2.0;
         yc = y * c;
         pk = pkm1 * z - pkm2 * yc;
         qk = qkm1 * z - qkm2 * yc;
         if( qk != 0 )
              {
              r = pk/qk;
              t = fabs( (ans - r)/r );
              ans = r;
              }
         else
              t = 1.0;
         pkm2 = pkm1;
         pkm1 = pk;
         qkm2 = qkm1;
         qkm1 = qk;
         if( fabs(pk) > big )
```

```
                {
                pkm2 /= big;
                pkm1 /= big;
                qkm2 /= big;
                qkm1 /= big;
                }
        }
    while( t > MACHEP );
    return( ans * ax );
    }
```

5.3.2 igam.c

Left tail of incomplete gamma integral

```
double igam( a, x )
double a, x;
    {
    double ans, ax, c, r, t;
    double lgam(), exp(), log();
    double igamc();
    static double big = BIG;

    if( (x <= 0) || ( a <= 0) )
        return( 0.0 );
    if( (x > 1.0) && (x > a ) )
        return( 1.0 - igamc(a,x) );
```

Compute $x^a e^{-x}/\Gamma(a)$

```
    ax = a * log(x) - x - lgam(a);
    if( ax < -MAXLOG )
        {
        mtherr( "igam", UNDERFLOW );
        return( 0.0 );
        }
    ax = exp(ax);
```

Power series

```
    r = a;
    c = 1.0;
    ans = 1.0;
    do
        {
        r += 1.0;
        c *= x/r;
        ans += c;
```

```
        }
    while( c/ans > MACHEP );
    return( ans * ax/a );
    }
```

5.3.3 Functional Inverse of Incomplete Gamma Integral

This program determines the value of the random variable x such that $1 - P(a, x)$ takes on a given value y. The parameter a is also regarded as given.

Starting with an approximate value of y, successively refined estimates are calculated by a Newton-Raphson iteration.

5.3.4 igami.c

```
#include "mconf.h"

extern double MACHEP, MAXNUM, MAXLOG, MINLOG;

double igami( a, y0 )
double a, y0;
    {
    double d, y, x0, lgm;
    int i;
    double igamc();
    double ndtri(), exp(), fabs(), log(), sqrt(), lgam();
```

Approximation to inverse function (AMS55 Chap. 26)

```
    d = 1.0/(9.0*a);
    y = ( 1.0 - d - ndtri(y0) * sqrt(d) );
    x0 = a * y * y * y;
    lgm = lgam(a);
```

Refine the approximation by a Newton-Raphson iteration.

```
    for( i=0; i<10; i++ )
        {
        if( x0 <= 0.0 )
            {
            mtherr( "igami", UNDERFLOW );
            return(0.0);
            }
        y = igamc(a,x0);
```

Compute the derivative of the function at this point.

```
        d = (a - 1.0) * log(x0) - x0 - lgm;
        if( d < -MAXLOG )
```

```
                {
                mtherr( "igami", UNDERFLOW );
                goto done;
                }
            d = -exp(d);
```

Step to the next approximation of x.

```
            if( d == 0.0 )
                goto done;
            d = (y - y0)/d;
            x0 = x0 - d;
            if( i < 3 )
                continue;
            if( fabs(d/x0) < 2.0 * MACHEP )
                break;
            }
done:
    return( x0 );
    }
```

5.4 Gamma Distribution

The distribution function of the gamma density is

$$\begin{aligned} \text{gdtr}(a,b,x) &= \frac{a^b}{\Gamma(b)} \int_0^x t^{b-1} e^{-at} dt \\ &= P(b, ax) \ . \end{aligned}$$

It is the same as the incomplete gamma integral except for the addition of the scale parameter a. The complementary function is

$$\text{gdtrc}(a,b,x) = 1 - \text{gdtr}(a,b,x) = P^*(b,ax) \ .$$

5.4.1 gdtr.c

Gamma distribution.

```
#include "mconf.h"

double gdtr( a, b, x )
double a, b, x;
    {
    double igam();

    if( x < 0.0 )
        {
        mtherr( "gdtr", DOMAIN );
```

```
        return( 0.0 );
        }
    return( igam( b, a * x ) );
    }
```

5.4.2 gdtrc.c

Complemented gamma distribution 1 – **gdtrc**().

```
double gdtrc( a, b, x )
double a, b, x;
    {
    double igamc();

    if( x < 0.0 )
        {
        mtherr( "gdtrc", DOMAIN );
        return( 0.0 );
        }
    return( igamc( b, a * x ) );
    }
```

5.5 χ^2 Distribution

The area under the left hand tail, from 0 to x, of the Chi-square probability density function with ν degrees of freedom is the integral

$$\begin{aligned} \text{chdtr}(\nu, \chi^2) &= \frac{1}{2^{\nu/2}\Gamma(\nu/2)} \int_0^{\chi^2} t^{\nu/2-1} e^{-t/2} dt \\ &= P(\nu/2, \chi^2/2) \ . \end{aligned}$$

This is a gamma distribution, and is therefore another special case of the incomplete gamma integral $P(a, x)$. The complementary function, the same integral from χ^2 to ∞, is

$$\text{chdtrc}(\nu, \chi^2) = \text{igamc}(\nu/2, \chi^2/2) \ .$$

The functional inverse of the complemented distribution is

$$\text{chdtri}(\nu, y) = 2 \ \text{igami}(\nu/2, y)$$

where $0 \le y \le 1$. The program returns an error condition if $\chi^2 < 0$ or $\nu < 1$.

5.5.1 chdtrc.c

Right tail of χ^2 distribution

```
#include "mconf.h"

double chdtrc(df,x)
double df, x;
    {
    double igamc();

    if( (x < 0.0) || (df < 1.0) )
        {
        mtherr( "chdtrc", DOMAIN );
        return(0.0);
        }
    return( igamc( df/2.0, x/2.0 ) );
    }
```

5.5.2 chdtr.c

Left tail of χ^2 distribution

```
double chdtr(df,x)
double df, x;
    {
    double igam();

    if( (x < 0.0) || (df < 1.0) )
        {
        mtherr( "chdtr", DOMAIN );
        return(0.0);
        }
    return( igam( df/2.0, x/2.0 ) );
    }
```

5.5.3 chdtri.c

Functional inverse of the right tail of χ^2 distribution

```
double chdtri( df, y )
double df, y;
    {
    double x;
    double igami();

    if( (y < 0.0) || (y > 1.0) || (df < 1.0) )
```

```
        {
        mtherr( "chdtri", DOMAIN );
        return(0.0);
        }

    x = igami( 0.5 * df, y );
    return( 2.0 * x );
    }
```

5.6 Poisson Distribution

The sum of the first $k+1$ terms of the Poisson distribution is

$$\begin{aligned} \mathrm{pdtr}(k,\lambda) &= \sum_{j=0}^{k} \frac{e^{-\lambda}\lambda^j}{j!} \\ &= 1 - P(k+1,\lambda) \ . \end{aligned}$$

It is a special case of the incomplete gamma integral, as can be seen by inspection of the power series expansion of $P(a,x)$.

Both k and λ are positive. The complementary function is

$$\begin{aligned} \mathrm{pdtrc}(k,\lambda) &= \sum_{j=k+1}^{\infty} \frac{e^{-\lambda}\lambda^j}{j!} \\ &= P(k+1,\lambda) \ . \end{aligned}$$

The functional inverse of $y = \mathrm{pdtr}(k,\lambda)$ can be found from the inverse gamma integral by

$$\lambda = \mathrm{pdtri}(k,y) = \mathrm{igami}(k+1,y)$$

if $0 \le y \le 1$.

5.6.1 pdtrc.c

Area under the right tail of the Poisson distribution

```
#include "mconf.h"

double pdtrc( k, m )
int k;
double m;
    {
    double v;
    double igam();
```

```
    if( (k < 0) || (m <= 0.0) )
        {
        mtherr( "pdtrc", DOMAIN );
        return( 0.0 );
        }
    v = k+1;
    return( igam( v, m ) );
    }
```

5.6.2 pdtr.c

Area under the left tail of the Poisson distribution

```
double pdtr( k, m )
int k;
double m;
    {
    double v;
    double igamc();

    if( (k < 0) || (m <= 0.0) )
        {
        mtherr( "pdtr", DOMAIN );
        return( 0.0 );
        }
    v = k+1;
    return( igamc( v, m ) );
    }
```

5.6.3 pdtri.c

Functional inverse of Poisson distribution

```
double pdtri( k, y )
int k;
double y;
    {
    double v;
    double igami();

    if( (k < 0) || (y < 0.0) || (y >= 1.0) )
        {
        mtherr( "pdtri", DOMAIN );
        return( 0.0 );
        }
    v = k+1;
```

Arithmetic	Domain	Trials	Peak	RMS
DEC	0, 30	1700	$7.7 \cdot 10^{-15}$	$1.5 \cdot 10^{-15}$
IEEE	0, 30	30000	$8.1 \cdot 10^{-14}$	$1.1 \cdot 10^{-14}$

Table 5.4: `beta`(a,b), relative error

```
v = igami( v, y );
return( v );
}
```

5.7 Beta Function

The beta function

$$B(a,b) = \int_0^1 t^{a-1}(1-t)^{b-1}dt = \frac{\Gamma(a)\Gamma(b)}{\Gamma(a+b)}$$

is, among other things, a normalizing factor for the beta distribution (see the next two sections). It has singularites in case either a or b is a negative integer.

To avoid unnecessary overflow, the gamma function arguments a, b, and $a+b$ must be tested against the maximum allowable input to the gamma function subroutine. If any of these inputs is too large, then it is necessary to use the routine for $\ln \Gamma(x)$ to find

$$w = \ln B(a,b) = \ln \Gamma(a) + \ln \Gamma(b) - \ln \Gamma(a+b)$$
$$B(a,b) = e^w.$$

The value of w must also be tested to ensure that there will not be overflow or underflow in the exponential function. The integer sign variable returned by the log gamma subroutine makes it easy to calculate the sign of $B(a,b)$.

Accuracy figures are presented in Table 5.4.

5.7.1 beta.c

Beta function program

```
#include "mconf.h"

#ifdef UNK
#define MAXGAM 34.84425627277176174
#endif
#ifdef DEC
#define MAXGAM 34.84425627277176174
#endif
#ifdef IBMPC
#define MAXGAM 171.624376956302725
#endif
#ifdef MIEEE
#define MAXGAM 171.624376956302725
#endif

extern double MAXLOG, MAXNUM;
extern int sgngam;

double beta( a, b )
double a, b;
    {
    double y, c;
    int sign;
    double fabs(), gamma(), lgam(), exp(), floor();

    if( a <= 0.0 )
        {
        if( a == floor(a) )
            goto over;
        }
    if( b <= 0.0 )
        {
        if( b == floor(b) )
            goto over;
        }
    sign = 1;
    y = a + b;
    if( fabs(y) > MAXGAM )
        {
        y = lgam(y);
        sign *= sgngam;
        y = lgam(b) - y;
        sign *= sgngam;
        y = lgam(a) + y;
        sign *= sgngam;
```

```
        if( y > MAXLOG )
                {
over:
                mtherr( "beta", OVERFLOW );
                return( sign * MAXNUM );
                }
        return( sign * exp(y) );
        }
y = gamma(y);
if( y == 0.0 )
        goto over;

if( a > b )
        {
        y = gamma(a)/y;
        y *= gamma(b);
        }
else
        {
        y = gamma(b)/y;
        y *= gamma(a);
        }
return(y);
}
```

5.8 Incomplete Beta Integral

The incomplete beta integral is defined for $0 \le x \le 1$ as

$$
\begin{aligned}
I_x(a,b) &= \frac{\Gamma(a+b)}{\Gamma(a)\Gamma(b)} \int_0^x t^{a-1}(1-t)^{b-1} dt \\
&= \frac{x^a \Gamma(a+b)}{a\Gamma(a)\Gamma(b)} \, {}_2F_1(a, 1-b; a+1; x)
\end{aligned}
$$

where ${}_2F_1$ is the Gauss hypergeometric function (see Section 7.1). Note that the constant expression is the beta function $1/B(a,b)$. It normalizes the integral so that $I_1(a,b) = 1$.

The connection between the incomplete beta function and some discrete probability distributions becomes apparent through integration by parts:

$$
\begin{aligned}
I_x(a,b) &= \frac{\Gamma(a+b)}{\Gamma(a)\Gamma(b)} \left\{ \frac{1}{a} t^a (1-t)^{b-1} + \frac{b-1}{a(a+1)} t^{a+1} (1-t)^{b-2} \right. \\
&\quad \left. + \frac{(b-1)(b-2)}{a(a+1)(a+2)} t^{a+2} (1-t)^{b-3} + \ldots \right\}_0^x .
\end{aligned}
$$

When b is a positive integer the partial integrations eventually reduce the exponent of $1-t$ to zero, leaving an elementary integral. The resulting finite sum has binomial terms. In fact, putting $b = n - a + 1$ gives

$$I_p(k+1, n-k) = \sum_{j=k+1}^{n} \binom{n}{j} p^j (1-p)^{n-j}$$

which is the (complemented) binomial distribution function. The binomial coefficients are denoted by the symbol

$$\begin{aligned} \binom{n}{j} &= \frac{n!}{j!\,(n-j)!} \\ &= \frac{I_x(j, n-j+1) - I_x(j+1, n-j)}{x^j (1-x)^{n-j}} \,. \end{aligned}$$

They satisfy, among many other combinatorial identities, the formula

$$\sum_{i=0}^{k} (-1)^{k-i} \binom{n}{i} = \binom{n-1}{k} \,.$$

This, together with the identity

$$p^n (1-p)^k = p^{n-1}(1-p)^k - p^{n-1}(1-p)^{k+1} \,,$$

can be used to show that

$$I_p(n, k) = \sum_{j=0}^{k-1} \binom{n+j-1}{j} p^n (1-p)^j$$

is the negative binomial distribution.

Two continuous distributions, the F distribution and Student's t distribution, can be obtained from the incomplete beta integral by change of variables. Programs that use $I_x(a, b)$ to compute each of the above cited distributions are given in the following sections.

The integral obeys a symmetry relation

$$I_x(a, b) = 1 - I_{1-x}(b, a) \,.$$

The computer program may test for the special cases $I_{x=0}(a, b) = 0$ and $I_{x=1}(a, b) = 1$ and should declare a domain error for x outside the interval [0,1]. A recurrence relation (AMS55 #26.5.16)

$$I_x(a, b) = \frac{\Gamma(a+b)}{\Gamma(a+1)\Gamma(b)} x^a (1-x)^b + I_x(a+1, b)$$

may be used to increase a if $0 < a \leq 1$.

If the integral is considered as a probability distribution (the beta distribution), the mean or expected value of x is $a/(a+b)$. If x is greater than this expected value, then the symmetry relation should be used to decrease the value of x for computation.

After these adjustments the function may be computed by one of the following continued fractions, depending on the value of

$$t = x\,\frac{a+b-2}{a-1}\,.$$

If $t < 1$, then use

$$I_x(a,b) = \frac{\Gamma(a+b)x^a(1-x)^b}{a\Gamma(a)\Gamma(b)}\left(\frac{1}{1+}\,\frac{a_1}{1+}\,\frac{a_2}{1+}\cdots\right)$$

where

$$\begin{aligned} a_{2k+1} &= -\frac{(a+k)(a+b+k)}{(a+2k)(a+2k+1)}x \\ a_{2k} &= -\frac{k(b-k)}{(a+2k-1)(a+2k)}x\,. \end{aligned}$$

If $t \geq 1$, then use (AMS55 #26.5.9)

$$I_x(a,b) = \frac{\Gamma(a+b)x^a(1-x)^{b-1}}{a\Gamma(a)\Gamma(b)}\left(\frac{a_1}{1+}\,\frac{a_2}{1+}\,\frac{a_3}{1+}\cdots\right)$$

in which

$$\begin{aligned} a_{2k} &= -\frac{(a+k-1)(b-k)}{(a+2k-2)(a+2k-1)}\,\frac{x}{1-x} \\ a_{2k+1} &= -\frac{k(a+b-1+k)}{(a+2k-1)(a+2k)}\,\frac{x}{1-x}\,. \end{aligned}$$

Whichever expansion is used, there should be a program escape for nonconvergence after a reasonable number of terms. There is a possibility of underflow in the exponential function while calculating the normalizing factor. If this happens, an underflow error should be flagged and the function should return 0 (or 1, if the symmetry relation was invoked).

Typical test results are shown in Table 5.5. For `ibeti.c`, a and b range from 0 to 100, x from 0 to 1. The procedures as described have a somewhat unsatisfactory level of error for extreme ratios of a and b. Larger errors than those shown may occur in that case.

5.8.1 ibet.c

Incomplete beta integral

Function	Arithmetic	Domain	Trials	Peak	RMS
ibet	DEC	$0, 100$	3300	$3.5 \cdot 10^{-14}$	$5.0 \cdot 10^{-15}$
ibet	IEEE	$0, 100$	10000	$3.9 \cdot 10^{-13}$	$5.2 \cdot 10^{-14}$
ibeti	DEC	$0, 1$	2500	$2.9 \cdot 10^{-13}$	$5.9 \cdot 10^{-15}$
ibeti	IEEE	$0, 1$	5000	$8.3 \cdot 10^{-13}$	$1.7 \cdot 10^{-14}$

Table 5.5: ibet(a, b, x), relative error; a, b, range from 0 to 100, x from 0 to 1.

```
#include "mconf.h"

#define BIG 1.44115188075855872E+17
extern double MACHEP, MINLOG, MAXLOG;

double ibet( aa, bb, xx )
double aa, bb, xx;
    {
    double ans, a, b, t, x, onemx;
    double lgam(), exp(), log(), fabs();
    double incbd(), incbcf();
    short flag;

    if( (xx <= 0.0) || ( xx >= 1.0) )
        {
        if( xx == 0.0 )
            return(0.0);
        if( xx == 1.0 )
            return( 1.0 );
        mtherr( "ibet", DOMAIN );
        return( 0.0 );
        }
    onemx = 1.0 - xx;
```

See if x is greater than the mean.

```
    if( xx > (aa/(aa+bb)) )
        {
        flag = 1;
        a = bb;
        b = aa;
        t = xx;
```

```
		x = onemx;
		}
	else
		{
		flag = 0;
		a = aa;
		b = bb;
		t = onemx;
		x = xx;
		}
```

Choose expansion for optimal convergence.

```
	ans = x * (a+b-2.0)/(a-1.0);
	if( ans < 1.0 )
		{
		ans = incbcf( a, b, x );
		t = b * log( t );
		}
	else
		{
		ans = incbd( a, b, x );
		t = (b-1.0) * log(t);
		}
adone:
	t += a*log(x) + lgam(a+b) - lgam(a) - lgam(b);
	t += log( ans/a );

	if( t < MINLOG )
		{
		if( flag == 0 )
			{
			mtherr( "ibet", UNDERFLOW );
			return( 0.0 );
			}
		else
			return(1.0);
		}

	t = exp(t);

	if( flag == 1 )
		t = 1.0 - t;

	return( t );
	}
```

Continued fraction expansion for $x(a+b-2) < a-1$

```
static double incbcf( a, b, x )
double a, b, x;
	{
	double xk, pk, pkm1, pkm2, qk, qkm1, qkm2;
	double k1, k2, k3, k4, k5, k6, k7, k8;
	double r, t, ans;
	static double big = BIG;
	double fabs();
	int n;

	k1 = a;
	k2 = a + b;
	k3 = a;
	k4 = a + 1.0;
	k5 = 1.0;
	k6 = b - 1.0;
	k7 = k4;
	k8 = a + 2.0;

	pkm2 = 0.0;
	qkm2 = 1.0;
	pkm1 = 1.0;
	qkm1 = 1.0;
	ans = 1.0;

	n = 0;
	do
		{

		xk = -( x * k1 * k2 )/( k3 * k4 );
		pk = pkm1 + pkm2 * xk;
		qk = qkm1 + qkm2 * xk;
		pkm2 = pkm1;
		pkm1 = pk;
		qkm2 = qkm1;
		qkm1 = qk;

		xk = ( x * k5 * k6 )/( k7 * k8 );
		pk = pkm1 + pkm2 * xk;
		qk = qkm1 + qkm2 * xk;
		pkm2 = pkm1;
		pkm1 = pk;
```

```
          qkm2 = qkm1;
          qkm1 = qk;

          if( qk != 0 )
               r = pk/qk;
          if( r != 0 )
               {
               t = fabs( (ans - r)/r );
               ans = r;
               }
          else
               t = 1.0;

          if( t < MACHEP )
               goto cdone;

          k1 += 1.0;
          k2 += 1.0;
          k3 += 2.0;
          k4 += 2.0;
          k5 += 1.0;
          k6 -= 1.0;
          k7 += 2.0;
          k8 += 2.0;

          if( (fabs(qk) + fabs(pk)) > big )
               {
               pkm2 /= big;
               pkm1 /= big;
               qkm2 /= big;
               qkm1 /= big;
               }
          if((fabs(qk) < MACHEP)
               || (fabs(pk) < MACHEP))
               {
               pkm2 *= big;
               pkm1 *= big;
               qkm2 *= big;
               qkm1 *= big;
               }
          }
     while( ++n < 100 );
cdone:
     return(ans);
     }
```

Continued fraction expansion for $x(a+b-2) > a-1$

```
static double incbd( a, b, x )
double a, b, x;
    {
    double xk, pk, pkm1, pkm2, qk, qkm1, qkm2;
    double k1, k2, k3, k4, k5, k6, k7, k8;
    double r, t, ans, z;
    static double big = BIG;
    double fabs();
    int n;

    k1 = a;
    k2 = b - 1.0;
    k3 = a;
    k4 = a + 1.0;
    k5 = 1.0;
    k6 = a + b;
    k7 = a + 1.0;;
    k8 = a + 2.0;

    pkm2 = 0.0;
    qkm2 = 1.0;
    pkm1 = 1.0;
    qkm1 = 1.0;
    z = x / (1.0-x);
    ans = 1.0;

    n = 0;
    do
        {

        xk = -( z * k1 * k2 )/( k3 * k4 );
        pk = pkm1 + pkm2 * xk;
        qk = qkm1 + qkm2 * xk;
        pkm2 = pkm1;
        pkm1 = pk;
        qkm2 = qkm1;
        qkm1 = qk;

        xk = ( z * k5 * k6 )/( k7 * k8 );
        pk = pkm1 + pkm2 * xk;
        qk = qkm1 + qkm2 * xk;
```

```
        pkm2 = pkm1;
        pkm1 = pk;
        qkm2 = qkm1;
        qkm1 = qk;

        if( qk != 0 )
             r = pk/qk;
        if( r != 0 )
             {
             t = fabs( (ans - r)/r );
             ans = r;
             }
        else
             t = 1.0;

        if( t < MACHEP )
             goto cdone;

        k1 += 1.0;
        k2 -= 1.0;
        k3 += 2.0;
        k4 += 2.0;
        k5 += 1.0;
        k6 += 1.0;
        k7 += 2.0;
        k8 += 2.0;

        if( (fabs(qk) + fabs(pk)) > big )
             {
             pkm2 /= big;
             pkm1 /= big;
             qkm2 /= big;
             qkm1 /= big;
             }
        if( (fabs(qk) < MACHEP)
             || (fabs(pk) < MACHEP) )
             {
             pkm2 *= big;
             pkm1 *= big;
             qkm2 *= big;
             qkm1 *= big;
             }
        }
    while( ++n < 100 );
cdone:
```

```
    return(ans);
    }
```

5.8.2 Functional Inverse of Incomplete Beta Integral

This computer routine determines the value of the random variable x such that $I_x(a, b)$ takes on a given value y. The parameters a and b are considered to be given, also.

Starting with a rough approximation to y, successively refined estimates are calculated by a Newton-Raphson iteration. The derivative required for the iteration is just the beta probability density function itself.

In part of the domain the initial approximation is so inaccurate that the Newton-Raphson method diverges. In this case a slower interval halving technique is used to find a sufficiently close starting point for the faster iteration. The program thus illustrates a general method of inverting a function numerically.

```
#include "mconf.h"

extern double MACHEP, MAXNUM, MAXLOG, MINLOG;

double ibeti( aa, bb, yy0 )
double aa, bb, yy0;
    {
    double a, b, y0;
    double d, y, x, x0, x1, lgm, yp, di;
    int i, n, rflg;
    double ibet();
    double ndtri(), exp(), fabs(), log(), sqrt(), lgam();

    if( yy0 <= 0 )
        return(0.0);
    if( yy0 >= 1.0 )
        return(1.0);
Approximation to inverse function (AMS55 #26.5.22)
    yp = -ndtri(yy0);
    if( yy0 > 0.5 )
        {
        rflg = 1;
        a = bb;
        b = aa;
        y0 = 1.0 - yy0;
        yp = -yp;
        }
```

```
else
     {
     rflg = 0;
     a = aa;
     b = bb;
     y0 = yy0;
     }
if( (aa <= 1.0) || (bb <= 1.0) )
     {
     y = yp * yp / 2.0;
     }
else
     {
     lgm = (yp * yp - 3.0)/6.0;
     x0 = 2.0/( 1.0/(2.0*a-1.0)
          + 1.0/(2.0*b-1.0) );
     y = yp * sqrt( x0 + lgm ) / x0
     - ( 1.0/(2.0*b-1.0) - 1.0/(2.0*a-1.0) )
     * (lgm + 5.0/6.0 - 2.0/(3.0*x0));
     y = 2.0 * y;
     if( y < MINLOG )
          {
          x0 = 1.0;
          goto under;
          }
     }
x = a/( a + b * exp(y) );
y = ibet( a, b, x );
yp = (y - y0)/y0;
if( fabs(yp) < 1.0e-1 )
     goto newt;
```

Resort to interval halving if not close enough

```
x0 = 0.0;
x1 = 1.0;
di = 0.5;
for( i=0; i<20; i++ )
     {
     if( i != 0 )
          {
          x = di * x1 + (1.0-di) * x0;
          y = ibet( a, b, x );
          yp = (y - y0)/y0;
          if( fabs(yp) < 1.0e-3 )
               goto newt;
          }
```

```
            if( y < y0 )
                {
                x0 = x;
                di = 0.5;
                }
            else
                {
                x1 = x;
                di *= di;
                if( di == 0.0 )
                    di = 0.5;
                }
            }
        if( x0 == 0.0 )
            {
under:
            mtherr( "ibeti", UNDERFLOW );
            x0 = 0.0;
            goto done;
            }
newt:
        x0 = x;
        lgm = lgam(a+b) - lgam(a) - lgam(b);
        for( i=0; i<10; i++ )
            {
```
Compute the function at this point.
```
            if( i != 0 )
                y = ibet(a,b,x0);
```
Compute the derivative of the function at this point.
```
            d = (a - 1.0) * log(x0)
                + (b - 1.0) * log(1.0-x0)
                + lgm;
            if( d < MINLOG )
                goto under;
            d = exp(d);
```
Step to the next approximation of x.
```
            d = (y - y0)/d;
            x = x0;
            x0 = x0 - d;
            if( (x0 <= 0.0) || (x0 >= 1.0) )
                goto under;
            if( i < 2 )
                continue;
            if( fabs(d/x0) < 256.0 * MACHEP )
                break;
```

```
        }
done:
    if( rflg )
        x0 = 1.0 - x0;
    return( x0 );
    }
```

5.9 Beta Distribution

The beta density function is given by

$$p_x(x) = \frac{\Gamma(a+b)}{\Gamma(a)\Gamma(b)} \, x^{a-1}(1-x)^{b-1} \; .$$

The area from zero to x under the beta density is the integral

$$P(x) = \frac{\Gamma(a+b)}{\Gamma(a)\Gamma(b)} \int_0^x t^{a-1}(1-t)^{b-1} dt \; .$$

This distribution function is identical to the incomplete beta integral, so

$$P(x) = I_x(a,b) \; .$$

The complemented function gives the area under the right tail of the density:

$$1 - P(1-x) = 1 - I_x(a,b) = I_{1-x}(b,a) \; .$$

5.9.1 btdtr.c

Beta distribution program

```
double btdtr( a, b, x )
double a, b, x;
    {
    double ibet();

    return( ibet( a, b, x ) );
    }
```

5.10 Binomial Distribution

The sum of the terms 0 through k of the binomial probability density is

$$\begin{aligned} \mathrm{bdtr}(k,n,p) &= \sum_{j=0}^{k} \binom{n}{j} p^j (1-p)^{n-j} \\ &= I_{1-p}(n-k, k+1) \; . \end{aligned}$$

As shown earlier, this is a special case of the incomplete beta integral. If $k = 0$, then

$$\text{bdtr}(0, n, p) = (1 - p)^n \; .$$

If $k = n$, the integral is 1. These should be tested as special cases. An error condition exists if $k < 0$, $n < k$, $p < 0$, or $p > 1$.

The complementary function is the sum of the terms $k + 1$ through n of the binomial probability density:

$$\begin{aligned} \text{bdtrc}(k, n, p) &= \sum_{j=k+1}^{n} \binom{n}{j} p^j (1 - p)^{n-j} \\ &= I_p(k + 1, n - k) \; . \end{aligned}$$

This has special cases

$$\begin{aligned} \text{bdtrc}(n, n, p) &= 0 \\ \text{bdtrc}(0, n, p) &= 1 - (1 - p)^n \; . \end{aligned}$$

The inverse function gives the event probability p such that the area under the density is equal to a given cumulative probability y. This can be found using the inverse beta integral by the relation

$$\text{bdtri}(n, k, y) = 1 - \text{ibeti}(n - k, k + 1, y) \; .$$

A special case is

$$\text{bdtri}(n, 0, y) = 1 - y^{\frac{1}{n}} .$$

5.10.1 bdtrc.c

Right tail of binomial distribution

```
#include "mconf.h"

double bdtrc( k, n, p )
int k, n;
double p;
	{
	double dk, dn;
	double ibet(), pow();

	if( (p < 0.0) || (p > 1.0) )
		goto domerr;
	if( k < 0 )
		return( 1.0 );
	if( n < k )
```

```
        {
domerr:
        mtherr( "bdtrc", DOMAIN );
        return( 0.0 );
        }
    if( k == n )
        return( 0.0 );
    dn = n - k;
    if( k == 0 )
        {
        dk = 1.0 - pow( 1.0-p, dn );
        }
    else
        {
        dk = k + 1;
        dk = ibet( dk, dn, p );
        }
    return( dk );
    }
```

5.10.2 bdtr.c

Left tail of binomial distribution

```
double bdtr( k, n, p )
int k, n;
double p;
    {
    double dk, dn;
    double ibet(), pow();

    if( (p < 0.0) || (p > 1.0) )
        goto domerr;
    if( (k < 0) || (n < k) )
        {
domerr:
        mtherr( "bdtr", DOMAIN );
        return( 0.0 );
        }
    if( k == n )
        return( 1.0 );
    dn = n - k;
    if( k == 0 )
        {
        dk = pow( 1.0-p, dn );
```

```
        }
    else
        {
        dk = k + 1;
        dk = ibet( dn, dk, 1.0 - p );
        }
    return( dk );
    }
```

5.10.3 bdtri.c

Functional inverse of binomial distribution

```
double bdtri( k, n, y )
int k, n;
double y;
    {
    double dk, dn, p;
    double ibeti(), pow();

    if( (y < 0.0) || (y > 1.0) )
        goto domerr;
    if( (k < 0) || (n <= k) )
        {
domerr:
        mtherr( "bdtri", DOMAIN );
        return( 0.0 );
        }
    dn = n - k;
    if( k == 0 )
        {
        p = 1.0 - pow( y, 1.0/dn );
        }
    else
        {
        dk = k + 1;
        p = 1.0 - ibeti( dn, dk, y );
        }
    return( p );
    }
```

5.11 Negative Binomial Distribution

The negative binomial density function gives the probability that in a sequence of Bernoulli trials the nth success occurs at the $n+j$th trial — i.e., that the nth success is preceded by j failures.

The sum of the terms 0 through k of the negative binomial distribution is the probability that the nth success is preceded by k or fewer failures. It is given by

$$\begin{aligned} \mathrm{nbdtr}(k, n, p) &= \sum_{j=0}^{k} \binom{n+j-1}{j} p^n (1-p)^j \\ &= I_p(n, k+1) \end{aligned}$$

which again involves the incomplete beta integral I. See Section 5.8 for additional information. The arguments are positive, with p ranging from 0 to 1.

The complemented function

$$\begin{aligned} \mathrm{nbdtrc}(k, n, p) &= \sum_{j=k+1}^{\infty} \binom{n+j-1}{j} p^n (1-p)^j \\ &= 1 - I_p(n, k+1) \\ &= I_{1-p}(k+1, n) \end{aligned}$$

is the probability that the nth success is preceded by more than k failures.

5.11.1 nbdtr.c

Left tail of negative binomial distribution

```
double nbdtr( k, n, p )
int k, n;
double p;
	{
	double dk, dn;
	double ibet();

	if( (p < 0.0) || (p > 1.0) )
		|| (k < 0) )
		{
		mtherr( "nbdtr", DOMAIN );
		return( 0.0 );
		}
	dk = k + 1;
	dn = n;
	return( ibet( dn, dk, p ) );
	}
```

5.11.2 nbdtrc.c

Right tail of negative binomial distribution

```
#include "mconf.h"

double nbdtrc( k, n, p )
int k, n;
double p;
    {
    double dk, dn;
    double ibet();

    if( (p < 0.0) || (p > 1.0) )
        goto domerr;
    if( k < 0 )
        {
domerr:
        mtherr( "nbdtr", DOMAIN );
        return( 0.0 );
        }
    dk = k+1;
    dn = n;
    return( ibet( dk, dn, 1.0 - p ) );
    }
```

5.12 *F* Distribution

Snedecor's density, or the variance ratio density, is the probability density function of

$$F = \frac{u_1/d_1}{u_2/d_2}$$

where u_1 and u_2 are random variables having Chi square distributions with d_1 and d_2 degrees of freedom, respectively. The distribution function is another special case of the incomplete beta integral. The area from 0 to F under the F density is

$$\text{fdtr}(d_1, d_2, F) = I_w(d_1/2, d_2/2)$$

where

$$w = \frac{d_1 F}{d_2 + d_1 F} .$$

An error exists if $F < 0$, or if either $d_1 < 1$ or $d_2 < 1$.

The complementary function is the area from F to ∞ under the F density. This is equal to

$$\text{fdtrc}(d_1, d_2, F) = I_v(d_2/2, d_1/2)$$

where

$$v = \frac{d_2}{d_2 + d_1 F} .$$

The inverse function, fdtrci(), can be obtained from the inverse of the incomplete beta integral. This finds the F density argument F such that the integral from F to ∞ of the F density is equal to the given probability P:

$$\begin{aligned} s &= \text{ibeti}(d_2/2, d_1/2, P) \\ \text{fdtrci}(d_1, d_2, P) &= \frac{d_2(1-s)}{d_1 s} . \end{aligned}$$

For the uncomplemented F distribution, the relation for the inverse, fdtri(), is

$$\begin{aligned} s &= \text{ibeti}(d_1/2, d_2/2, P) \\ \text{fdtri}(d_1, d_2, P) &= \frac{d_2 s}{d_1(1-s)} . \end{aligned}$$

An error exists if $P < 0$, $P > 1$, or if either $d_1 < 1$ or $d_2 < 1$.

5.12.1 fdtrc.c

Right tail of F distribution

```
#include "mconf.h"

double fdtrc( ia, ib, x )
int ia, ib;
double x;
	{
	double a, b, w;
	double ibet();

	if( (ia < 1) || (ib < 1) || (x < 0.0) )
		{
		mtherr( "fdtrc", DOMAIN );
		return( 0.0 );
		}
	a = ia;
	b = ib;
	w = b / (b + a * x);
	return( ibet( b/2.0, a/2.0, w ) );
	}
```

5.12.2 fdtr.c

Left tail of F distribution

```
double fdtr( ia, ib, x )
int ia, ib;
double x;
    {
    double a, b, w;
    double ibet();

    if( (ia < 1) || (ib < 1) || (x < 0.0) )
        {
        mtherr( "fdtr", DOMAIN );
        return( 0.0 );
        }
    a = ia;
    b = ib;
    w = a * x;
    w = w / (b + w);
    return( ibet(a/2.0, b/2.0, w) );
    }
```

5.12.3 fdtrci.c

Functional inverse of right tail of F distribution

```
double fdtrci( ia, ib, y )
int ia, ib;
double y;
    {
    double a, b, w, x;
    double ibeti();

    if( (ia < 1) || (ib < 1) || (y <= 0.0)
        || (y > 1.0) )
        {
        mtherr( "fdtri", DOMAIN );
        return( 0.0 );
        }
    a = ia;
    b = ib;
    w = ibeti( 0.5*b, 0.5*a, y );
    x = (b - b*w)/(a*w);
    return(x);
    }
```

5.13 Student's t distribution

The integral from minus infinity to t of the Student t density with integer $k > 0$ degrees of freedom is

$$\text{stdtr}(k,t) = \frac{\Gamma\left(\frac{k+1}{2}\right)}{\sqrt{k\pi}\,\Gamma\left(\frac{k}{2}\right)} \int_{-\infty}^{t} \left(1 + \frac{x^2}{k}\right)^{-\frac{k+1}{2}} dx \; .$$

The integral from $-\infty$ to $-t$ is equal to

$$\frac{1}{2}\, I_w(k/2, 1/2) \; ,$$

where

$$w = \frac{k}{k+t^2} \; .$$

Though this is a special case of the incomplete beta integral, a more direct computation for $t \geq -1$ can be derived easily through integration by parts. The integral from $-t$ to t is found by first setting

$$u = 1 + \frac{x^2}{k} \; .$$

If k is odd, find

$$\begin{aligned} v &= \frac{x}{\sqrt{k}} \\ s &= 1 + \frac{2}{3u} + \frac{2 \cdot 4}{3 \cdot 5u^2} + \ldots + \frac{2 \cdot 4 \cdots (k-3)}{3 \cdot 5 \cdots (k-2)u^{(k-3)/2}} \\ p &= \frac{2}{\pi}\left(\tan^{-1} v + \frac{vs}{u}\right) \; . \end{aligned}$$

If k is even,

$$\begin{aligned} s &= 1 + \frac{1}{2u} + \frac{1 \cdot 3}{2 \cdot 4u^2} + \ldots + \frac{1 \cdot 3 \cdots (k-3)}{2 \cdot 4 \cdots (k-2)u^{(k-2)/2}} \\ p &= \frac{st}{\sqrt{ku}} \; . \end{aligned}$$

For either odd or even k the integral is equal to

$$\text{stdtr}(k,t) = \frac{1+p}{2} \; .$$

Since the function is symmetric about $t = 0$, the area under the right tail of the density is found by calling the function with $-t$ instead of t.

Arithmetic	Domain	Trials	Peak	RMS
DEC	$0, 24$	12000	$4.7 \cdot 10^{-17}$	$8.9 \cdot 10^{-18}$
DEC	$-24, 0$	11000	$2.3 \cdot 10^{-15}$	$2.7 \cdot 10^{-16}$
IEEE	$0, 24$	30000	$4.5 \cdot 10^{-16}$	$8.0 \cdot 10^{-17}$
IEEE	$-24, 0$	30000	$1.9 \cdot 10^{-14}$	$2.3 \cdot 10^{-15}$

Table 5.6: `stdtr`(k, x), relative error

The functional inverse of the distribution can be found, from the relation to the incomplete beta integral indicated above, through

$$\begin{aligned} \mathbf{stdtr}(k, -t) &= y \ \text{(given)} \\ w &= \mathbf{ibeti}(\tfrac{k}{2}, \tfrac{1}{2}, 2y) \\ t &= \sqrt{\frac{k}{w} - k}\ . \end{aligned}$$

Table 5.6 gives relative error of the program for the Student distribution with k varying randomly between 1 and 25. The "Domain" column in the table refers to t.

5.13.1 stdtr.c

Student's t distribution function

```
#include "mconf.h"

extern double PI, MACHEP;

double stdtr( k, t )
int k;
double t;
    {
    double x, rk, z, f, tz, p, xsqk;
    double sqrt(), atan(), ibet();
    int j;

    if( k <= 0 )
        {
        mtherr( "stdtr", DOMAIN );
```

```
          return(0.0);
          }
     if( t == 0 )
          return( 0.5 );
     if( t < -1.0 )
          {
          rk = k;
          z = rk / (rk + t * t);
          p = 0.5 * ibet( 0.5*rk, 0.5, z );
          return( p );
          }
```

Compute integral from $-t$ to $+t$

```
     if( t < 0 )
          x = -t;
     else
          x = t;
     rk = k;
     z = 1.0 + ( x * x )/rk;
     if( (k & 1) != 0)
          {
```

Computation for odd k

```
          xsqk = x/sqrt(rk);
          p = atan( xsqk );
          if( k > 1 )
               {
               f = 1.0;
               tz = 1.0;
               j = 3;
               while( (j<=(k-2))
                    && ( (tz/f) > MACHEP ) )
                    {
                    tz *= (j-1)/( z * j );
                    f += tz;
                    j += 2;
                    }
               p += f * xsqk/z;
               }
          p *= 2.0/PI;
          }
     else
          {
```

Computation for even k

```
          f = 1.0;
          tz = 1.0;
          j = 2;
```

```
        while( ( j <= (k-2) )
            && ( (tz/f) > MACHEP ) )
            {
            tz *= (j - 1)/( z * j );
            f += tz;
            j += 2;
            }
        p = f * x/sqrt(z*rk);
        }
    if( t < 0 )
        p = -p;
```

Note destruction of relative accuracy if $p < 0$.

```
    p = 0.5 + 0.5 * p;
    return(p);
    }
```

5.14 Gaussian Distribution

In terms of the error function

$$\text{erf } x = \frac{2}{\sqrt{\pi}} \int_0^x e^{-t^2} dt$$

the Gaussian or normal probability distribution function is

$$\begin{aligned} N(x) &= \frac{1}{\sqrt{2\pi}} \int_{-\infty}^x e^{t^2/2} dt \\ &= \frac{1 + \text{erf}(x/\sqrt{2})}{2} \\ &= 1 - \frac{\text{erfc}(x/\sqrt{2})}{2} \end{aligned}$$

where erfc $x = 1 - \text{erf } x$.

To compute erfc x, the domain is partitioned into several segments. If $|x| < 1.0$, then erfc is found via erf x as above. For $1.0 < |x| < 8.0$,

$$\text{erfc } |x| \approx e^{-x^2} \frac{P(|x|)}{Q(|x|)}$$

where

$P(t) =$
$+2.4619698147353052524 \cdot 10^{-10}t^8$
$+0.56418956483106882197 7t^7$
$+7.4632105644226991268 7t^6$
$+48.637197098568136661 4t^5$
$+196.52083295607709824 2t^4$
$+526.44519499547735863 1t^3$
$+934.52852717195760754 0t^2$
$+1027.5518868951571027 2t$
$+557.53533536939932752 6$,

$Q(t) =$
$+1.0t^8$
$+13.228195115474499250 8t^7$
$+86.707214088598974232 9t^6$
$+354.93777888781989106 2t^5$
$+975.70850174320548975 3t^4$
$+1823.9091668790973628 9t^3$
$+2246.3376081871098179 2t^2$
$+1656.6630919416135018 2t$
$+557.53534081772767554 6$.

The rational form approximates $e^{-x^2}\text{erfc}(x)$ with a theoretical relative error of $3.1 \cdot 10^{-18}$. A separate rational approximation is required for larger arguments. If $x \geq 8.0$, then

$$\text{erfc } |x| \approx e^{-x^2} \frac{R(|x|)}{S(|x|)}$$

where

$R(t) =$
$+0.56418958354775507398 4t^5$
$+1.2753667075997810441 6t^4$
$+5.0190504225118047741 4t^3$
$+6.1602109799305358519 5t^2$
$+7.4097426995044893916 0t$
$+2.9788666537210024067 0$,

$S(t) =$
$+1.0t^6$
$+2.2605286322011727659 0t^5$
$+9.3960352493800143467 3t^4$
$+12.048953980809665660 5t^3$
$+17.081445074756589722 2t^2$
$+9.6089680906328587819 8t$
$+3.3690764510008151605 0$.

The theoretical relative error is $3.2 \cdot 10^{-18}$. For sufficiently large $x > 0$ there will be underflow in the exponential function. This occurs for $x > 26.6$ in IEEE arithmetic; the exact point of underflow will depend on whether denormalized tiny numbers are implemented. In DEC arithmetic the underflow threshold is $x > 9.231948545$. The computer routine should return zero and signal an underflow error in this case. If x is negative, then use

$$\text{erfc } x = 2 - \text{erfc}(-x) .$$

The function approaches +2 for x large and negative, so there is no underflow error.

To compute erf x there are two cases. If $|x| > 1.0$ then

$$\text{erf } x = 1 - \text{erfc } x .$$

For smaller x,

$$\text{erf } x \approx x \frac{T(x^2)}{U(x^2)}$$

where

Function	Arith.	Domain	Trials	Peak	RMS
ndtr	DEC	$-13, 0$	8000	$2.1 \cdot 10^{-15}$	$4.8 \cdot 10^{-16}$
ndtr	IEEE	$-13, 0$	30000	$3.4 \cdot 10^{-14}$	$6.7 \cdot 10^{-15}$
erf	DEC	$0, 1$	14000	$4.7 \cdot 10^{-17}$	$1.5 \cdot 10^{-17}$
erf	IEEE	$0, 1$	30000	$3.7 \cdot 10^{-16}$	$1.0 \cdot 10^{-16}$
erfc	DEC	$0, 9.23$	12000	$5.1 \cdot 10^{-16}$	$1.2 \cdot 10^{-16}$
erfc	IEEE	$0, 26.6$	30000	$5.7 \cdot 10^{-14}$	$1.5 \cdot 10^{-14}$
ndtri	DEC	0.125, 1	11000	$7.6 \cdot 10^{-17}$	$1.6 \cdot 10^{-17}$
ndtri	DEC	$6 \cdot 10^{-39}$, .135	10000	$6.0 \cdot 10^{-17}$	$1.3 \cdot 10^{-17}$
ndtri	IEEE	0.125, 1	20000	$7.2 \cdot 10^{-16}$	$1.3 \cdot 10^{-16}$
ndtri	IEEE	$3 \cdot 10^{-308}$,.135	50000	$4.6 \cdot 10^{-16}$	$9.8 \cdot 10^{-17}$

Table 5.7: Error functions, relative error

$$
\begin{array}{ll}
T(t) = & U(t) = \\
+9.6049737398705161638749t^4 & +1.0t^5 \\
+90.026019720384268921 7t^3 & +33.561714164750309964 7t^4 \\
+2232.0053459468431922 6t^2 & +521.35794978015267979 5t^3 \\
+7003.3251411280507547 3t & +4594.3238297098012798 7t^2 \\
+55592.301301039496276 8 \ , & +22629.000061389093424 6t \\
 & +49267.394260863592108 6 \ .
\end{array}
$$

This has relative error $2.2 \cdot 10^{-18}$. With these functions to call upon, the Gaussian distribution is calculated in one of two ways. If $|x| < \sqrt{2}/2$ then

$$
N(x) = \frac{1 + \operatorname{erf}(x/\sqrt{2})}{2} \ .
$$

For larger x, compute

$$
y = \frac{\operatorname{erfc}(x/\sqrt{2})}{2} \ .
$$

If $x < 0$, $N(x) = y$; else $N(x) = 1 - y$. Experimental results for the error functions and the functional inverse `ndtri.c` are given in Table 5.7. The figures for `ndtri.c` and DEC arithmetic apply to a slightly different program than the one shown here.

5.14.1 ndtr.c

Gaussian distribution function

```
#include "mconf.h"

#ifdef DEC
static short P[] = {
0030207,0054445,0011173,0021706,
0040020,0067272,0030661,0122075,
0040756,0151236,0173053,0067042,
0041502,0106175,0062555,0151457,
0042104,0102525,0047401,0003667,
0042403,0116176,0011446,0075303,
0042551,0120723,0061641,0123275,
0042600,0070651,0007264,0134516,
0042413,0061102,0167507,0176625
};
Leading coefficient 1.0 omitted from Q.
static short Q[] = {
0041123,0123257,0165741,0017142,
0041655,0065027,0173413,0115450,
0042261,0074011,0021573,0004150,
0042563,0166530,0013662,0007200,
0042743,0176427,0162443,0105214,
0043014,0062546,0153727,0123772,
0042717,0012470,0006227,0067424,
0042413,0061103,0003042,0013254
};
static short R[] = {
0040020,0067272,0101024,0155421,
0040243,0037467,0056706,0026462,
0040640,0116017,0120665,0034315,
0040705,0020162,0143350,0060137,
0040755,0016234,0134304,0130157,
0040476,0122700,0051070,0015473
};
Leading 1.0 omitted from S.
static short S[] = {
0040420,0126200,0044276,0070413,
0041026,0053051,0007302,0063746,
0041100,0144203,0174051,0061151,
0041210,0123314,0126343,0177646,
0041031,0137125,0051431,0033011,
0040527,0117362,0152661,0066201
};
static short T[] = {
0041031,0126770,0170672,0166101,
0041664,0006522,0072360,0031770,
0043013,0100025,0162641,0126671,
0043332,0155231,0161627,0076200,
0044131,0024115,0021020,0117343
};
```

```
#ifdef IIEEE
static short P[] = {
0x6479,0xa24f,0xeb24,0x3df0,
0x3488,0x4636,0x0dd7,0x3fe2,
0x6dc4,0xdec5,0xda53,0x401d,
0xba66,0xacad,0x518f,0x4048,
0x20f7,0xa9e0,0x90aa,0x4068,
0xcf58,0xc264,0x738f,0x4080,
0x34d8,0x6c74,0x343a,0x408d,
0x972a,0x21d6,0x0e35,0x4090,
0xffb3,0x5de8,0x6c48,0x4081
};
static short Q[] = {
0x23cc,0xfd7c,0x74d5,0x402a,
0x7365,0xfee1,0xad42,0x4055,
0x610d,0x246f,0x2f01,0x4076,
0x41d0,0x02f6,0x7dab,0x408e,
0x7151,0xfca4,0x7fa2,0x409c,
0xf4ff,0xdafa,0x8cac,0x40a1,
0xede2,0x0192,0xe2a7,0x4099,
0x42d6,0x60c4,0x6c48,0x4081
};
static short R[] = {
0x9b62,0x5042,0x0dd7,0x3fe2,
0xc5a6,0xebb8,0x67e6,0x3ff4,
0xa71a,0xf436,0x1381,0x4014,
0x0c0c,0x58dd,0xa40e,0x4018,
0x960e,0x9718,0xa393,0x401d,
0x0367,0x0a47,0xd4b8,0x4007
};
static short S[] = {
0xce21,0x0917,0x1590,0x4002,
0x4cfd,0x21d8,0xcac5,0x4022,
0x2c4d,0x7f05,0x1910,0x4028,
0x7ff5,0x959c,0x14d9,0x4031,
0x26c1,0xaa63,0x37ca,0x4023,
0x2d90,0x5ab6,0xf3de,0x400a
};
static short T[] = {
0x5d88,0x1e37,0x35bf,0x4023,
0x067f,0x4e9e,0x81aa,0x4056,
0x35b7,0xbcb4,0x7002,0x40a1,
0xef90,0x3c72,0x5b53,0x40bb,
0x13dc,0xa442,0x2509,0x40eb
};
```

Leading 1.0 omitted from U.

```
static short U[] = {
0041406,0037461,0177575,0032714,
0042402,0053350,0123061,0153557,
0043217,0111227,0032007,0164217,
0043660,0145000,0004013,0160114,
0044100,0071544,0167107,0125471
};
#define UTHRESH 14.0
#endif
```

```
static short U[] = {
0xa6ba,0x3fef,0xc7e6,0x4040,
0x3aee,0x14c6,0x4add,0x4080,
0xfd12,0xe680,0xf252,0x40b1,
0x7c0a,0x0101,0x1940,0x40d6,
0xf567,0x9dc8,0x0e6c,0x40e8
};
#define UTHRESH 37.519379347
#endif
```

```
#include "mconf.h"
extern double SQRTH, MAXLOG;

double ndtr(a)
double a;
    {
    double x, y, z;
    double fabs(), erf(), erfc();

    x = a * SQRTH;
    z = fabs(x);
    if( z < SQRTH )
        y = 0.5 + 0.5 * erf(x);
    else
        {
        y = 0.5 * erfc(z);
        if( x > 0 )
            y = 1.0 - y;
        }
    return(y);
    }
```

5.14.2 erfc.c

Complemented error function. See ndtr.c for the coefficients.

```
double erfc(a)
double a;
    {
    double p,q,x,y,z;
    double polevl(), p1evl(), exp(), log(), erf();

    if( a < 0.0 )
        x = -a;
```

```
	else
		x = a;
	if( x < 1.0 )
		return( 1.0 - erf(a) );
	z = -a * a;
	if( z < -MAXLOG )
		{
under:
		mtherr( "erfc", UNDERFLOW );
		return( 0.0 );
		}
	z = exp(z);
	if( x < 8.0 )
		{
		p = polevl( x, P, 8 );
		q = p1evl( x, Q, 8 );
		}
	else
		{
		p = polevl( x, R, 5 );
		q = p1evl( x, S, 6 );
		}
	y = (z * p)/q;
	if( a < 0 )
		y = 2.0 - y;
	if( y == 0.0 )
		goto under;
	return(y);
	}
```

5.14.3 erf.c

Error function. See ndtr.c for the coefficients.

```
double erf(x)
double x;
	{
	double y, z;
	double fabs(), erfc(), polevl(), p1evl();

	if( fabs(x) > 1.0 )
		return( 1.0 - erfc(x) );
	z = x * x;
	y = x * polevl( z, T, 4 ) / p1evl( z, U, 5 );
```

```
return( y );
}
```

5.14.4 Functional Inverse of Gaussian Distribution

The functional inverse of the normal distribution function finds the argument, x, for which the area

$$\frac{1}{\sqrt{2\pi}} \int_{-\infty}^{x} e^{-t^2/2} dt$$

under the Gaussian probability density function is equal to y. A typical application of this function is to generate normal deviates from a sequence of uniformly distributed random numbers.

For small values $0 < y < e^{-2}$ of the area, a form of approximation is suggested by the asymptotic formula

$$1 - y \approx \frac{1}{x\sqrt{2\pi}} e^{-x^2/2} \left(1 - \frac{1}{x^2} + \cdots\right) .$$

A first approximation to the inverse is the quantity

$$z = \sqrt{-2.0 \ln y} .$$

Substituting this back into the asymptotic formula and examining the residual yields a somewhat refined estimate

$$x \approx z - \frac{\ln z - \ln \sqrt{2\pi}}{z} .$$

The form chosen for the rational approximation is

$$x \approx z - \frac{\ln z}{z} - 1/z \, \frac{P(1/z)}{Q(1/z)} .$$

There are two rational functions P/Q. The first, P_1/Q_1, is for $1.27 \cdot 10^{-14} < y < 0.135$ (or $2 < z < 8$):

$P_1(t) =$	$Q_1(t) =$
$4.0554489230596241992 3 \cdot 10^0 t^8$	$1.0t^8$
$+3.1525109459989386615 4 \cdot 10^1 t^7$	$+1.5779988325646674973 1 \cdot 10^1 t^7$
$+5.7162819224642128816 2 \cdot 10^1 t^6$	$+4.5390763512887921058 4 \cdot 10^1 t^6$
$+4.4080507389320083470 0 \cdot 10^1 t^5$	$+4.1317203825467203044 0 \cdot 10^1 t^5$
$+1.4684956192885802401 4 \cdot 10^1 t^4$	$+1.5042538569290750340 8 \cdot 10^1 t^4$
$+2.1866330685079026753 9 \cdot 10^0 t^3$	$+2.5046494620830941597 9 \cdot 10^0 t^3$
$-1.4025607917135449587 5 \cdot 10^{-1} t^2$	$-1.4218292285478778857 4 \cdot 10^{-1} t^2$
$-3.5042462682784820341 8 \cdot 10^{-2} t$	$-3.8080640769157827719 4 \cdot 10^{-2} t$
$-8.5745678515468541361 1 \cdot 10^{-4}$,	$-9.3325948089545742737 2 \cdot 10^{-4}$.

This approximates the residual with an absolute accuracy of $3.6 \cdot 10^{-18}$. The second version of P/Q, P_2/Q_2 is for $e^{-2048} = 3.67 \cdot 10^{-890} < y < 1.27 \cdot 10^{-14} = e^{-32}$. For this interval the coefficients are

$P_2(t) =$	$Q_2(t) =$
$3.2377489177694603597 0 \cdot 10^0 t^8$	$1.0t^8$
$+6.91522889068984211695 \cdot 10^0 t^7$	$+6.02427039364742014255 \cdot 10^0 t^7$
$+3.93881025292474443415 \cdot 10^0 t^6$	$+3.67983563856160859403 \cdot 10^0 t^6$
$+1.33303460815807542389 \cdot 10^0 t^5$	$+1.37702099489081330271 \cdot 10^0 t^5$
$+2.01485389549179081538 \cdot 10^{-1} t^4$	$+2.16236993594496635890 \cdot 10^{-1} t^4$
$+1.23716634817820021358 \cdot 10^{-2} t^3$	$+1.34204006088543189037 \cdot 10^{-2} t^3$
$+3.01581553508235416007 \cdot 10^{-4} t^2$	$+3.28014464682127739104 \cdot 10^{-4} t^2$
$+2.65806974686737550832 \cdot 10^{-6} t$	$+2.89247864745380683936 \cdot 10^{-6} t$
$+6.23974539184983293730 \cdot 10^{-9}$,	$+6.79019408009981274425 \cdot 10^{-9}$.

Over its approximation interval, this form has an absolute accuracy of $1.1 \cdot 10^{-17}$. For larger arguments the program computes

$$w = y - 0.5$$

in terms of which the approximation is

$$\frac{x}{\sqrt{2\pi}} = w + w^3 \frac{P_0(w^2)}{Q_0(w^2)}$$

where the polynomials are

	$Q_0(t) =$
	$1.0t^8$
$P_0(t) =$	$+1.95448858338141759834 \cdot 10^0 t^7$
$-5.99633501014107895267 \cdot 10^1 t^4$	$+4.67627912898881538453 \cdot 10^0 t^6$
$+9.80010754185999661536 \cdot 10^1 t^3$	$+8.63602421390890590575 \cdot 10^1 t^5$
$-5.66762857469070293439 \cdot 10^1 t^2$	$-2.25462687854119370527 \cdot 10^2 t^4$
$+1.39312609387279679503 \cdot 10^1 t$	$+2.00260212380060660359 \cdot 10^2 t^3$
-1.23916583867381258016 ,	$-8.20372256168333339912 \cdot 10^1 t^2$
	$+1.59056225126211695515 \cdot 10^1 t$
	-1.18331621121330003142 .

The interval of this approximation is $0 \leq w \leq 0.375$. Relative accuracy is $5.9 \cdot 10^{-18}$.

5.14.5 ndtri.c

Functional inverse of Gaussian distribution

```
#include "mconf.h"
```

```
#include "mconf.h"
extern double MAXNUM;
#if DEC                              #if IIEEE
```

approximation for $0 \leq |y - 0.5| \leq 3/8$

```
static short P0[20] = {
0141557,0155170,0071360,0120550,
0041704,0000214,0172417,0067307,
0141542,0132204,0040066,0156723,
0041136,0163161,0157276,0007747,
0140236,0116374,0073666,0051764,
};
```

```
static short P0[20] = {
0x142d,0x0e5e,0xfb4f,0xc04d,
0xedd9,0x9ea1,0x8011,0x4058,
0xdbba,0x8806,0x5690,0xc04c,
0xc1fd,0x3bd7,0xdcce,0x402b,
0xca7e,0x8ef6,0xd39f,0xbff3,
};
```

Leading coefficient 1.0 is omitted from Q0.

```
static short Q0[32] = {
0040372,0026256,0110403,0123707,
0040625,0122024,0020277,0026661,
0041654,0134161,0124134,0007244,
0142141,0073162,0133021,0131371,
0042110,0041235,0043516,0057767,
0141644,0011417,0036155,0137305,
0041176,0076556,0004043,0125430,
0140227,0073347,0152776,0067251,
};
```

```
static short Q0[36] = {
0x74f9,0xd220,0x4595,0x3fff,
0xe5b6,0x8417,0xb482,0x4012,
0x81d4,0x350b,0x970e,0x4055,
0x365f,0x56c2,0x2ece,0xc06c,
0xcbff,0xa8e9,0x0853,0x4069,
0xb7d9,0xe78d,0x8261,0xc054,
0x7563,0xc104,0xcfad,0x402f,
0xcdd5,0xfabf,0xeedc,0xbff2,
};
```

```
static short P1[36] = {
0040601,0143074,0150744,0073326,
0041374,0031554,0113253,0146016,
0041544,0123272,0012463,0176771,
0041460,0051160,0103560,0156511,
0041152,0172624,0117772,0030755,
0040413,0170713,0151545,0176413,
0137417,0117512,0022154,0131671,
0137017,0104257,0071432,0007072,
0135540,0143363,0063137,0036166,
};
```

```
static short P1[36] = {
0x8edb,0x9a3c,0x38c7,0x4010,
0x7982,0x92d5,0x866d,0x403f,
0x7fbf,0x42a6,0x94d7,0x404c,
0x1ba9,0x10ee,0x0a4e,0x4046,
0x463e,0x93ff,0x5eb2,0x402d,
0xbfa1,0x7a6c,0x7e39,0x4001,
0x9677,0x448d,0xf3e9,0xbfc1,
0x41c7,0xee63,0xf115,0xbfa1,
0xe78f,0x6ccb,0x18de,0xbf4c,
};
```

Leading coefficient 1.0 is omitted from Q1.

```
static short Q1[32] = {
0041174,0075325,0004736,0120326,
0041465,0110044,0047561,0045567,
0041445,0042321,0012142,0030340,
0041160,0127074,0166076,0141051,
0040440,0046055,0040745,0150400,
0137421,0114146,0067330,0010621,
0137033,0175162,0025555,0114351,
0135564,0122773,0145750,0030357,
};
```

```
static short Q1[32] = {
0xd41b,0xa13b,0x8f5a,0x402f,
0x296f,0x89ee,0xb204,0x4046,
0x461c,0x228c,0xa89a,0x4044,
0xd845,0x9d87,0x15c7,0x402e,
0xba20,0xa83c,0x0985,0x4004,
0x0232,0xcddb,0x330c,0xbfc2,
0xb31d,0x456d,0x7f4e,0xbfa3,
0x061e,0x797d,0x94bf,0xbf4e,
};
```

```
static short P2[36] = {
0040517,0033507,0036236,0125641,
0040735,0044616,0014473,0140133,
0040574,0012567,0114535,0102541,
0040252,0120340,0143474,0150135,
0037516,0051057,0115361,0031211,
0036512,0131204,0101511,0125144,
0035236,0016627,0043160,0140216,
```

```
static short P2[36] = {
0xd574,0xe793,0xe6e8,0x4009,
0x780b,0xc327,0xa931,0x401b,
0xb0ac,0xf32b,0x82ae,0x400f,
0x9a0c,0x18e7,0x541c,0x3ff5,
0x2651,0xf35e,0xca45,0x3fc9,
0x354d,0x9069,0x5650,0x3f89,
0x1812,0xe8ce,0xc3b2,0x3f33,
```

```
0033462,0060512,0060141,0010641,
0031326,0062541,0101304,0077706,
};
```

```
0x2234,0x4c0c,0x4c29,0x3ec6,
0x8ff9,0x3058,0xccac,0x3e3a,
};
```

Leading coefficient 1.0 is omitted from Q2.

```
static short Q2[32] = {
0040700,0143322,0132137,0040501,
0040553,0101155,0053221,0140257,
0040260,0041071,0052573,0010004,
0037535,0066472,0177261,0162330,
0036533,0160475,0066666,0036132,
0035253,0174533,0027771,0044027,
0033502,0016147,0117666,0063671,
0031351,0047455,0141663,0054751,
};
static short s2p[] = {
0040440,0066230,0177661,0034055
};
#define s2pi *(double *)s2p
#endif
```

```
static short Q2[32] = {
0xe828,0x568b,0x18da,0x4018,
0x3816,0xaad2,0x704d,0x400d,
0x6200,0x2aaf,0x0847,0x3ff6,
0x3c9b,0x5fd6,0xada7,0x3fcb,
0xc78b,0xadb6,0x7c27,0x3f8b,
0x2903,0x65ff,0x7f2b,0x3f35,
0xccf7,0xf3f6,0x438c,0x3ec8,
0x6b3d,0xb876,0x29e5,0x3e3d,
};
static short s2p[] = {
0x2706,0x1ff6,0x0d93,0x4004
};
#define s2pi *(double *)s2p
#endif
```

```
#include "mconf.h"
extern double MAXNUM;

double ndtri(y0)
double y0;
    {
    double x, y, z, y2, x0, x1;
    int code;
    double polevl(), p1evl(), log(), sqrt();

    if( y0 <= 0.0 )
        {
        mtherr( "ndtri", DOMAIN );
        return( -MAXNUM );
        }
    if( y0 >= 1.0 )
        {
        mtherr( "ndtri", DOMAIN );
        return( MAXNUM );
        }
    code = 1;
    y = y0;
```

$0.135\ldots = e^{-2}$

```
    if( y > (1.0 - 0.13533528323661269189) )
        {
        y = 1.0 - y;
```

```
        code = 0;
        }
    if( y > 0.13533528323661269189 )
        {
        y = y - 0.5;
        y2 = y * y;
        x = y2 * polevl( y2, P0, 4)/p1evl( y2, Q0, 8);
        x = ( y + y * x ) * s2pi;
        return(x);
        }
    x = sqrt( -2.0 * log(y) );
    x0 = x - log(x)/x;
    z = 1.0/x;
```

$y > e^{-32} = 1.2664165549 \cdot 10^{-14}$

```
    if( x < 8.0 )
        x1 = z * polevl( z, P1, 8 )/p1evl( z, Q1, 8 );
    else
        x1 = z * polevl( z, P2, 8 )/p1evl( z, Q2, 8 );
    x = x0 - x1;
    if( code != 0 )
        x = -x;
    return( x );
    }
```

6

Bessel Functions

6.1 $J_0(x)$

Bessel functions of the first and second kind have an infinite number of zeros located at nonuniform spacing. Relative accuracy is likely to be limited by cancellation error near the zeros. Considering the order ν and the argument x as two independent variables, the general case is difficult to program. Several different expansions are needed to cover different intervals of ν and x. Good quality routines for fixed order can be developed using rational approximations. Programs to compute J_0, Y_0, J_1 and Y_1 will therefore be discussed first.

The fundamental expansion is

$$J_\nu(x) = (\tfrac{1}{2}x)^\nu \sum_{k=0}^{\infty} \frac{(-\frac{1}{4}x^2)^k}{\Gamma(\nu+k+1)\ k!} \ .$$

A program for J_0 may begin by replacing x with $|x|$, since $J_0(-x) = J_0(x)$. In double precision arithmetic, when $x < 10^{-6}$, terms of order x^3 may be neglected and $J_0(x) \approx 1 - x^2/4$ from the first terms of the ascending power series.

For $x \leq 5.0$ an effective approach is to develop a rational approximation in which the first two zeros are factored out. Thus,

$$J_0(x) \approx (x^2 - r_1^2)(x^2 - r_2^2)\frac{RP(x^2)}{RQ(x^2)} \ .$$

The squares of the first two zeros of $J_0(x)$ are

$$\begin{aligned} r_1^2 &= 5.78318596294678452117599575845 58 \\ r_2^2 &= 30.4712623436620863990781631750 23 \ . \end{aligned}$$

Including the zeros explicitly yields a considerable accuracy iprovement. However, it is impractical to include more than the first few zeros in this way; furthermore the Hankel expansion or other asymptotic forms become applicable at larger values of x. The approximation polynomials for $|x| \leq 5$ are

$$
\begin{aligned}
RP(t) = \\
&-4.79443220978201773821 \cdot 10^9 t^3 \\
&+1.95617491946556577543 \cdot 10^{12} t^2 \\
&-2.49248344360967716204 \cdot 10^{14} t \\
&+9.70862251047306323952 \cdot 10^{15} ,
\end{aligned}
$$

$$
\begin{aligned}
RQ(t) = \\
&+1.0 t^8 \\
&+4.99563147152651017219 \cdot 10^2 t^7 \\
&+1.73785401676374683123 \cdot 10^5 t^6 \\
&+4.84409658339962045305 \cdot 10^7 t^5 \\
&+1.11855537045356834862 \cdot 10^{10} t^4 \\
&+2.11277520115489217587 \cdot 10^{12} t^3 \\
&+3.10518229857422583814 \cdot 10^{14} t^2 \\
&+3.18121955943204943306 \cdot 10^{16} t \\
&+1.71086294081043136091 \cdot 10^{18} .
\end{aligned}
$$

The rational form, regarded as an approximation to

$$J_0(x)/(x^2 - r_1^2)(x^2 - r_2^2) ,$$

has a theoretical relative error of $8.6 \cdot 10^{-18}$. This and the other minimax approximations for J_0, J_1, Y_0, and J_1 were calculated using 100-decimal arithmetic, with which the Bessel functions were evaluated to 43 decimal accuracy.

If $|x| > 5.0$ an approximation may be used that has a form similar to Hankel's asymptotic expansion

$$
\begin{aligned}
J_\nu(x) &= \sqrt{\frac{2}{\pi x}} \, [P(\nu, x) \cos\theta - Q(\nu, x) \sin\theta] \\
P(\nu, x) &\approx 1 - \frac{(u-1)(u-9)}{2!(8x)^2} + \frac{(u-1)(u-9)(u-25)(u-49)}{4!(8x)^4} - \dots \\
Q(\nu, x) &\approx \frac{u-1}{8x} - \frac{(u-1)(u-9)(u-25)}{3!(8x)^3} + \dots \\
\theta &= x - (\tfrac{1}{2}\nu + \tfrac{1}{4})\pi \\
u &= 4\nu^2 .
\end{aligned}
$$

For $J_0(x)$ the parameters are $u = 0$, $\theta = x - \pi/4$. To obtain a best approximation of this form in the interval $[5, \infty]$, $P(\nu, x)$ and $Q(\nu, x)$ are replaced by rational functions in $w = 5/x$. Setting $q = w^2 = 25/x^2$, the approximation is

$$J_0(x) \approx \sqrt{\frac{2}{\pi x}} \left(\frac{PP(q)}{PQ(q)} \cos\theta - w \frac{QP(q)}{QQ(q)} \sin\theta \right) .$$

Owing to the imprecise determination of θ and the summing of two trigonometric functions, it is impractical to maintain high relative accuracy near the zeros with Hankel's formula. The approximation polynomials are

$PP(t) =$
$7.96936729297347051624 \cdot 10^{-4}t^6$
$+8.28352392107440799803 \cdot 10^{-2}t^5$
$+1.23953371646414299388t^4$
$+5.44725003058768775090t^3$
$+8.74716500199817011941t^2$
$+5.30324038235394892183t$
$+0.99999999999999997821$,

$PQ(t) =$
$9.24408810558863637013 \cdot 10^{-4}t^6$
$+8.56288474354474431428 \cdot 10^{-2}t^5$
$+1.25352743901058953537t^4$
$+5.47097740330417105182t^3$
$+8.76190883237069594232t^2$
$+5.30605288235394617618t$
$+1.00000000000000000218$,

$QP(t) =$
$-1.13663838898469149931 \cdot 10^{-2}t^7$
$-1.28252718670509318512 \cdot 10^{0}t^6$
$-1.95539544257735972385 \cdot 10^{1}t^5$
$-9.32060152123768231369 \cdot 10^{1}t^4$
$-1.77681167980488050595 \cdot 10^{2}t^3$
$-1.47077505154951170175 \cdot 10^{2}t^2$
$-5.14105326766599330220 \cdot 10^{1}t$
$-6.05014350600728481186 \cdot 10^{0}$,

$QQ(t) =$
$1.0t^7$
$+6.43178256118178023184 \cdot 10^{1}t^6$
$+8.56430025976980587198 \cdot 10^{2}t^5$
$+3.88240183605401609683 \cdot 10^{3}t^4$
$+7.24046774195652478189 \cdot 10^{3}t^3$
$+5.93072701187316984827 \cdot 10^{3}t^2$
$+2.06209331660327847417 \cdot 10^{3}t$
$+2.42005740240291393179 \cdot 10^{2}$.

Theoretical relative error of the PP/PQ form is $4.4 \cdot 10^{-18}$; of the QP/QQ form, $2.9 \cdot 10^{-18}$.

6.1.1 j0.c

```
#include "mconf.h"

#if DEC
static short PP[28] = {
0035520,0164604,0140733,0054470,
0037251,0122605,0115356,0107170,
0040236,0124412,0071500,0056303,
0040656,0047737,0045720,0045263,
0041013,0172143,0045004,0142103,
0040651,0132045,0026241,0026406,
0040200,0000000,0000000,0000000,
};
static short PQ[28] = {
0035562,0052006,0070034,0134666,
0037257,0057055,0055242,0123424,
0040240,0071626,0046630,0032371,
0040657,0011077,0032013,0012731,
0041014,0030307,0050331,0006414,
0040651,0145457,0065021,0150304,
0040200,0000000,0000000,0000000,
};
static short QP[32] = {
0136472,0035021,0142451,0141115,
0140244,0024731,0150620,0105642,
```

```
#if IIEEE
static short PP[28] = {
0x6b27,0x983b,0x1d30,0x3f4a,
0xd1cf,0xb35d,0x34b0,0x3fb5,
0x0b98,0x4e68,0xd521,0x3ff3,
0x0956,0xe97a,0xc9fb,0x4015,
0x9888,0x6940,0x7e8c,0x4021,
0x25a1,0xa594,0x3684,0x4015,
0x0000,0x0000,0x0000,0x3ff0,
};
static short PQ[28] = {
0x9737,0xce03,0x4a80,0x3f4e,
0x54e3,0xab54,0xebc5,0x3fb5,
0x069f,0xc9b3,0x0e72,0x3ff4,
0x62bb,0xe681,0xe247,0x4015,
0x21a1,0xea1b,0x8618,0x4021,
0x3a19,0xed42,0x3965,0x4015,
0x0000,0x0000,0x0000,0x3ff0,
};
static short QP[32] = {
0x384a,0x38a5,0x4742,0xbf87,
0x1174,0x3a32,0x853b,0xbff4,
```

```
0141234,0067177,0124161,0060141,
0141672,0064572,0151557,0043036,
0142061,0127141,0003127,0043517,
0142023,0011727,0060271,0144544,
0141515,0122142,0126620,0143150,
0140701,0115306,0106715,0007344,
};
```

Leading 1.0 omitted from QQ.

```
static short QQ[28] = {
0041600,0121272,0004741,0026544,
0042526,0015605,0105654,0161771,
0043162,0123155,0165644,0062645,
0043342,0041675,0167576,0130756,
0043271,0052720,0165631,0154214,
0043000,0160576,0034614,0172024,
0042162,0000570,0030500,0051235,
};
static short YP[32] = {
0043563,0120677,0042264,0046166,
0146137,0140371,0113444,0042260,
0050241,0175707,0100502,0063344,
0152144,0125737,0007265,0164526,
0053637,0051621,0163035,0060546,
0155105,0004416,0107306,0060023,
0056035,0045133,0030132,0000024,
0155603,0065132,0144061,0131732,
};
```

Leading 1.0 omitted from YQ.

```
static short YQ[28] = {
0042602,0024422,0135557,0162663,
0045030,0155665,0044075,0160135,
0047200,0035432,0105446,0104005,
0051240,0167331,0056063,0022743,
0053223,0127746,0025764,0012160,
0055064,0044206,0177532,0145545,
0056536,0111375,0163715,0127201,
};
#endif
```

```
0x2c0c,0xf50e,0x8dcf,0xc033,
0xe8c4,0x5a6d,0x4d2f,0xc057,
0xe8ea,0x20ca,0x35cc,0xc066,
0x392d,0xec17,0x627a,0xc062,
0x18cd,0x55b2,0xb48c,0xc049,
0xa1dd,0xd1b9,0x3358,0xc018,
};

static short QQ[28] = {
0x25ac,0x413c,0x1457,0x4050,
0x9c7f,0xb175,0xc370,0x408a,
0x8cb5,0xbd74,0x54cd,0x40ae,
0xd63e,0xbdef,0x4877,0x40bc,
0x3b11,0x1d73,0x2aba,0x40b7,
0x9e82,0xc731,0x1c2f,0x40a0,
0x0a54,0x0628,0x402f,0x406e,
};
static short YP[32] = {
0x898f,0xe896,0x7437,0x40ce,
0x8896,0x32e4,0xf81f,0xc16b,
0x4cdd,0xf028,0x3f78,0x41f4,
0xbd2b,0xe1d6,0x957b,0xc26c,
0xac2d,0x3cc3,0xea72,0x42d3,
0xcc02,0xd1d8,0xa121,0xc328,
0x4003,0x660b,0xa94b,0x4363,
0x367b,0x5906,0x6d4b,0xc350,
};

static short YQ[28] = {
0xfcb6,0x576d,0x4522,0x4090,
0xbc0c,0xa907,0x1b76,0x4123,
0xd101,0x5164,0x0763,0x41b0,
0x64bc,0x2b86,0x1ddb,0x4234,
0x828e,0xc57e,0x75fc,0x42b2,
0x596d,0xdfeb,0x8910,0x4326,
0xb5d0,0xbcf9,0xd25f,0x438b,
};
#endif
```

```
#ifdef UNK
static double DR1 = 5.78318596294678452118E0;
static double DR2 = 3.04712623436620863991E1;
#endif
```

```
#ifdef DEC
static short R1[] = {
0040671,0007734,0001061,0056734};
#define DR1 *(double *)R1
static short R2[] = {
```

```
#ifdef IIEEE
static short R1[] = {
0x2bbb,0x8046,0x21fb,0x4017};
#define DR1 *(double *)R1
static short R2[] = {
```

```
0041363,0142445,0030416,0165567};
#define DR2 *(double *)R2
static short RP[16] = {
0150216,0161235,0064344,0014450,
0052343,0135216,0035624,0144153,
0154142,0130247,0003310,0003667,
0055411,0173703,0047772,0176635,
};
```

Leading 1.0 omitted from RQ.

```
static short RQ[32] = {
0042371,0144025,0032265,0136137,
0044451,0133131,0132420,0151466,
0046470,0144641,0072540,0030636,
0050446,0126600,0045042,0044243,
0052365,0172633,0110301,0071063,
0054215,0032424,0062272,0043513,
0055742,0005013,0171731,0072335,
0057275,0170646,0036663,0013134,
};
#endif
```

```
0xdd6f,0xa621,0x78a4,0x403e};
#define DR2 *(double *)R2
static short RP[16] = {
0x8325,0xad1c,0xdc53,0xc1f1,
0x990d,0xc772,0x7751,0x427c,
0x00f7,0xe0d9,0x5614,0xc2ec,
0x5fb4,0x69ff,0x3ef8,0x4341,
};

static short RQ[32] = {
0xb78c,0xa696,0x3902,0x407f,
0x1a67,0x36a2,0x36cb,0x4105,
0x0634,0x2eac,0x1934,0x4187,
0x4914,0x0944,0xd5b0,0x4204,
0x2e46,0x7218,0xbeb3,0x427e,
0x48e9,0x8c97,0xa6a2,0x42f1,
0x2e9c,0x7e7b,0x4141,0x435c,
0x62cc,0xc7b6,0xbe34,0x43b7,
};
#endif
```

```
double j0(x)
double x;
    {
    double polevl(), p1evl();
    double w, z, p, q, xn;
    double sin(), cos(), sqrt();
    extern double PIO4, SQ2OPI;


    if( x < 0 )
        x = -x;

    if( x <= 5.0 )
        {
        z = x * x;
        if( x < 1.0e-5 )
            return( 1.0 - z/4.0 );

        p = (z - DR1) * (z - DR2);
        p = p * polevl( z, RP, 3)/p1evl( z, RQ, 8 );
        return( p );
        }

    w = 5.0/x;
    q = 25.0/(x*x);
```

```
p = polevl( q, PP, 6)/polevl( q, PQ, 6 );
q = polevl( q, QP, 7)/p1evl( q, QQ, 7 );
xn = x - PIO4;
p = p * cos(xn) - w * q * sin(xn);
return( p * SQ2OPI / sqrt(x) );
}
```

6.2 $Y_0(x)$

Bessel's function of the second kind of order zero has the ascending power series expansion

$$Y_0(x) = \frac{2}{\pi}\left[\ln\frac{x}{2}+\gamma\right]J_0(x) + \frac{2}{\pi}\left[\frac{u}{(1!)^2}-(1+\frac{1}{2})\frac{u^2}{(2!)^2}+(1+\frac{1}{2}+\frac{1}{3})\frac{u^3}{(3!)^2}-\ldots\right]$$

where

$$u = \frac{1}{4}x^2$$

and γ is Euler's constant. As x approaches zero, $Y_0(x)$ exhibits a logarithmic singularity. The program must test for $y \le 0$ and report an error in this event. For purposes of computing the approximation coefficients, the function

$$f(x) = Y_0(x) - \frac{2}{\pi}J_0(x)\ln x \approx \frac{YP(x^2)}{YQ(x^2)}$$

was employed. This removed the singularity at $x = 0$, as

$$f(0) = \frac{2}{\pi}(\ln 1/2 + \gamma) = -0.073804295108687225 \; .$$

The approximation polynomials are

$YP(t) =$	$YQ(t) =$
$1.55924367855235737965 \cdot 10^4 t^7$	$1.0t^7$
$-1.46639295903971606143 \cdot 10^7 t^6$	$+1.04128353664259848412 \cdot 10^3 t^6$
$+5.43526477051876500413 \cdot 10^9 t^5$	$+6.26107330137134956842 \cdot 10^5 t^5$
$-9.82136065717911466409 \cdot 10^{11} t^4$	$+2.68919633393814121987 \cdot 10^8 t^4$
$+8.75906394395366999549 \cdot 10^{13} t^3$	$+8.64002487103935000337 \cdot 10^{10} t^3$
$-3.46628303384729719441 \cdot 10^{15} t^2$	$+2.02979612750105546709 \cdot 10^{13} t^2$
$+4.42733268572569800351 \cdot 10^{16} t$	$+3.17157752842975028269 \cdot 10^{15} t$
$-1.84950800436986690637 \cdot 10^{16}$,	$+2.50596256172653059228 \cdot 10^{17}$.

These satisfy the absolute error criterion, whereas all the other rational approximations in this section satisfy the relative error criterion. The theoretical accuracy of this rational form is $7.6 \cdot 10^{-22}$.

Using $f(x)$ the approximation for $x \le 5.0$ is

$$Y_0(x) \approx \frac{YP(x^2)}{YQ(x^2)} + \frac{2}{\pi} J_0(x) \ln x \;.$$

For $x > 5.0$, the Hankel expansion is again used:

$$Y_0(x) \approx \sqrt{\frac{2}{\pi x}} \left(\frac{PP(q)}{PQ(q)} \sin\theta + w \frac{QP(q)}{QQ(q)} \cos\theta \right)$$

with the same definitions as for $J_0(x)$.

The following constants are required:

$$\begin{aligned} \pi/4 &= 0.78539816339744830962 \\ \sqrt{2/\pi} &= 0.79788456080286535588 \;. \end{aligned}$$

6.2.1 y0.c

Bessel function of second kind, order zero. See `j0.c` for the coefficients.

```
extern double MAXNUM, TWOOPI, SQ2OPI, PIO4;

double y0(x)
double x;
    {
    double polevl(), p1evl();
    double w, z, p, q, xn;
    double j0(), log(), sin(), cos(), sqrt();

    if( x <= 5.0 )
        {
        if( x <= 0.0 )
            {
            mtherr( "y0", DOMAIN );
            return( -MAXNUM );
            }
        z = x * x;
        w = polevl( z, YP, 7) / p1evl( z, YQ, 7 );
        w += TWOOPI * log(x) * j0(x);
        return( w );
        }
    w = 5.0/x;
    z = 25.0 / (x * x);
    p = polevl( z, PP, 6)/polevl( z, PQ, 6 );
    q = polevl( z, QP, 7)/p1evl( z, QQ, 7 );
```

```
xn = x - PIO4;
p = p * sin(xn) + w * q * cos(xn);
return( p * SQ2OPI / sqrt(x) );
}
```

6.3 Modulus and Phase

An alternative expansion of Bessel functions for large x is in terms of a modulus $M(x)$ and phase $\theta(x)$ such that

$$\begin{aligned} \tan(\theta_n(x)) &= \frac{Y_n(x)}{J_n(x)} \\ M_n^2(x) &= J_n^2(x) + Y_n^2(x) \, . \end{aligned}$$

For $n = 0$, $x \geq 5$ an approximation for M_0 is

$$\begin{aligned} w &= 5/x \\ M_0^2(x) &\approx w \, \frac{P(w^2)}{Q(w^2)} \end{aligned}$$

where

$$\begin{aligned} P(t) = {} & 9.85992596732645506170 \cdot 10^{-2} t^6 \\ & +1.05103033565098609508 \cdot 10^1 t^5 \\ & +1.56315803159483429826 \cdot 10^2 t^4 \\ & +6.81919056479882997314 \cdot 10^2 t^3 \\ & +1.08821029978292105736 \cdot 10^3 t^2 \\ & +6.56406822629655523114 \cdot 10^2 t \\ & +1.23258123334451107419 \cdot 10^2 \, , \end{aligned}$$

$$\begin{aligned} Q(t) = {} & 1.0 t^6 \\ & +8.74553293398745562438 \cdot 10^1 t^5 \\ & +1.25212679410254107858 \cdot 10^3 t^4 \\ & +5.39696752023882233061 \cdot 10^3 t^3 \\ & +8.57225822309516355115 \cdot 10^3 t^2 \\ & +5.16024746453339017062 \cdot 10^3 t \\ & +9.68067036906940679580 \cdot 10^2 \, . \end{aligned}$$

The phase is approximated by

$$\begin{aligned} w &= 5/x \\ \theta_0(x) &\approx x - \frac{\pi}{4} + w \, \frac{P(w^2)}{Q(w^2)} \end{aligned}$$

where

$$\begin{aligned} P(t) = {} & -1.36197755003940839285 \cdot 10^{-2} t^6 \\ & -1.17984575715066233497 \cdot 10^0 t^5 \\ & -1.45181090519066236709 \cdot 10^1 t^4 \\ & -5.42453877822018247077 \cdot 10^1 t^3 \\ & -7.61837168510915882777 \cdot 10^1 t^2 \\ & -4.12833908194942236882 \cdot 10^1 t \\ & -7.07494447054639240247 \, , \end{aligned}$$

$$\begin{aligned} Q(t) = {} & 1.0 t^6 \\ & +5.60440101469414274852 \cdot 10^1 t^5 \\ & +6.20167231193037134789 \cdot 10^2 t^4 \\ & +2.22979723476514463817 \cdot 10^3 t^3 \\ & +3.08111518709149716368 \cdot 10^3 t^2 \\ & +1.65723141983841478014 \cdot 10^3 t \\ & +2.82997778821856322525 \cdot 10^2 \, . \end{aligned}$$

This approximation for the phase satisfies the absolute error criterion. In terms of M and θ,

$$\begin{aligned} J_n(x) &= M_n(x)\cos(\theta_n(x)) \\ Y_n(x) &= M_n(x)\sin(\theta_n(x)) \ . \end{aligned}$$

6.4 $J_1(x)$

Bessel's function of the first kind, of order 1, has the power series

$$J_1(x) = \frac{x}{2}\sum_{k=0}^{\infty}\frac{(-x^2/4)^k}{k!\,\Gamma(k+2)} \ .$$

It is an odd function:

$$J_1(-x) = -J_1(x) \ .$$

An approximation form similar to the one used for J_0 is applicable. For $x \leq 5$,

$$J_1(x) \approx x(x^2 - r_1^2)(x^2 - r_2^2)\frac{RP(x^2)}{RQ(x^2)}$$

where

$$\begin{aligned} r_1^2 &= 14.68197064212389 32572 \\ r_2^2 &= 49.21845632169460 36703 \end{aligned}$$

and the approximation polynomials are

$RP(t) =$
static double RP[4] = {
$-8.99971225705559398224 \cdot 10^8 t^3$
$+4.52228297998194034323 \cdot 10^{11} t^2$
$-7.27494245221818276015 \cdot 10^{13} t$
$+3.68295732863852883286 \cdot 10^{15}$,

$RQ(t) =$
$+1.0t^8$
$+6.20836478118054335476 \cdot 10^2 t^7$
$+2.56987256757748830383 \cdot 10^5 t^6$
$+8.35146791431949253037 \cdot 10^7 t^5$
$+2.21511595479792499675 \cdot 10^{10} t^4$
$+4.74914122079991414898 \cdot 10^{12} t^3$
$+7.84369607876235854894 \cdot 10^{14} t^2$
$+8.95222336184627338078 \cdot 10^{16} t$
$+5.32278620332680085395 \cdot 10^{18}$.

When regarded as an approximation to $J_1(x)/(x^2 - r_1)(x^2 - r_2)$, this rational form has a relative accuracy of $1.1 \cdot 10^{-18}$.

Hankel's asymptotic expansion is used for $x > 5$, in the form

$$J_1(x) \approx \sqrt{\frac{2}{\pi x}}\left(\frac{PP(w^2)}{PQ(w^2)}\cos\theta - w\frac{QP(w^2)}{QQ(w^2)}\sin\theta\right)$$

where $\theta = x - 3\pi/4$ and $w = 5/x$. The polynomial coefficients are

$$PP(t) = 7.62125616208173112003 \cdot 10^{-4}t^6 +7.31397056940917570436 \cdot 10^{-2}t^5 +1.12719608129684925192t^4 +5.11207951146807644818t^3 +8.42404590141772420927t^2 +5.21451598682361504063t +1.00000000000000000254 ,$$

$$PQ(t) = 5.71323128072548699714 \cdot 10^{-4}t^6 +6.88455908754495404082 \cdot 10^{-2}t^5 +1.10514232634061696926t^4 +5.07386386128601488557t^3 +8.39985554327604159757t^2 +5.20982848682361821619t +0.99999999999999997461,$$

$$QP(t) = 5.10862594750176621635 \cdot 10^{-2}t^7 +4.98213872951233449420 \cdot 10^{0}t^6 +7.58238284132545283818 \cdot 10^{1}t^5 +3.66779609360150777800 \cdot 10^{2}t^4 +7.10856304998926107277 \cdot 10^{2}t^3 +5.97489612400613639965 \cdot 10^{2}t^2 +2.11688757100572135698 \cdot 10^{2}t +2.52070205858023719784 \cdot 10^{1} ,$$

$$QQ(t) = +1.0t^7 +7.42373277035675149943 \cdot 10^{1}t^6 +1.05644886038262816351 \cdot 10^{3}t^5 +4.98641058337653607651 \cdot 10^{3}t^4 +9.56231892404756170795 \cdot 10^{3}t^3 +7.99704160447350683650 \cdot 10^{3}t^2 +2.82619278517639096600 \cdot 10^{3}t +3.36093607810698293419 \cdot 10^{2} .$$

Theoretical relative error of PP/PQ is $5.1 \cdot 10^{-18}$; that of QP/QQ is $1.1 \cdot 10^{-18}$.

6.4.1 j1.c

```
#include "mconf.h"

#if DEC
static short RP[16] = {
0147526,0110742,0063322,0077052,
0051722,0112720,0065034,0061530,
0153604,0052227,0033147,0105650,
0055121,0055025,0032276,0022015,
};
Leading 1.0 omitted from RQ.
static short RQ[32] = {
0042433,0032610,0155604,0033473,
0044572,0173320,0067270,0006616,
0046637,0045246,0162225,0006606,
0050645,0004773,0157577,0053004,
0052612,0033734,0001667,0176501,
0054462,0054121,0173147,0121367,
0056237,0002777,0121451,0176007,
0057623,0136253,0131601,0044710,
};
static short PP[28] = {
0035507,0144542,0061543,0024326,
0037225,0145105,0017766,0022661,
0040220,0043766,0010254,0133255,
0040643,0113047,0142611,0151521,
```

```
#if IIEEE
static short RP[16] = {
0x4fc5,0x4cda,0xd23c,0xc1ca,
0x8c6b,0x0d43,0x52ba,0x425a,
0xf175,0xe6cc,0x8a92,0xc2d0,
0xc482,0xa697,0x2b42,0x432a,
};

static short RQ[32] = {
0x86e7,0x1b70,0x66b1,0x4083,
0x01b2,0x0dd7,0x5eda,0x410f,
0xa1b1,0xdc92,0xe954,0x4193,
0xeac1,0x7bef,0xa13f,0x4214,
0xffa8,0x8076,0x46fb,0x4291,
0xf45f,0x3ecc,0x4b0a,0x4306,
0x3f81,0xf465,0xe0bf,0x4373,
0x2939,0x7670,0x7795,0x43d2,
};
static short PP[28] = {
0x651b,0x4c6c,0xf92c,0x3f48,
0xc4b6,0xa3fe,0xb948,0x3fb2,
0x96d6,0xc215,0x08fe,0x3ff2,
0x3a6a,0xf8b1,0x72c4,0x4014,
```

```
0041006,0144344,0055351,0074261,
0040646,0156520,0120574,0006416,
0040200,0000000,0000000,0000000,
};
static short PQ[28] = {
0035425,0142330,0115041,0165514,
0037214,0177352,0145105,0052026,
0040215,0072515,0141207,0073255,
0040642,0056427,0137222,0106405,
0041006,0062716,0166427,0165450,
0040646,0133352,0035425,0123304,
0040200,0000000,0000000,0000000,
};
static short QP[32] = {
0037121,0037723,0055605,0151004,
0040637,0066656,0031554,0077264,
0041627,0122714,0153170,0161466,
0042267,0061712,0036520,0140145,
0042461,0133315,0131573,0071176,
0042425,0057525,0147500,0013201,
0042123,0130122,0061245,0154131,
0041311,0123772,0064254,0172650,
};
```

Leading 1.0 omitted from QQ.

```
static short QQ[28] = {
0041624,0074603,0002112,0101670,
0042604,0007135,0010162,0175565,
0043233,0151510,0157757,0172010,
0043425,0064506,0112006,0104276,
0043371,0164125,0032271,0164242,
0043060,0121425,0122750,0136013,
0042250,0005773,0053472,0146267,
};
static short YP[24] = {
0047626,0112763,0013715,0133045,
0152026,0134552,0142033,0024411,
0053720,0045245,0102210,0077565,
0155347,0000321,0136415,0102031,
0056463,0146550,0055633,0032605,
0157054,0171012,0167361,0054265,
};
```

Leading 1.0 omitted from YQ.

```
static short YQ[32] = {
0042424,0111515,0044773,0153014,
0044546,0005405,0171307,0075774,
0046614,0023575,0047105,0063556,
0050613,0143034,0101533,0156026,
0052541,0175367,0166514,0114257,
0054415,0014466,0134350,0171154,
```

```
0x2f16,0x8b5d,0xd91c,0x4020,
0x81a2,0x142f,0xdbaa,0x4014,
0x0000,0x0000,0x0000,0x3ff0,
};
static short PQ[28] = {
0x3d69,0x1344,0xb89b,0x3f42,
0xaa83,0x5948,0x9fdd,0x3fb1,
0xeed6,0xb850,0xaea9,0x3ff1,
0x51a1,0xf7d2,0x4ba2,0x4014,
0xfd65,0xdda2,0xccb9,0x4020,
0xb4d9,0x4762,0xd6dd,0x4014,
0x0000,0x0000,0x0000,0x3ff0,
};
static short QP[32] = {
0xba40,0x6b70,0x27fa,0x3faa,
0x8fd6,0xc66d,0xedb5,0x4013,
0x1c67,0x9acf,0xf4b9,0x4052,
0x180d,0x47aa,0xec79,0x4076,
0x6e50,0xb66f,0x36d9,0x4086,
0x02d0,0xb9e8,0xabea,0x4082,
0xbb0b,0x4c54,0x760a,0x406a,
0x9eb5,0x4d15,0x34ff,0x4039,
};
static short QQ[28] = {
0x5077,0x6089,0x8f30,0x4052,
0x5f6f,0xa20e,0x81cb,0x4090,
0xfe81,0x1bfd,0x7a69,0x40b3,
0xd118,0xd280,0xad28,0x40c2,
0x3d14,0xa697,0x3d0a,0x40bf,
0x1781,0xb4bd,0x1462,0x40a6,
0x5997,0x6ae7,0x017f,0x4075,
};
static short YP[24] = {
0xb6c5,0x62f9,0xd2be,0x41d2,
0x6521,0x5883,0xd72d,0xc262,
0x0fef,0xb091,0x0954,0x42da,
0xb083,0x37a1,0xe01a,0xc33c,
0x66b1,0x0b73,0x79ad,0x4386,
0x2b17,0x5dde,0x9e41,0xc3a5,
};
static short YQ[32] = {
0x7ac2,0xa93f,0x9269,0x4082,
0xef7f,0xbe58,0xc160,0x410c,
0xacee,0xa9c8,0x84ef,0x4191,
0x7b83,0x906b,0x78c3,0x4211,
0x9316,0xfda9,0x3f5e,0x428c,
0x1e4e,0xd71d,0xa326,0x4301,
```

```
0056164,0017436,0025075,0022101,
0057534,0103614,0103663,0121772,
};
#endif
```

```
0xa488,0xc547,0x83e3,0x436e,
0x747f,0x90f6,0x90f1,0x43cb,
};
#endif
```

```
#ifdef UNK
static double Z1 = 1.46819706421238932572E1;
static double Z2 = 4.92184563216946036703E1;
#endif
```

```
#ifdef DEC
static short DZ1[] = {
0041152,0164532,0006114,0010540};
static short DZ2[] = {
0041504,0157663,0001625,0020621};
#define Z1 (*(double *)DZ1)
#define Z2 (*(double *)DZ2)
#endif
```

```
#ifdef IIEEE
static short DZ1[] = {
0x822c,0x4189,0x5d2b,0x402d};
static short DZ2[] = {
0xa432,0x6072,0x9bf6,0x4048};
#define Z1 (*(double *)DZ1)
#define Z2 (*(double *)DZ2)
#endif
```

```
double j1(x)
double x;
	{
	extern double PIO4, THPIO4, SQ2OPI;
	double polevl(), p1evl();
	double w, z, p, q, xn;
	double sin(), cos(), sqrt();
	w = x;
	if( x < 0 )
		w = -x;
	if( w <= 5.0 )
		{
		z = x * x;
		w = polevl( z, RP, 3 ) / p1evl( z, RQ, 8 );
		w = w * x * (z - Z1) * (z - Z2);
		return( w );
		}
	w = 5.0/x;
	z = w * w;
	p = polevl( z, PP, 6)/polevl( z, PQ, 6 );
	q = polevl( z, QP, 7)/p1evl( z, QQ, 7 );
	xn = x - THPIO4;
	p = p * cos(xn) - w * q * sin(xn);
	return( p * SQ2OPI / sqrt(x) );
	}
```

6.5 $Y_1(x)$

For $Y_1(x)$, the Bessel function of the second kind of order 1, the procedure is similar to the one for Y_0. There is again a singularity at $x = 0$, which must be provided for as an overflow error. For $0 < x \leq 5$,

$$Y_1(x) \approx x\frac{YP(x^2)}{YQ(x^2)} + \frac{2}{\pi}\left[J_1(x)\ln x - 1/x\right]$$

where

$$\begin{array}{ll} & YQ(t) = +1.0t^8 \\ YP(t) = & +5.94301592346128195359 \cdot 10^2 t^7 \\ 1.26320474790178026440 \cdot 10^9 t^5 & +2.35564092943068577943 \cdot 10^5 t^6 \\ -6.47355876379160291031 \cdot 10^{11} t^4 & +7.34811944459721705660 \cdot 10^7 t^5 \\ +1.14509511541823727583 \cdot 10^{14} t^3 & +1.87601316108706159478 \cdot 10^{10} t^4 \\ -8.12770255501325109621 \cdot 10^{15} t^2 & +3.88231277496238566008 \cdot 10^{12} t^3 \\ +2.02439475713594898196 \cdot 10^{17} t & +6.20557727146953693363 \cdot 10^{14} t^2 \\ -7.78877196265950026825 \cdot 10^{17}\ , & +6.87141087355300489866 \cdot 10^{16} t \\ & +3.97270608116560655612 \cdot 10^{18}\ . \end{array}$$

The theoretical absolute error for this rational form is $1.2 \cdot 10^{-19}$. For $x > 5$,

$$Y_1(x) = \sqrt{\frac{2}{\pi x}}\left(\frac{PP(w^2)}{PQ(w^2)}\sin\theta + w\frac{QP(w^2)}{QQ(w^2)}\cos\theta\right)$$

where again $\theta = x - 3\pi/4$, $w = 5/x$, and the polynomials are exactly the same as for J_1.

Required constants are

$$\begin{aligned} 3\pi/4 &= 2.35619449019234492885 \\ \sqrt{2/\pi} &= 0.79788456080286535588\ . \end{aligned}$$

6.5.1 y1.c

Bessel function of second kind, order one. See `j1.c` for the octal and hexadecimal values of coefficients.

```
extern double MAXNUM, TWOOPI, THPIO4, SQ2OPI;

double y1(x)
double x;
    {
    double polevl(), p1evl();
    double w, z, p, q, xn;
    double j1(), log(), sin(), cos(), sqrt();
```

```
if( x <= 5.0 )
	{
	if( x <= 0.0 )
		{
		mtherr( "y1", DOMAIN );
		return( -MAXNUM );
		}
	z = x * x;
	w = x * (polevl( z, YP, 5 ) / p1evl( z, YQ, 8 ));
	w += TWOOPI * ( j1(x) * log(x) - 1.0/x );
	return( w );
	}
w = 5.0/x;
z = w * w;
p = polevl( z, PP, 6)/polevl( z, PQ, 6 );
q = polevl( z, QP, 7)/p1evl( z, QQ, 7 );
xn = x - THPIO4;
p = p * sin(xn) + w * q * cos(xn);
return( p * SQ2OPI / sqrt(x) );
}
```

6.6 $J_n(x)$

In the section on continued fractions the following expression for the ratio of two Bessel functions was derived:

$$\frac{J_n(x)}{J_{n-1}(x)} = \frac{1}{2n/x -} \; \frac{1}{2(n+1)/x -} \; \frac{1}{2(n+2)/x - \cdots} .$$

Starting with $k = n$, the recurrence

$$J_{k-1}(x) = \frac{2k}{x} J_k(x) - J_{k+1}(x)$$

may be used to compute successively

$$\begin{aligned}
\frac{J_{n-2}(x)}{J_n(x)} &= \frac{2n-2}{x} \frac{J_{n-1}(x)}{J_n(x)} - 1 \\
\frac{J_{n-3}(x)}{J_n(x)} &= \frac{2n-4}{x} \frac{J_{n-2}(x)}{J_n(x)} - \frac{J_{n-1}(x)}{J_n(x)} \\
&\cdots \\
\frac{J_0(x)}{J_n(x)} &= \frac{2}{x} \frac{J_1(x)}{J_n(x)} - \frac{J_2(x)}{J_n(x)} \\
&= f(x) .
\end{aligned}$$

$J_0(x)$ may be found by the expansions given in the previous section. Hence

$$J_n(x) = \frac{J_0(x)}{f(x)} .$$

This method of computation is not very suitable for large x or n. If x is large, numerical convergence of the continued fraction may be unsatisfactory. If n is large, reduction of the order to $k = 0$ by recurrence will be time consuming and probably inaccurate. For these cases the expansions given later in the section on $J_\nu(x)$ may be preferred.

While computing the recurrence, a refinement by Blanch[1] may be used to avoid amplification of cancellation error. Each new term J_m is tested relative to the two items subtracted. If its magnitude is small then the next term may be computed by

$$\begin{aligned} J_{m-1}(x) &= \frac{2m}{x} J_m(x) - J_{m+1}(x) \\ &= \frac{2m}{x}\left[\frac{2(m+1)}{x} J_{m+1}(x) - J_{m+2}(x)\right] - J_{m+1} \\ &= \left[\frac{4m(m+1)}{x^2} - 1\right] J_{m+1}(x) - \frac{2m}{x} J_{m+2}(x) . \end{aligned}$$

The next following term is computed in the normal way by

$$J_{m-2}(x) = \frac{2(m-1)}{x} J_{m-1}(x) - J_m(x).$$

In the first instance, the absolute error in computing J_m would have been multiplied by the factor $2m/x$, whereas in the modified procedure it is not amplified.

A useful simplification of the above procedure is to examine $f(x) = J_0(x)/J_n$ and $f_1(x) = J_1(x)/J_n$ at the end of the sequence of recurrence steps. If f_1 is larger in magnitude than f, then find J_n by

$$J_n(x) = \frac{J_1(x)}{f_1(x)} .$$

For a program, see the subroutine **recur**() in Section 6.13.

6.7 $I_0(x)$

The modified Bessel function of order zero is the even function

$$I_0(x) = J_0(ix) .$$

[1] Blanch, G. (1964) "Numerical evaluation of continued fractions," *SIAM Review* 6, 383-421.

Coefficients in Table 6.1 are for insertion into the Chebyshev expansion

$$f(x) = \sum_{i=0}^{\infty}{}' \; c_i T_i(x) \; .$$

The primed summation indicates that the zeroth degree Chebyshev coefficient is to be divided by two; this has already been done in the tables. If $x \le 8$, the expansion is (see Section 2.2)

$$I_0(x) = e^x \sum_{i=0}^{\infty}{}' \; A_i \; T_i(\tfrac{x}{4} - 1) \; .$$

If $x > 8$,

$$I_0(x) = \frac{e^x}{\sqrt{x}} \sum_{i=0}^{\infty}{}' \; B_i \; T_i(\tfrac{16}{x} - 1) \; .$$

6.7.1 i0.c

```
#include "mconf.h"
```

Chebyshev coefficients for $e^{-x} I_0(x)$

```
#ifdef DEC
static short A[] = {
0121642,0162671,0004646,0103567,
0022431,0115424,0135755,0026104,
0123214,0023533,0110365,0156635,
0023767,0033304,0117662,0172716,
0124522,0100426,0012277,0157531,
0025254,0155062,0054461,0030465,
0126010,0131143,0013560,0153604,
0026517,0170577,0006336,0114437,
0127227,0162253,0152243,0052734,
0027724,0142766,0061641,0160200,
0130416,0123760,0116564,0125262,
0031066,0144035,0021246,0054641,
0131537,0053664,0060131,0102530,
0032201,0155664,0165153,0020652,
0132617,0061434,0074423,0176145,
0033225,0174444,0136147,0122542,
0133624,0031576,0056453,0020470,
0034211,0175305,0172321,0041314,
0134561,0054462,0147040,0165315,
0035105,0124333,0120203,0162532,
0135427,0013750,0174257,0055221,
0035726,0161654,0050220,0100162,
0136215,0131361,0000325,0041110,
```

```
#ifdef IIEEE
static short A[] = {
0xd0ef,0x2134,0x5cb7,0xbc54,
0xa589,0x977d,0x3362,0x3c83,
0xbbb4,0x721e,0x84eb,0xbcb1,
0x5eba,0x93f6,0xe6d8,0x3cde,
0xfbeb,0xc297,0x5022,0xbd0a,
0x2627,0x4b26,0x9b46,0x3d35,
0x1af0,0x62ee,0x164c,0xbd61,
0xd324,0xe19b,0xfe2f,0x3d89,
0x6abc,0x7a94,0xfc95,0xbdb2,
0x3c10,0xcc74,0x98be,0x3dda,
0x9556,0x13ae,0xd4fe,0xbe01,
0xcb34,0xa454,0xd903,0x3e26,
0x30ab,0x8c0b,0xeaf6,0xbe4b,
0x6435,0x9d4d,0x3b76,0x3e70,
0x7f8d,0x8f22,0xec63,0xbe91,
0xf4ac,0x978c,0xbf24,0x3eb2,
0x6427,0xcba5,0x866f,0xbed2,
0x2859,0xbe9a,0x3f58,0x3ef1,
0x1d5a,0x59c4,0x2b26,0xbf0e,
0x7cab,0x7410,0xb51b,0x3f28,
0xeb52,0x1f15,0xe2fd,0xbf42,
0x100e,0x8a12,0xdc75,0x3f5a,
0xa849,0x201a,0xb65e,0xbf71,
```

Table 6.1: Chebyshev Coefficients for $I_0(x)$

i	A_i	B_i
29	$-4.415 \cdot 10^{-18}$	
28	$+3.3308 \cdot 10^{-17}$	
27	$-2.43128 \cdot 10^{-16}$	
26	$+1.715391 \cdot 10^{-15}$	
25	$-1.1685333 \cdot 10^{-14}$	
24	$+7.6761855 \cdot 10^{-14}$	$-7.233 \cdot 10^{-18}$
23	$-4.85644678 \cdot 10^{-13}$	$-4.831 \cdot 10^{-18}$
22	$+2.955052663 \cdot 10^{-12}$	$+4.4656 \cdot 10^{-17}$
21	$-1.7268262914 \cdot 10^{-11}$	$+3.4612 \cdot 10^{-17}$
20	$+9.6758090354 \cdot 10^{-11}$	$-2.82762 \cdot 10^{-16}$
19	$-5.18979560164 \cdot 10^{-10}$	$-3.42549 \cdot 10^{-16}$
18	$+2.659823724682 \cdot 10^{-9}$	$+1.772560 \cdot 10^{-15}$
17	$-1.3000250099862 \cdot 10^{-8}$	$+3.811681 \cdot 10^{-15}$
16	$+6.0469950225419 \cdot 10^{-8}$	$-9.554847 \cdot 10^{-15}$
15	$-2.67079385394061 \cdot 10^{-7}$	$-4.1505693 \cdot 10^{-14}$
14	$+1.117387539120104 \cdot 10^{-6}$	$+1.5400862 \cdot 10^{-14}$
13	$-4.416738358458751 \cdot 10^{-6}$	$+3.85277838 \cdot 10^{-13}$
12	$+1.6448448070728897 \cdot 10^{-5}$	$+7.18012445 \cdot 10^{-13}$
11	$-5.7541950100821037 \cdot 10^{-5}$	$-1.794178532 \cdot 10^{-12}$
10	$+1.88502885095841656 \cdot 10^{-4}$	$-1.3215811840 \cdot 10^{-11}$
9	$-5.76375574538582366 \cdot 10^{-4}$	$-3.1499165280 \cdot 10^{-11}$
8	$+1.639475616941335798 \cdot 10^{-3}$	$+1.1889147108 \cdot 10^{-11}$
7	$-4.324309995050575944 \cdot 10^{-3}$	$+4.94060238822 \cdot 10^{-10}$
6	$+1.0546460394594998318 \cdot 10^{-2}$	$+3.396232025708 \cdot 10^{-9}$
5	$-2.3737414805899468816 \cdot 10^{-2}$	$+2.2666689904982 \cdot 10^{-8}$
4	$+4.9305284239670708488 \cdot 10^{-2}$	$+2.04891858946906 \cdot 10^{-7}$
3	$-9.4901097048047644421 \cdot 10^{-2}$	$+2.891370520834756 \cdot 10^{-6}$
2	$+0.17162090152220877534$9	$+6.8897583469168240 \cdot 10^{-5}$
1	$-0.30468267234319839868$3	$+3.3691164782556940\cdot 10^{-3}$
0	$+0.67679527440947608499$5	$+0.80449041101410883160$8

```
0036454,0145417,0117357,0017352,
0136702,0072367,0104415,0133574,
0037111,0172126,0072505,0014544,
0137302,0055601,0120550,0033523,
0037457,0136543,0136544,0043002,
0137633,0177536,0001276,0066150,
0040055,0041164,0100655,0010521
};
```

Chebyshev coefficients for $e^{-x}\sqrt{x}I_0(x)$

```
static short B[] = {
0122005,0066672,0123124,0054311,
0121662,0033323,0030214,0104602,
0022515,0170300,0113314,0020413,
0022437,0117350,0035402,0007146,
0123243,0000135,0057220,0177435,
0123305,0073476,0144106,0170702,
0023777,0071755,0017527,0154373,
0024211,0052214,0102247,0033270,
0124454,0017763,0171453,0012322,
0125072,0166316,0075505,0154616,
0024612,0133770,0065376,0025045,
0025730,0162143,0056036,0001632,
0026112,0015077,0150464,0063542,
0126374,0101030,0014274,0065457,
0127150,0077271,0125763,0157617,
0127412,0104350,0040713,0120445,
0027121,0023765,0057500,0001165,
0030407,0147146,0003643,0075644,
0031151,0061445,0044422,0156065,
0031702,0132224,0003266,0125551,
0032534,0000076,0147153,0005555,
0033502,0004536,0004016,0026055,
0034620,0076433,0142314,0171215,
0036134,0146145,0013454,0101104,
0040115,0171425,0062500,0047133
};
#endif
```

```
0xe3dd,0xf3dd,0x9961,0x3f85,
0xb6f0,0xf121,0x4e9e,0xbf98,
0xa32d,0xcea8,0x3e8a,0x3fa9,
0x06ea,0x342d,0x4b70,0xbfb8,
0x88c0,0x77ac,0xf7ac,0x3fc5,
0xcd8d,0xc057,0x7feb,0xbfd3,
0xa22a,0x9035,0xa84e,0x3fe5,
};

static short B[] = {
0x8b19,0x54ca,0xadb7,0xbc60,
0x9130,0x6611,0x46da,0xbc56,
0x8421,0x12d9,0xbe18,0x3c89,
0x41cd,0x0760,0xf3dd,0x3c83,
0x1fe4,0xabd2,0x600b,0xbcb4,
0xde38,0xd908,0xaee7,0xbcb8,
0xfb1f,0xa3ea,0xee7d,0x3cdf,
0xe6d7,0x9094,0x2a91,0x3cf1,
0x629a,0x7e65,0x83fe,0xbd05,
0xbb32,0xcf68,0x5d99,0xbd27,
0xc545,0x0d5f,0x56ff,0x3d11,
0xc073,0x6b83,0x1c8c,0x3d5b,
0x8cec,0xfa26,0x4347,0x3d69,
0x8d66,0x0317,0x9043,0xbd7f,
0x7bf2,0x357e,0x0fd7,0xbdad,
0x7425,0x0839,0x511d,0xbdc1,
0x004f,0xabe8,0x24fe,0x3daa,
0x6f75,0xc0f4,0xf9cc,0x3e00,
0x5b87,0xa922,0x2c64,0x3e2d,
0xd56d,0x80d6,0x5692,0x3e58,
0x616e,0xd9cd,0x8007,0x3e8b,
0xc586,0xc101,0x412b,0x3ec8,
0x9e52,0x7899,0x0fa3,0x3f12,
0x9049,0xa2e5,0x998c,0x3f6b,
0x09cb,0xaca8,0xbe62,0x3fe9
};
#endif
```

```
double i0(x)
double x;
	{
	double y;
	double chbevl(), exp(), sqrt();
	int i;

	if( x < 0 )
		x = -x;
```

```
    if( x <= 8.0 )
        {
        y = (x/2.0) - 2.0;
        return( exp(x) * chbevl( y, A, 30 ) );
        }
    return( exp(x)
            * chbevl( 32.0/x - 2.0, B, 25 ) / sqrt(x) );
    }
```

Program for $e^{-x} I_0(x)$

```
double i0e( x )
double x;
    {
    double y;
    int i;
    double chbevl(), exp(), sqrt();

    if( x < 0 )
        x = -x;
    if( x <= 8.0 )
        {
        y = (x/2.0) - 2.0;
        return( chbevl( y, A, 30 ) );
        }
    return( chbevl( 32.0/x - 2.0, B, 25 ) / sqrt(x) );
    }
```

6.8 $I_1(x)$

The modified Bessel function of order one is the odd function

$$I_1(x) = -i \; J_1(ix) \; .$$

Table 6.2 gives Chebyshev coefficients for two expansions. If $x \leq 8$,

$$I_1(x) = xe^x {\sum_{i=0}^{\infty}}' \; A_i \; T_i(\tfrac{x}{4} - 1) \; .$$

As x approaches zero,

$$\lim_{x \to 0} \frac{I_1(x)}{x} = \frac{1}{2} \; .$$

Table 6.2: Chebyshev Coefficients for $I_1(x)$

i	A_i	B_i
28	$+2.778 \cdot 10^{-18}$	
27	$-2.1114 \cdot 10^{-17}$	
26	$+1.55363 \cdot 10^{-16}$	
25	$-1.105597 \cdot 10^{-15}$	
24	$+7.600684 \cdot 10^{-15}$	$+7.517 \cdot 10^{-18}$
23	$-5.0421855 \cdot 10^{-14}$	$+4.414 \cdot 10^{-18}$
22	$+3.22379337 \cdot 10^{-13}$	$-4.6503 \cdot 10^{-17}$
21	$-1.983974398 \cdot 10^{-12}$	$-3.2095 \cdot 10^{-17}$
20	$+1.1736186299 \cdot 10^{-11}$	$+2.96263 \cdot 10^{-16}$
19	$-6.6634897235 \cdot 10^{-11}$	$+3.30820 \cdot 10^{-16}$
18	$+3.62559028155 \cdot 10^{-10}$	$-1.880355 \cdot 10^{-15}$
17	$-1.887249751723 \cdot 10^{-9}$	$-3.814403 \cdot 10^{-15}$
16	$+9.381537386496 \cdot 10^{-9}$	$+1.0420277 \cdot 10^{-14}$
15	$-4.4450591287963 \cdot 10^{-8}$	$+4.2724400 \cdot 10^{-14}$
14	$+2.00329475355214 \cdot 10^{-7}$	$-2.1015418 \cdot 10^{-14}$
13	$-8.56872026469545 \cdot 10^{-7}$	$-4.08355111 \cdot 10^{-13}$
12	$+3.470251308137678 \cdot 10^{-6}$	$-7.19855178 \cdot 10^{-13}$
11	$-1.3273163656039436 \cdot 10^{-5}$	$+2.035628544 \cdot 10^{-12}$
10	$+4.7815651075500542 \cdot 10^{-5}$	$+1.4125807437 \cdot 10^{-11}$
9	$-1.61760815825896746 \cdot 10^{-4}$	$+3.2526035830 \cdot 10^{-11}$
8	$+5.12285956168575773 \cdot 10^{-4}$	$-1.8974958124 \cdot 10^{-11}$
7	$-1.513572450631253149 \cdot 10^{-3}$	$-5.58974346220 \cdot 10^{-10}$
6	$+4.156422944312888157 \cdot 10^{-3}$	$-3.835380385964 \cdot 10^{-9}$
5	$-1.056408489462619816 \cdot 10^{-2}$	$-2.6314688468895 \cdot 10^{-8}$
4	$+2.472644903062651682 \cdot 10^{-2}$	$-2.51223623787021 \cdot 10^{-7}$
3	$-5.294598120809499142 \cdot 10^{-2}$	$-3.882564808877690 \cdot 10^{-6}$
2	$+0.102643658689847095384$	$-1.10588938762623716 \cdot 10^{-4}$
1	-0.176416518357834055153	$-9.761097491361468408 \cdot 10^{-3}$
0	$+0.252587186443633654823$	$+0.778576235018280120474$

If $x > 8$,

$$I_1(x) = \frac{e^x}{\sqrt{x}} \sum_{i=0}^{\infty}{}' \; B_i \; T_i(\tfrac{16}{x} - 1) \; .$$

As x increases,

$$\lim_{x \to \infty} e^{-x} \sqrt{x} I_1(x) = \frac{1}{\sqrt{2\pi}} \; .$$

6.8.1 i1.c

Programs for modified Bessel function of order 1

```
#include "mconf.h"
```

Chebyshev coefficients for $e^{-x} I_1(x)$

```
#ifdef DEC
static short A[] = {
0021514,0174520,0060742,0000241,
0122302,0137206,0016120,0025663,
0023063,0017437,0026235,0176536,
0123637,0052523,0170150,0125632,
0024410,0165770,0030251,0044134,
0125143,0012160,0162170,0054727,
0025665,0075702,0035716,0145247,
0126413,0116032,0176670,0015462,
0027116,0073425,0110351,0105242,
0127622,0104034,0137530,0037364,
0030307,0050645,0120776,0175535,
0131001,0130331,0043523,0037455,
0031441,0026160,0010712,0100174,
0132076,0164761,0022706,0017500,
0032527,0015045,0115076,0104076,
0133146,0001714,0015434,0144520,
0033550,0161166,0124215,0077050,
0134136,0127715,0143365,0157170,
0034510,0106652,0013070,0064130,
0135051,0117126,0117264,0123761,
0035406,0045355,0133066,0175751,
0135706,0061420,0054746,0122440,
0036210,0031232,0047235,0006640,
0136455,0012373,0144235,0011523,
0036712,0107437,0036731,0015111,
0137130,0156742,0115744,0172743,
0037322,0033326,0124667,0124740,
0137464,0123210,0021510,0144556,
0037601,0051433,0111123,0177721
};
```

```
#ifdef IBMPC
static short A[] = {
0x4014,0x0c3c,0x9f2a,0x3c49,
0x0576,0xc38a,0x57d0,0xbc78,
0xbfac,0xe593,0x63e3,0x3ca6,
0x1573,0x7e0d,0xeaaa,0xbcd3,
0x290c,0x0615,0x1d7f,0x3d01,
0x0b3b,0x1c8f,0x628e,0xbd2c,
0xd955,0x4779,0xaf78,0x3d56,
0x0366,0x5fb7,0x7383,0xbd81,
0x3154,0xb21d,0xcee2,0x3da9,
0x07de,0x97eb,0x5103,0xbdd2,
0xdf6c,0xb43f,0xea34,0x3df8,
0x67e6,0x28ea,0x361b,0xbe20,
0x5010,0x0239,0x258e,0x3e44,
0xc3e8,0x24b8,0xdd3e,0xbe67,
0xd108,0xb347,0xe344,0x3e8a,
0x992a,0x8363,0xc079,0xbeac,
0xafc5,0xd511,0x1c4e,0x3ecd,
0xbbcf,0xb8de,0xd5f9,0xbeeb,
0x0d0b,0x42c7,0x11b5,0x3f09,
0x94fe,0xd3d6,0x33ca,0xbf25,
0xdf7d,0xb6c6,0xc95d,0x3f40,
0xd4a4,0x0b3c,0xcc62,0xbf58,
0xa1b4,0x49d3,0x0653,0x3f71,
0xa26a,0x7913,0xa29f,0xbf85,
0x2349,0xe7bb,0x51e3,0x3f99,
0x9ebc,0x537c,0x1bbc,0xbfab,
0xf53c,0xd536,0x46da,0x3fba,
0x192e,0x0469,0x94d1,0xbfc6,
0x7ffa,0x724a,0x2a63,0x3fd0
};
```

Chebyshev coefficients $e^{-x} \sqrt{x} I_1(x)$

```
static short B[] = {
0022012,0125555,0115227,0043456,
0021642,0156127,0052075,0145203,
0122526,0072435,0111231,0011664,
0122424,0001544,0161671,0114403,
0023252,0144257,0163532,0142121,
0023276,0132162,0174045,0013204,
0124007,0077154,0057046,0110517,
0124211,0066650,0116127,0157073,
0024473,0133413,0130551,0107504,
0025100,0064741,0032631,0040364,
0124675,0045101,0071551,0012400,
0125745,0161054,0071637,0011247,
0126112,0117410,0035525,0122231,
0026417,0037237,0131034,0176427,
0027170,0100373,0024742,0025725,
0027417,0006417,0105303,0141446,
0127246,0163716,0121202,0060137,
0130431,0123122,0120436,0166000,
0131203,0144134,0153251,0124500,
0131742,0005234,0122732,0033006,
0132606,0157751,0072362,0121031,
0133602,0043372,0047120,0015626,
0134747,0165774,0001125,0046462,
0136437,0166402,0117746,0155137,
0040107,0050305,0125330,0124241
};
#endif
```

```
static short B[] = {
0xe8e6,0xb352,0x556d,0x3c61,
0xb950,0xea87,0x5b8a,0x3c54,
0x2277,0xb253,0xcea3,0xbc8a,
0x3320,0x9c77,0x806c,0xbc82,
0x588a,0xfceb,0x5915,0x3cb5,
0xa2d1,0x5f04,0xd68e,0x3cb7,
0xd22a,0x8bc4,0xefcd,0xbce0,
0xfbc7,0x138a,0x2db5,0xbcf1,
0x31e8,0x762d,0x76e1,0x3d07,
0x281e,0x26b3,0x0d3c,0x3d28,
0x22a0,0x2e6d,0xa948,0xbd17,
0xe255,0x8e73,0xbc45,0xbd5c,
0xb493,0x076a,0x53e1,0xbd69,
0x9fa3,0xf643,0xe7d3,0x3d81,
0x457b,0x653c,0x101f,0x3daf,
0x7865,0xf158,0xe1a1,0x3dc1,
0x4c0c,0xd450,0xdcf9,0xbdb4,
0xdd80,0x5423,0x34ca,0xbe03,
0x3528,0x9ad5,0x790b,0xbe30,
0x46c1,0x94bb,0x4153,0xbe5c,
0x5443,0x2e9e,0xdbfd,0xbe90,
0x0373,0x49ca,0x48df,0xbed0,
0xa9a6,0x804a,0xfd7f,0xbf1c,
0xdb4c,0x53fc,0xfda0,0xbf83,
0x1514,0xb55b,0xea18,0x3fe8
};
#endif
```

```
double i1(x)
double x;
     {
     double y, z;
     double chbevl(), exp(), sqrt(), fabs();
     int i;

     z = fabs(x);
     if( z <= 8.0 )
          {
          y = (z/2.0) - 2.0;
          z = chbevl( y, A, 29 ) * z * exp(z);
          }
     else
          {
          z = exp(z) * chbevl( 32.0/z - 2.0, B, 25 )
               / sqrt(z);
```

```
        }
    if( x < 0.0 )
        z = -z;
    return( z );
    }
```

Program for $e^{-x}I_1(x)$

```
double i1e( x )
double x;
    {
    double y, z;
    double chbevl(), exp(), sqrt(), fabs();
    int i;

    z = fabs(x);
    if( z <= 8.0 )
        {
        y = (z/2.0) - 2.0;
        z = chbevl( y, A, 29 ) * z;
        }
    else
        {
        z = chbevl( 32.0/z - 2.0, B, 25 ) / sqrt(z);
        }
    if( x < 0.0 )
        z = -z;
    return( z );
    }
```

6.9 $I_\nu(x)$

The modified Bessel function of general order is defined as

$$I_\nu(x) = (-i)^\nu J_\nu(ix) \ .$$

It may be computed with moderately good accuracy in terms of the confluent hypergeometric function (see Section 7.1), according to the formula

$$I_\nu(x) = \frac{(\frac{1}{2}x)^\nu e^{-x}}{\Gamma(\nu+1)} \ {}_1F_1(\nu + \tfrac{1}{2}; 2\nu + 1; 2x) \ .$$

If x is negative, x^ν must be real for I_ν to be real. When ν is a negative integer n,

$$I_{-n}(x) = I_n(x) \ .$$

Accuracy is diminished if ν is near, but not equal to, a negative integer.

6.9.1 iv.c

Modified Bessel function of noninteger order

```
extern double MACHEP, MAXNUM;

double iv( v, x )
double v, x;
	{
	double hyp1f1(), exp(), gamma(), log();
	double floor(), fabs();
	int k, sign;
	double t, ax;

	t = floor(v);
	if( v < 0.0 )
		{
		if( t == v )
			{
			v = -v;
			t = -t;
			}
		}
	sign = 1;
	if( x < 0.0 )
		{
		if( t != v )
			{
			mtherr( "iv", DOMAIN );
			return( 0.0 );
			}
		if( v != 2.0 * floor(v/2.0) )
			sign = -1;
		}
```

Avoid logarithm singularity

```
	if( x == 0.0 )
		{
		if( v == 0.0 )
			return( 1.0 );
		if( v < 0.0 )
			{
			mtherr( "iv", OVERFLOW );
			return( MAXNUM );
			}
		else
			return( 0.0 );
```

```
        }
    ax = fabs(x);
    t = v * log( 0.5 * ax ) - x;
    t = sign * exp(t) / gamma( v + 1.0 );
    ax = v + 0.5;
    return( t * hyp1f1( ax, 2.0 * ax, 2.0 * x ) );
    }
```

6.10 $K_0(x)$

The modified Bessel function of the third kind, order zero, has power series expansions that are special cases of the ones given below in the section on $K_n(x)$. As x approaches zero,

$$\lim_{x\to 0} K_0(x) + \ln \frac{x}{2} = -\gamma \ ,$$

where γ is Euler's constant. As x increases,

$$\lim_{x\to\infty} e^x \sqrt{x} K_0(x) = \sqrt{\frac{\pi}{2}} \ .$$

Chebyshev coefficients are given in Table 6.3. They are used in the same way as the ones for $I_0(x)$. The odd numbered Chebyshev coefficients C_i are equal to zero; the tabulated values refer to C_{2i}. The expansion works as written when the argument of the Chebyshev polynomials is proportional to x^2. If $x \le 2$,

$$K_0(x) = \sum_{i=0}^{\infty}{}' \ C_{2i} \ T_i(\tfrac{x^2}{2} - 1) - I_0(x) \ \ln \frac{x}{2} \ .$$

If $x > 2$,

$$K_0(x) = \frac{e^{-x}}{\sqrt{x}} \sum_{i=0}^{\infty}{}' \ D_i \ T_i(\tfrac{4}{x} - 1) \ .$$

6.10.1 k0.c

```
#include "mconf.h"
```

Chebyshev coefficients for $K_0(x) + \ln \frac{1}{2}x \ I_0(x)$

```
#ifdef DEC
static short A[] = {
0023036,0073417,0032477,0165673,
0025077,0154126,0016046,0012517,
0027066,0011342,0035211,0005041,
```

```
#ifdef IIEEE
static short A[] = {
0xfd77,0xe6a7,0xcee1,0x3ca3,
0xc2aa,0xc384,0xfb0a,0x3d27,
0x2144,0x4751,0xc25c,0x3da6,
```

Table 6.3: Chebyshev Coefficients for $K_0(x)$

i	C_{2i}	D_i
24		$+5.300 \cdot 10^{-18}$
23		$-1.6476 \cdot 10^{-17}$
22		$+5.2104 \cdot 10^{-17}$
21		$-1.67823 \cdot 10^{-16}$
20		$+5.51206 \cdot 10^{-16}$
19		$-1.848593 \cdot 10^{-15}$
18		$+6.340076 \cdot 10^{-15}$
17		$-2.2275133 \cdot 10^{-14}$
16		$+8.0328908 \cdot 10^{-14}$
15		$-2.98009692 \cdot 10^{-13}$
14		$+1.140340588 \cdot 10^{-12}$
13		$-4.514597883 \cdot 10^{-12}$
12		$+1.8559491150 \cdot 10^{-11}$
11		$-7.9574892445 \cdot 10^{-11}$
10		$+3.57739728140 \cdot 10^{-10}$
9	$+1.37447 \cdot 10^{-16}$	$-1.697534509389 \cdot 10^{-9}$
8	$+4.2598161 \cdot 10^{-14}$	$+8.574034017414 \cdot 10^{-9}$
7	$+1.0349695258 \cdot 10^{-11}$	$-4.6604898976879 \cdot 10^{-8}$
6	$+1.904516377220 \cdot 10^{-9}$	$+2.76681363944502 \cdot 10^{-7}$
5	$+2.53479107902615 \cdot 10^{-7}$	$-1.831755522719119 \cdot 10^{-6}$
4	$+2.2862121031194518 \cdot 10^{-5}$	$+1.3949813718876499 \cdot 10^{-5}$
3	$+1.264615411446925923 \cdot 10^{-3}$	$-1.28495495816278026 \cdot 10^{-4}$
2	$+3.5979936515361501627 \cdot 10^{-2}$	$+1.5698838857300533751 \cdot 10^{-3}$
1	$+0.34428989999246284868 86$	$-3.1448101311964500543 \cdot 10^{-2}$
0	$-0.53532739323390276872 0$	$+2.4403030820659554546 8$

```
0031002,0160233,0037454,0050224,
0032610,0012747,0037712,0173741,
0034277,0144007,0172147,0162375,
0035645,0140563,0125431,0165626,
0037023,0057662,0125124,0102051,
0037660,0043304,0004411,0166707,
0140011,0005467,0047227,0130370
};
```

Chebyshev coefficients for $e^x\sqrt{x}K_0(x)$

```
static short B[] = {
0021703,0106456,0076144,0173406,
0122227,0173144,0116011,0030033,
0022560,0044562,0006506,0067642,
0123101,0076243,0123273,0131013,
0023436,0157713,0056243,0141331,
0124005,0032207,0063726,0164664,
0024344,0066342,0051756,0162300,
0124710,0121365,0154053,0077022,
0025264,0161166,0066246,0077420,
0125647,0141671,0006443,0103212,
0026240,0076431,0077147,0160445,
0126636,0153741,0174002,0105031,
0027243,0040102,0035375,0163073,
0127656,0176256,0113476,0044653,
0030304,0125544,0006377,0130104,
0130751,0047257,0110537,0127324,
0031423,0046400,0014772,0012164,
0132110,0025240,0155247,0112570,
0032624,0105314,0007437,0021574,
0133365,0155243,0174306,0116506,
0034152,0004776,0061643,0102504,
0135006,0136277,0036104,0175023,
0035715,0142217,0162474,0115022,
0137000,0147671,0065177,0134356,
0040434,0026754,0175163,0044070
};
#endif
```

```
0x8a13,0x67e5,0x5c13,0x3e20,
0x5efc,0xe7f9,0x02bc,0x3e91,
0xfca0,0xfe8c,0xf900,0x3ef7,
0x3d73,0x7563,0xb82e,0x3f54,
0x9085,0x554a,0x6bf6,0x3fa2,
0x3db9,0x8121,0x08d8,0x3fd6,
0xf61f,0xe9d2,0x2166,0xbfe1
};

static short B[] = {
0x9ee1,0xcf8c,0x71a5,0x3c58,
0x2603,0x9381,0xfecc,0xbc72,
0xcdf4,0x41a8,0x092e,0x3c8e,
0x7641,0x74d7,0x2f94,0xbca8,
0x785b,0x6b94,0xdbf9,0x3cc3,
0xdd36,0xecfa,0xa690,0xbce0,
0xdc98,0x4a7d,0x8d9c,0x3cfc,
0x6fc2,0xbb05,0x145e,0xbd19,
0xcfe2,0xcd94,0x9c4e,0x3d36,
0x70d1,0x21a4,0xf877,0xbd54,
0xfc25,0x2fcc,0x0fa3,0x3d74,
0x5143,0x3f00,0xdafc,0xbd93,
0xbcc7,0x475f,0x6808,0x3db4,
0xc935,0xd2e7,0xdf95,0xbdd5,
0xf608,0x819f,0x956c,0x3df8,
0xf5db,0xf22b,0x29d5,0xbe1d,
0x428e,0x033f,0x69a0,0x3e42,
0xf2af,0x1b54,0x0554,0xbe69,
0xe46f,0x81e3,0x9159,0x3e92,
0xd3a9,0x7f18,0xbb54,0xbebe,
0x70a9,0xcc74,0x413f,0x3eed,
0x9f42,0xe788,0xd797,0xbf20,
0x9342,0xfca7,0xb891,0x3f59,
0xf71e,0x2d4f,0x19f7,0xbfa0,
0x6907,0x9f4e,0x85bd,0x4003
};
#endif
```

```
extern double PI;
extern double MAXNUM;

double k0(x)
double x;
	{
	double y, z;
	double chbevl(), exp(), i0(), log(), sqrt();
```

```
    if( x <= 0.0 )
        {
        mtherr( "k0", DOMAIN );
        return( MAXNUM );
        }
    if( x <= 2.0 )
        {
        y = x * x - 2.0;
        y = chbevl( y, A, 10 ) - log( 0.5 * x ) * i0(x);
        return( y );
        }
    z = 8.0/x - 2.0;
    y = exp(-x) * chbevl( z, B, 25 ) / sqrt(x);
    return(y);
    }
```

Program for $e^x K_0(x)$

```
double k0e( x )
double x;
    {
    double y;
    double chbevl(), exp(), i0(), log(), sqrt();
    int i;

    if( x <= 0.0 )
        {
        mtherr( "k0e", DOMAIN );
        return( MAXNUM );
        }

    if( x <= 2.0 )
        {
        y = x * x - 2.0;
        y = chbevl( y, A, 10 ) - log( 0.5 * x ) * i0(x);
        return( y * exp(x) );
        }
    y = chbevl( 8.0/x - 2.0, B, 25 ) / sqrt(x);
    return(y);
    }
```

6.11 $K_1(x)$

Expansions for the modified Bessel function of the third kind, order one are special cases of the ones given below for $K_n(x)$. As x approaches zero,

$$\lim_{x \to 0} x K_1(x) = \tfrac{1}{4} \; .$$

As x increases,

$$\lim_{x \to \infty} e^x \sqrt{x} K_1(x) = \sqrt{\frac{\pi}{2}} \; .$$

With coefficients from Table 6.4, the expansion for $x \leq 2$ is

$$K_1(x) = I_1(x) \, \ln \frac{x}{2} + \frac{1}{x} {\sum_{i=0}^{\infty}}' \; A_{2i} \; T_i(\tfrac{x^2}{2} - 1) \; .$$

The expansion for $x > 2$ is

$$K_1(x) = \frac{e^{-x}}{\sqrt{x}} {\sum_{i=0}^{\infty}}' \; B_i \; T_i(\tfrac{4}{x} - 1) \; .$$

The tabulated coefficients for $i = 0$ have already been divided by two.

6.11.1 k1.c

```
#include "mconf.h"
```

Chebyshev coefficients for $x(K_1(x) - \ln(x/2) I_1(x))$

```
#ifdef DEC                                  #ifdef IIEEE
#define MINNUM 6.0e-39                      #define MINNUM 1.0e-308
static short A[] = {                        static short A[] = {
0122001,0110501,0164746,0151255,            0xda56,0x3d3c,0x3228,0xbc60,
0124056,0165213,0150034,0147377,            0x99e0,0x7a03,0xdd51,0xbce5,
0126073,0124026,0167207,0001044,            0xe045,0xddd0,0x7502,0xbd67,
0130033,0030735,0141061,0033116,            0x26ca,0xb846,0x663b,0xbde3,
0131676,0020350,0121341,0107175,            0x31d0,0x145c,0xc41d,0xbe57,
0133443,0046631,0062031,0070716,            0x2e3a,0x2c83,0x69b3,0xbec4,
0135065,0067427,0026435,0164022,            0xbd02,0xe5a3,0xade2,0xbf26,
0136344,0112234,0165752,0006222,            0x4192,0x9d7d,0x9293,0xbf7c,
0137373,0015622,0017016,0155636,            0xdb74,0x43c1,0x6372,0xbfbf,
0137664,0150333,0125730,0067240,            0x0dd4,0x757b,0x9a1b,0xbfd6,
0040303,0036411,0130200,0043120             0x08ca,0x3610,0x67a1,0x3ff8
};                                          };
```

Chebyshev coefficients for $e^x \sqrt{x} K_1(x)$

```
static short B[] = {                        static short B[] = {
0121724,0061352,0013041,0150076,            0x3a08,0x42c4,0x8c5d,0xbc5a,
0022245,0074324,0016172,0173232,            0x5ed3,0x838f,0xaf1a,0x3c74,
```

Table 6.4: Chebyshev Coefficients for $K_1(x)$

i	A_{2i}	B_i
24		$-5.757 \cdot 10^{-18}$
23		$+1.7941 \cdot 10^{-17}$
22		$-5.6895 \cdot 10^{-17}$
21		$+1.83809 \cdot 10^{-16}$
20		$-6.05705 \cdot 10^{-16}$
19		$+2.038703 \cdot 10^{-15}$
18		$-7.019837 \cdot 10^{-15}$
17		$+2.4771544 \cdot 10^{-14}$
16		$-8.9767052 \cdot 10^{-14}$
15		$+3.34841967 \cdot 10^{-13}$
14		$-1.289173961 \cdot 10^{-12}$
13		$+5.139639673 \cdot 10^{-12}$
12		$-2.1299678384 \cdot 10^{-11}$
11		$+9.2183151876 \cdot 10^{-11}$
10	$-7.024 \cdot 10^{-18}$	$-4.19035475934 \cdot 10^{-10}$
9	$-2.427450 \cdot 10^{-15}$	$+2.015049755197 \cdot 10^{-9}$
8	$-6.66690169 \cdot 10^{-13}$	$-1.0345762465678 \cdot 10^{-8}$
7	$-1.41148839263 \cdot 10^{-10}$	$+5.7410841254500 \cdot 10^{-8}$
6	$-2.2133876307347 \cdot 10^{-8}$	$-3.50196060308781 \cdot 10^{-7}$
5	$-2.433406141565968 \cdot 10^{-6}$	$+2.406484947837217 \cdot 10^{-6}$
4	$-1.73028895751305206 \cdot 10^{-4}$	$-1.9361979741660830 \cdot 10^{-5}$
3	$-6.975723859639864350 \cdot 10^{-3}$	$+1.95215518471351631 \cdot 10^{-4}$
2	-0.122611180822657148235	$-2.857816859622779387 \cdot 10^{-3}$
1	-0.353155960776544875667	$+0.103923736576817238437$
0	$+1.525300227338947770 53$	$+2.720626190484442669 45$

```
0122603,0030250,0135670,0165221,
0023123,0165362,0023561,0060124,
0123456,0112436,0141654,0073623,
0024022,0163557,0077564,0006753,
0124374,0165221,0131014,0026524,
0024737,0017512,0144250,0175451,
0125312,0021456,0123136,0076633,
0025674,0077720,0020125,0102607,
0126265,0067543,0007744,0043701,
0026664,0152702,0033002,0074202,
0127273,0055234,0120016,0071733,
0027712,0133200,0042441,0075515,
0130346,0057000,0015456,0074470,
0031012,0074441,0051636,0111155,
0131461,0136444,0177417,0002101,
0032166,0111743,0032176,0021410,
0132674,0001224,0076555,0027060,
0033441,0077430,0135226,0106663,
0134242,0065610,0167155,0113447,
0035114,0131304,0043664,0102163,
0136073,0045065,0171465,0122123,
0037324,0152767,0147401,0017732,
0040456,0017275,0050061,0062120,
};
#endif
```

```
0x1d52,0x1777,0x6615,0xbc90,
0x2c0b,0x44ee,0x7d5e,0x3caa,
0x8ef2,0xd875,0xd2a3,0xbcc5,
0x81bd,0xefee,0x5ced,0x3ce2,
0x85ab,0x3641,0x9d52,0xbcff,
0x1f65,0x5915,0xe3e9,0x3d1b,
0xcfb3,0xd4cb,0x4465,0xbd39,
0xb0b1,0x040a,0x8ffa,0x3d57,
0x88f8,0x61fc,0xadec,0xbd76,
0x4f10,0x46c0,0x9ab8,0x3d96,
0xce7b,0x9401,0x6b53,0xbdb7,
0x2f6a,0x08a4,0x56d0,0x3dd9,
0xcf27,0x0365,0xcbc0,0xbdfc,
0xd24e,0x2a73,0x4f24,0x3e21,
0xe088,0x9fe1,0x37a4,0xbe46,
0xc461,0x668f,0xd27c,0x3e6e,
0xa5c6,0x8fad,0x8052,0xbe97,
0xd1b6,0x1752,0x2fe3,0x3ec4,
0xb2e5,0x1dcd,0x4d71,0xbef4,
0x908e,0x88f6,0x9658,0x3f29,
0xb48a,0xbe66,0x6946,0xbf67,
0x23fb,0xf9e0,0x9abe,0x3fba,
0x2c8a,0xaa06,0xc3d7,0x4005
};
#endif
```

```
extern double PI;
extern double MAXNUM;

double k1(x)
double x;
	{
	double y, z;
	double chbevl(), exp(), i1(), log(), sqrt();
	int i;

	if( x <= MINNUM )
		{
		mtherr( "k1", DOMAIN );
		return( MAXNUM );
		}
	if( x <= 2.0 )
		{
		y = x * x - 2.0;
		y = log( 0.5 * x ) * i1(x)
			+ chbevl( y, A, 11 ) / x;
```

```
        return( y );
        }
    return( exp(-x)
        * chbevl( 8.0/x - 2.0, B, 25 ) / sqrt(x) );
    }
```

Program for $e^x K_1(x)$

```
double k1e( x )
double x;
    {
    double y;
    double chbevl(), exp(), i1(), log(), sqrt();
    int i;

    if( x <= 0.0 )
        {
        mtherr( "k1e", DOMAIN );
        return( MAXNUM );
        }
    if( x <= 2.0 )
        {
        y = x * x - 2.0;
        y = log( 0.5 * x ) * i1(x)
            + chbevl( y, A, 11 ) / x;
        return( y * exp(x) );
        }
    return( chbevl( 8.0/x - 2.0, B, 25 ) / sqrt(x) );
    }
```

6.12 $K_n(x)$

For integer $n \geq 0$, the modified Bessel function of the third kind can be computed by one of two expansions. If $x < 9.55$, for a double precision routine use (AMS55 #9.6.11)

$$\begin{aligned} K_n(x) &= \tfrac{1}{2}(\tfrac{1}{2}x)^{-n} \sum_{k=0}^{n-1} \frac{(n-k-1)!\,(-\tfrac{1}{4}x^2)^k}{k!} \\ &\quad + (-1)^n \tfrac{1}{2}(\tfrac{1}{2}x)^n \sum_{k=0}^{\infty} (\psi(k+1) + \psi(n+k+1)) \frac{(\tfrac{1}{4}x^2)^k}{k!\,(n+k)!} \\ &\quad + (-1)^{n+1} I_n(x) \ln \frac{x}{2} \end{aligned}$$

where

$$\psi(1) = -\gamma$$
$$\psi(m) = -\gamma + \sum_{k=1}^{m-1} \frac{1}{k}$$

and γ is Euler's constant. For $x > 9.55$,

$$u = 4\nu^2$$
$$K_\nu(x) \approx \sqrt{\frac{\pi}{2x}}\, e^{-x} \left[1 + \frac{u - 1^2}{1!\,(8x)^1} + \frac{(u-1^2)(u-3^2)}{2!\,(8x)^2} + \dots\right].$$

The following symmetry applies:

$$K_{-n}(x) = K_n(x).$$

The maximum value of n is limited by the largest representable value of $\Gamma(n)$. $K_n(0)$ is infinite. The function is complex valued for $x < 0$.

6.12.1 kn.c

```
#define EUL 5.772156649015328606065e-1
#define MAXFAC 31
extern double MACHEP, MAXNUM, MAXLOG, PI;

double kn( nn, x )
int nn;
double x;
    {
    double k, kf, nk1f, nkf, zn, t, s, z0, z;
    double ans, fn, pn, pk, zmn, tlg, tox;
    int i, n;
    double fabs(), exp(), log(), sqrt();

    if( nn < 0 )
        n = -nn;
    else
        n = nn;

    if( n > MAXFAC )
        {
overf:
        mtherr( "kn", OVERFLOW );
        return( MAXNUM );
        }
```

```
if( x <= 0.0 )
     {
     if( x < 0.0 )
          mtherr( "kn", DOMAIN );
     else
          mtherr( "kn", SING );
     return( MAXNUM );
     }


if( x > 9.55 )
     goto asymp;

ans = 0.0;
z0 = 0.25 * x * x;
fn = 1.0;
pn = 0.0;
zmn = 1.0;
tox = 2.0/x;

if( n > 0 )
     { compute factorial of n and psi(n)
     pn = -EUL;
     k = 1.0;
     for( i=1; i<n; i++ )
          {
          pn += 1.0/k;
          k += 1.0;
          fn *= k;
          }

     zmn = tox;

     if( n == 1 )
          {
          ans = 1.0/x;
          }
     else
          {
          nk1f = fn/n;
          kf = 1.0;
          s = nk1f;
          z = -z0;
          zn = 1.0;
          for( i=1; i<n; i++ )
```

```
                {
                nk1f = nk1f/(n-i);
                kf = kf * i;
                zn *= z;
                t = nk1f * zn / kf;
                if( (MAXNUM - fabs(t)) < fabs(s) )
                        goto overf;
                if( (tox > 1.0)
                        && ((MAXNUM/tox) < zmn) )
                        goto overf;
                s += t;
                zmn *= tox;
                }
        s *= 0.5;
        t = fabs(s);
        if( (zmn > 1.0) && ((MAXNUM/zmn) < t) )
                goto overf;
        if( (t > 1.0) && ((MAXNUM/t) < zmn) )
                goto overf;
        ans = s * zmn;
        }
    }


tlg = 2.0 * log( 0.5 * x );
pk = -EUL;
if( n == 0 )
        {
        pn = pk;
        t = 1.0;
        }
else
        {
        pn = pn + 1.0/n;
        t = 1.0/fn;
        }
s = (pk+pn-tlg)*t;
k = 1.0;
do
        {
        t *= z0 / (k * (k+n));
        pk += 1.0/k;
        pn += 1.0/(k+n);
        s += (pk+pn-tlg)*t;
        k += 1.0;
```

```
        }
    while( fabs(t/s) > MACHEP );

    s = 0.5 * s / zmn;
    if( n & 1 )
        s = -s;
    ans += s;
    return(ans);
```

Asymptotic expansion for Kn(x)

```
asymp:
    if( x > MAXLOG )
        {
        mtherr( "kn", UNDERFLOW );
        return(0.0);
        }
    k = n;
    pn = 4.0 * k * k;
    pk = 1.0;
    z0 = 8.0 * x;
    fn = 1.0;
    t = 1.0;
    s = t;
    nkf = MAXNUM;
    i = 0;
    do
        {
        z = pn - pk * pk;
        t = t * z /(fn * z0);
        nk1f = fabs(t);
        if( (i >= n) && (nk1f > nkf) )
            {
            goto adone;
            }
        nkf = nk1f;
        s += t;
        fn += 1.0;
        pk += 2.0;
        i += 1;
        }
    while( fabs(t/s) > MACHEP );
adone:
    ans = exp(-x) * sqrt( PI/(2.0*x) ) * s;
    return(ans);
    }
```

6.13 $J_\nu(x)$

Several alternative expansions must be considered when attempting to compute Bessel functions of arbitrary order over a wide domain. The boundaries of the optimum domain for each expansion depend on the precision of the arithmetic. The tests given here are suitable for a double precision program. Both the Hankel expansion and the ascending power series are subject to severe arithmetic cancellation error. For $x > 10$, loss of precision due to cancellation will be about 7 least significant bits or less if

$$x < 3.6\sqrt{\nu}$$

for the power series, or if

$$x > 0.075\nu^2$$

for the Hankel expansion. This loss of least significant bits is almost independent of the arithmetic precision. If neither of these inequalities is satisfied, the recurrence relation used in the previously described method for J_n may be used to reduce ν until either the power series or the Hankel expansion can be applied safely. In double precision arithmetic if $x < 17$ then ν may be reduced by recurrence to $2 < \nu < 4$. The largest absolute error occurs in this region, when $5 < x < 10$ and $10 < \nu < 13$. In double precision, the boundary between applicability of the Hankel expansion and the power series is given approximately by

$$t = 0.0083k^2 + 0.09k + 12.9$$

for $k < 26$, and

$$t = 0.95k$$

if $k \geq 26$. If $x > t$ then the Hankel expansion theoretically converges to double precision and has less cancellation error than the power series. If $x \leq t$ then the power series expansion is better.

As x increases, the labor of calculating the continued fraction and the recurrence to reduce ν grows. Eventually special expansions are required to accommodate large x and ν. The two expansions to be described give 9 or more significant figures if x and ν are greater than 200. A uniform asymptotic expansion (AMS55 9.3.35)[2] can be used unless x is very close to ν. It is applicable if

$$\left| \frac{x-\nu}{\sqrt[3]{\nu}} \right| > 0.7 \, .$$

The expansion is in terms of a variable ζ. Defining

$$w = 1 - \left(\frac{x}{\nu}\right)^2 \equiv 1 - z^2$$

[2]These expansions are given in AMS55, chapter 9.3. For additional information, consult W. G. Bickley, "Bessel functions and formulae" in British Association for the Advancement of Science, Bessel functions, Part II. Functions of positive integer order, Mathematical Tables, vol. VI (Cambridge Univ. Press, 1950).

then

$$\zeta = \left\{ \frac{3}{2} \ln \frac{1 + w^{\frac{1}{2}}}{z} - \frac{3}{2} w^{\frac{1}{2}} \right\}^{\frac{2}{3}}$$

if $w > 0$, or, if $w \le 0$,

$$\zeta = \left\{ \frac{3}{2} (-w)^{\frac{1}{2}} - \frac{3}{2} \cos^{-1} \frac{1}{z} \right\}^{\frac{2}{3}} .$$

In terms of this variable the uniform expansion is

$$J_\nu(\nu z) \approx \left(\frac{4\zeta}{1 - z^2} \right)^{\frac{1}{4}} \left\{ \frac{\mathrm{Ai}(\nu^{\frac{2}{3}}\zeta)}{\nu^{\frac{1}{3}}} \sum_{k=0}^{\infty} \frac{a_k(\zeta)}{\nu^{2k}} + \frac{\mathrm{Ai}'(\nu^{\frac{2}{3}}\zeta)}{\nu^{\frac{5}{3}}} \sum_{k=0}^{\infty} \frac{b_k(\zeta)}{\nu^{2k}} \right\}$$

where the Ai are Airy functions (see the next section). The coefficients a_k, b_k are found as follows. Define

$$\begin{aligned}
\lambda_0 &= 1 \\
\lambda_s &= \frac{(2s+1)(2s+3)\cdots(6s-1)}{144^s \; s!} \\
\mu_0 &= 1 \\
\mu_s &= -\lambda_s \frac{6s+1}{6s-1} \\
u_0(t) &= 1 \\
u_1(t) &= \frac{3t - 5t^3}{24} \\
u_{k+1}(t) &= \tfrac{1}{2} t^2 (1 - t^2) u_k'(t) + \frac{1}{8} \int_0^t (1 - 5t^2) u_k(t) dt \\
r &= \frac{1}{\sqrt{1 - z^2}} = 1/w^{1/2} .
\end{aligned}$$

Then the coefficients are

$$\begin{aligned}
a_k(\zeta) &= \sum_{s=0}^{2k} \mu_s \zeta^{-\frac{3s}{2}} u_{2k-s}(r) \\
b_k(\zeta) &= -\zeta^{-\frac{1}{2}} \sum_{s=0}^{2k+1} \lambda_s \zeta^{-\frac{3s}{2}} u_{2k-s+1}(r) .
\end{aligned}$$

The numerical values of λ and μ are given in the computer program listing below. In terms of polynomials P_1, P_2, etc., whose coefficients are shown in the computer program, the numerical coefficients u_k are

$$\begin{aligned}
u_0 &= 1 \\
u_k &= r^k P_k(1/w) .
\end{aligned}$$

For x very close to ν — specifically, if

$$\left|\frac{x-\nu}{\sqrt[3]{\nu}}\right| \leq 0.7,$$

then use the asymptotic expansion (AMS55 9.3.23)

$$J_\nu(\nu + z\nu^{\frac{1}{3}}) \approx \frac{2^{\frac{1}{3}}}{\nu^{\frac{1}{3}}}\mathrm{Ai}(-2^{\frac{1}{3}}z)\sum_{k=0}^{\infty}\frac{f_k(z)}{\nu^{\frac{2k}{3}}} + \frac{2^{\frac{2}{3}}}{\nu}\mathrm{Ai}'(-2^{\frac{1}{3}}z)\sum_{k=0}^{\infty}\frac{g_k(z)}{\nu^{\frac{2k}{3}}}.$$

The coefficients $f_i(z)$, $g_i(z)$ are polynomials in x as shown in the computer program.

6.13.1 jv.c

Bessel function program for noninteger order. The program rejects negative x unless n is an integer. Also, negative n is limited to $n > -200$ unless n is an integer.

```
extern double MAXNUM, MACHEP, MINLOG, MAXLOG;
#define BIG 1.44115188075855872E+17

double jv( n, x )
double n, x;
    {
    double k, q, t, y, an;
    int i, sign, nint;
    double fabs(), floor(), j0(), j1(), jnx(), jvs(), hankel();
    double recur(), sqrt();

    nint = 0; Flag for integer n
    sign = 1; Flag for sign inversion
    an = fabs( n );
    y = floor( an );
    if( y == an )
        {
        nint = 1;
        i = an - 16384.0 * floor( an/16384.0 );
        if( n < 0.0 )
            {
            if( i & 1 )
                sign = -sign;
            n = an;
            }
        if( x < 0.0 )
            {
```

```
                if( i & 1 )
                        sign = -sign;
                x = -x;
                }
        if( n == 0.0 )
                return( j0(x) );
        if( n == 1.0 )
                return( sign * j1(x) );
        }
if( (x < 0.0) && (y != an) )
        {
        mtherr( "Jv", DOMAIN );
        y = 0.0;
        goto done;
        }
y = fabs(x);
if( y < MACHEP )
        goto underf;
k = 3.6 * sqrt(y);
if( an > 17.0 )
        {
        t = 3.6 * sqrt(an);
        if( y < t )
                return( sign * jvs(n,x) );
        if( an < k )
                return( sign * hankel(n,x) );
        }
if( an < 200.0 )
        {
        if( nint != 0 )
                {
                k = 0.0;
                q = recur( &n, x, &k );
                if( k == 0.0 )
                        {
                        y = j0(x)/q;
                        goto done;
                        }
                if( k == 1.0 )
                        {
                        y = j1(x)/q;
                        goto done;
                        }
                }
        if( k <= 17.0 )
```

```
            {
            k = 2.0;
            }
        if( an > (k + 3.0) )
            {
            if( n < 0.0 )
                k = -k;
            q = n - floor(n);
            k = floor(k) + q;
            if( n > 0.0 )
                q = recur( &n, x, &k );
            else
                {
                t = k;
                k = n;
                q = recur( &t, x, &k );
                k = t;
                }
            if( q == 0.0 )
                {
underf:
                y = 0.0;
                goto done;
                }
            }
        else
            {
            k = n;
            q = 1.0;
            }
        y = fabs(k);
        if( y < 26.0 )
            t = (0.0083*y + 0.09)*y + 12.9;
        else
            t = 0.9 * y;
        if( x > t )
            y = hankel(k,x);
        else
            y = jvs(k,x);
        if( n > 0.0 )
            y /= q;
        else
            y *= q;
        }
    else
```

```
        {
```

For large n, use the uniform expansion or the transitional expansion unless x is of the order of n^2.

```
        if( n < 0.0 )
            {
            mtherr( "Jv", TLOSS );
            y = 0.0;
            goto done;
            }
        t = x/n;
        t /= n;
        if( t > 0.3 )
            y = hankel(n,x);
        else
            y = jnx(n,x);
        }
done:  return( sign * y);
    }
```

Reduce the order by backward recurrence. AMS55 #9.1.27 and 9.1.73.

```
static double recur( n, x, newn )
double *n;
double x;
double *newn;
    {
    double pkm2, pkm1, pk, pkp1, qkm2, qkm1;
    double k, ans, qk, xk, yk, r, t, k2, kf;
    static double big = BIG;
    int nflag, ctr;
    double fabs(), floor();
```

continued fraction for Jn(x)/Jn-1(x)

```
    if( *n < 0.0 )
        nflag = 1;
    else
        nflag = 0;
fstart:
    pkm2 = 0.0;
    qkm2 = 1.0;
    pkm1 = x;
```

```
    qkm1 = *n + *n;
    xk = -x * x;
    yk = qkm1;
    ans = 1.0;
    ctr = 0;
    do
        {
        yk += 2.0;
        pk = pkm1 * yk + pkm2 * xk;
        qk = qkm1 * yk + qkm2 * xk;
        pkm2 = pkm1;
        pkm1 = pk;
        qkm2 = qkm1;
        qkm1 = qk;
        if( qk != 0 )
            r = pk/qk;
        if( r != 0 )
            {
            t = fabs( (ans - r)/r );
            ans = r;
            }
        else
            t = 1.0;
        if( t < MACHEP )
            goto done;
        if( fabs(pk) > big )
            {
            pkm2 /= big;
            pkm1 /= big;
            qkm2 /= big;
            qkm1 /= big;
            }
        }
    while( t > MACHEP );
done:
```

Change n to $n-1$ if $n<0$ and the continued fraction is small.

```
    if( nflag > 0 )
        {
        if( fabs(ans) < 0.125 )
            {
            nflag = -1;
            *n = *n - 1.0;
            goto fstart;
            }
        }
```

```
    kf = *newn;
    pk = 1.0;
    pkm1 = 1.0/ans;
    k = *n - 1.0;
    r = 2 * k;
    do
        {
        pkm2 = (pkm1 * r - pk * x) / x;
        pkp1 = pk;
        pk = pkm1;
        pkm1 = pkm2;
        r -= 2.0;
        k -= 1.0;
        }
    while( k > (kf + 0.5) );
```

Take the larger of the last two iterates
on the theory that it may have less cancellation error.

```
    if( (kf >= 0.0) && (fabs(pk) > fabs(pkm1)) )
        {
        k += 1.0;
        pkm2 = pk;
        }
    *newn = k;
    return( pkm2 );
    }
```

Ascending power series for $J_v(x)$.
AMS55 #9.1.10.

```
extern double PI;
extern int sgngam;

static double jvs( n, x )
double n, x;
    {
    double t, u, y, z, k;
    double lgam(), exp(), log(), fabs(), pow(), gamma();

    z = -x * x / 4.0;
    u = 1.0;
    y = u;
    k = 1.0;
    t = 1.0;
```

```
while( t > MACHEP )
     {
     u *= z / (k * (n+k));
     y += u;
     k += 1.0;
     if( y != 0 )
          t = fabs( u/y );
     }
if( x < 0.0 )
     {
     y = y * pow( 0.5 * x, n ) / gamma( n + 1.0 );
     }
else
     {
     t = n * log(x/2.0) - lgam(n + 1.0);
     if( t < MINLOG )
          {
          return( 0.0 );
          }
     if( t > MAXLOG )
          {
          t = log(y) + t;
          if( t > MAXLOG )
               {
               mtherr( "Jv", OVERFLOW );
               return( MAXNUM );
               }
          else
               {
               y = sgngam * exp(t);
               return(y);
               }
          }
     y = sgngam * y * exp( t );
     }
return(y);
}
```

Hankel's asymptotic expansion for large x.
AMS55 #9.2.5.

```
static double hankel( n, x )
double n, x;
```

```
{
double t, u, z, k, sign, conv;
double p, q, j, m, pp, qq;
double fabs(), sqrt(), sin(), cos();
int flag;

m = 4.0*n*n;
j = 1.0;
z = 8.0 * x;
k = 1.0;
p = 1.0;
u = (m - 1.0)/z;
q = u;
sign = 1.0;
conv = 1.0;
flag = 0;
t = 1.0;
while( t > MACHEP )
	{
	k += 2.0;
	j += 1.0;
	sign = -sign;
	u *= (m - k * k)/(j * z);
	p += sign * u;
	k += 2.0;
	j += 1.0;
	u *= (m - k * k)/(j * z);
	q += sign * u;
	t = fabs(u/p);
	if( t < conv )
		{
		conv = t;
		qq = q;
		pp = p;
		flag = 1;
		}
```

Stop if the terms start getting larger.

```
	if( (flag != 0) && (t > conv) )
		{
		goto hank1;
		}
	}
hank1:
u = x - (0.5*n + 0.25) * PI;
t = sqrt( 2.0/(PI*x) ) * ( pp * cos(u) - qq * sin(u) );
```

```
    return( t );
    }
```

Asymptotic expansion for large n.
AMS55 #9.3.35.

```
static double lambda[] = {
    1.0,
    1.041666666666666666666667E-1,
    8.355034722222222222222222E-2,
    1.282265745563271604938272E-1,
    2.918490264641404642489712E-1,
    8.816272674437576524187671E-1,
    3.321408281862767544702647E+0,
    1.499576298686255465867237E+1,
    7.892301301158651813848139E+1,
    4.744515388682643231611949E+2,
    3.207490090890661934704328E+3
    };
static double mu[] = {
    1.0,
    -1.458333333333333333333333E-1,
    -9.874131944444444444444444E-2,
    -1.433120539158950617283951E-1,
    -3.172272026784135480967078E-1,
    -9.424291479571202491373028E-1,
    -3.511203040826354261542798E+0,
    -1.572726362036804512982712E+1,
    -8.228143909718594444224656E+1,
    -4.923553705236705240352022E+2,
    -3.316218568547972508762102E+3
    };
static double P1[] = {
    -2.083333333333333333333333E-1,
    1.250000000000000000000000E-1
    };
static double P2[] = {
    3.342013888888888888888889E-1,
    -4.010416666666666666666667E-1,
    7.031250000000000000000000E-2
    };
static double P3[] = {
    -1.025812596450617283950617E+0,
```

```
     1.846462673611111111111111E+0,
     -8.912109375000000000000000E-1,
     7.324218750000000000000000E-2
     };
static double P4[] = {
     4.669584423426247427983539E+0,
     -1.120700261622299382716049E+1,
     8.789123535156250000000000E+0,
     -2.364086914062500000000000E+0,
     1.121520996093750000000000E-1
     };
static double P5[] = {
     -2.8212072558200244877E1,
     8.4636217674600734632E1,
     -9.1818241543240017361E1,
     4.2534998745388454861E1,
     -7.3687943594796316964E0,
     2.27108001708984375E-1
     };
static double P6[] = {
     2.1257013003921712286E2,
     -7.6525246814118164230E2,
     1.0599904525279998779E3,
     -6.9957962737613254123E2,
     2.1819051174421159048E2,
     -2.6491430486951555525E1,
     5.7250142097473144531E-1
     };
static double P7[] = {
     -1.9194576623184069963E3,
     8.0617221817373093845E3,
     -1.3586550006434137439E4,
     1.1655393336864533248E4,
     -5.3056469786134031084E3,
     1.2009029132163524628E3,
     -1.0809091978839465550E2,
     1.7277275025844573975E0
     };

static double jnx( n, x )
double n, x;
     {
     double zeta, sqz, zz, zp, np;
     double cbn, n23, t, z, sz;
     double pp, qq, z32i, zzi;
```

```
double ak, bk, akl, bkl;
int sign, doa, dob, nflg, k, s, tk, tkp1, m;
static double u[8];
static double ai, aip, bi, bip;
double acos(), fabs(), sqrt(), cbrt(), log();
double airy(), polevl(), hankel(), jnt();
```

Test for x very close to n.
Use expansion for transition region if so.

```
cbn = cbrt(n);
z = (x - n)/cbn;
if( fabs(z) <= 0.7 )
     return( jnt(n,x) );
z = x/n;
zz = 1.0 - z*z;
if( zz == 0.0 )
     return(0.0);
if( zz > 0.0 )
     {
     sz = sqrt( zz );
     t = 1.5 * (log( (1.0+sz)/z ) - sz );
     zeta = cbrt( t * t );
     nflg = 1;
     }
else
     {
     sz = sqrt(-zz);
     t = 1.5 * (sz - acos(1.0/z));
     zeta = -cbrt( t * t );
     nflg = -1;
     }
z32i = fabs(1.0/t);
sqz = cbrt(t);
n23 = cbrt( n * n );
t = n23 * zeta;
airy( t, &ai, &aip, &bi, &bip );
```

Polynomials in the expansion

```
u[0] = 1.0;
zzi = 1.0/zz;
u[1] = polevl( zzi, P1, 1 )/sz;
u[2] = polevl( zzi, P2, 2 )/zz;
u[3] = polevl( zzi, P3, 3 )/(sz*zz);
pp = zz*zz;
u[4] = polevl( zzi, P4, 4 )/pp;
u[5] = polevl( zzi, P5, 5 )/(pp*sz);
```

```
    pp *= zz;
    u[6] = polevl( zzi, P6, 6 )/pp;
    u[7] = polevl( zzi, P7, 7 )/(pp*sz);
    pp = 0.0;
    qq = 0.0;
    np = 1.0;
```

Flags to stop when terms get larger

```
    doa = 1;
    dob = 1;
    akl = MAXNUM;
    bkl = MAXNUM;
    for( k=0; k<=3; k++ )
        {
        tk = 2 * k;
        tkp1 = tk + 1;
        zp = 1.0;
        ak = 0.0;
        bk = 0.0;
        for( s=0; s<=tk; s++ )
            {
            if( doa )
                {
                if( (s & 3) > 1 )
                    sign = nflg;
                else
                    sign = 1;
                ak += sign * mu[s] * zp * u[tk-s];
                }
            if( dob )
                {
                m = tkp1 - s;
                if( ((m+1) & 3) > 1 )
                    sign = nflg;
                else
                    sign = 1;
                bk += sign * lambda[s] * zp * u[m];
                }
            zp *= z32i;
            }
        if( doa )
            {
            ak *= np;
            t = fabs(ak);
            if( t < akl )
                {
```

```
                akl = t;
                pp += ak;
                }
            else
                doa = 0;
            }
        if( dob )
            {
            bk += lambda[tkp1] * zp * u[0];
            bk *= -np/sqz;
            t = fabs(bk);
            if( t < bkl )
                {
                bkl = t;
                qq += bk;
                }
            else
                dob = 0;
            }
        if( np < MACHEP )
            break;
        np /= n*n;
        }
```

Normalizing factor $(4\zeta/(1-z^2))^{1/4}$

```
    t = 4.0 * zeta/zz;
    t = sqrt( sqrt(t) );
    t *= ai*pp/cbrt(n) + aip*qq/(n23*n);
    return(t);
    }
```

Asymptotic expansion for transition region,
n large and x close to n.
AMS55 #9.3.23.

```
static double PF2[] = {
    -9.00000000000000000000e-2,
    8.57142857142857142857e-2
    };
static double PF3[] = {
    1.36714285714285714286e-1,
    -5.49206349206349206349e-2,
    -4.44444444444444444444e-3
    };
```

```
static double PF4[] = {
     1.35000000000000000000e-3,
     -1.60360544217687074830e-1,
     4.25901875901875901880e-2,
     2.73304473304473304470e-3
     };
static double PG1[] = {
     -2.42857142857142857140e-1,
     1.42857142857142857140e-2
     };
static double PG2[] = {
     -9.00000000000000000000e-3,
     1.93968253968253968250e-1,
     -1.17460317460317460320e-2
     };
static double PG3[] = {
     1.96071428571428571430e-2,
     -1.59836940836940836940e-1,
     6.38383838383838383840e-3
     };

static double jnt( n, x )
double n, x;
     {
     double cbrt(), airy(), fabs(), polevl();
     double z, zz, z3;
     double cbn, n23, cbtwo;
     double ai, aip, bi, bip;
     double nk, fk, gk, pp, qq;
     double F[5], G[4];
     int k;

     cbn = cbrt(n);
     z = (x - n)/cbn;
     cbtwo = cbrt( 2.0 );
     zz = -cbtwo * z;
     airy( zz, &ai, &aip, &bi, &bip );
     zz = z * z;
     z3 = zz * z;
     F[0] = 1.0;
     F[1] = -z/5.0;
     F[2] = polevl( z3, PF2, 1 ) * zz;
     F[3] = polevl( z3, PF3, 2 );
     F[4] = polevl( z3, PF4, 3 ) * z;
     G[0] = 0.3 * zz;
```

```
G[1] = polevl( z3, PG1, 1 );
G[2] = polevl( z3, PG2, 2 ) * z;
G[3] = polevl( z3, PG3, 2 ) * zz;
pp = 0.0;
qq = 0.0;
nk = 1.0;
n23 = cbrt( n * n );
for( k=0; k<=4; k++ )
	{
	fk = F[k]*nk;
	pp += fk;
	if( k != 4 )
		{
		gk = G[k]*nk;
		qq += gk;
		}
	nk /= n23;
	}
fk = cbtwo * ai * pp/cbn + cbrt(4.0) * aip * qq/n;
return(fk);
}
```

6.14 Airy Functions

The Airy functions $\mathrm{Ai}(x)$, $\mathrm{Bi}(x)$ are solutions of the differential equation $y''(x) = xy$. They may be written in terms of two auxiliary functions

$$\begin{aligned} f(x) &= 1 + \frac{x^3}{3!} + \frac{1 \cdot 4x^6}{6!} + \frac{1 \cdot 4 \cdot 7x^9}{9!} + \ldots \\ g(x) &= x + \frac{2x^4}{4!} + \frac{2 \cdot 5x^7}{7!} + \frac{2 \cdot 5 \cdot 8x^{10}}{10!} + \ldots \end{aligned}$$

and the constants

$$\begin{aligned} c_1 &= \frac{1}{3^{\frac{2}{3}}\Gamma(\frac{2}{3})} = \mathrm{Ai}(0) = 3^{-\frac{1}{2}}\mathrm{Bi}(0) \\ c_2 &= \frac{1}{3^{\frac{1}{3}}\Gamma(\frac{1}{3})} = -\mathrm{Ai}'(0) = 3^{-\frac{1}{2}}\mathrm{Bi}'(0) \; . \end{aligned}$$

In terms of the auxiliary functions, the Airy functions are

$$\begin{aligned} \mathrm{Ai}(x) &= c_1 f(x) - c_2 g(x) \\ \mathrm{Bi}(x) &= 3^{\frac{1}{2}} \left[c_1 f(x) + c_2 g(x) \right] . \end{aligned}$$

Power series for the derivatives Ai′ and Bi′ follow immediately by differentiating $f(x)$ and $g(x)$ term by term. The representation of $\mathrm{Ai}(x)$ is very ill-behaved for positive x. Both f and g increase rapidly, while $\mathrm{Ai}(x)$ tends to zero. At $x = 10$, f and g are on the order of 10^8 but $\mathrm{Ai}(x) \approx 10^{-8}$. The resulting cancellation error is essentially total in relative terms for double precision arithmetic, and the absolute accuracy of about 10^{-8} is also very poor. Accurate *a priori* computation using f and g therefore requires extended precision arithmetic using high precision values of the constants,

$$\begin{aligned}
\Gamma(\tfrac{1}{3}) &= 2.6789385347077476336556929409746776441286893779573 \\
\Gamma(\tfrac{2}{3}) &= 1.3541179394264004169452880281545137855193272660568 \\
c_1 &= 0.35502805388781723926006318600418317639797917419918 \\
c_2 &= 0.25881940379280679840518356018920396347909113835493 \,.
\end{aligned}$$

Asymptotic expansions are available for large x. These literal expansions provide double precision accuracy if $|x| > 8.5$.

$$\begin{aligned}
b_0 &= 1 \\
b_k &= \frac{(2k+1)(2k+3)\cdots(6k-1)}{216^k\ k!} \\
\zeta &= \tfrac{2}{3}x^{\frac{3}{2}} \\
\mathrm{Ai}(x) &\approx \frac{1}{2\pi^{\frac{1}{2}}x^{\frac{1}{4}}e^{\zeta}}\sum_{k=0}^{\infty}\frac{(-1)^k b_k}{\zeta^k} \,.
\end{aligned}$$

The following rational approximation may be used to achieve double precision accuracy when $2 \le \zeta \le 1024$.

$$\mathrm{Ai}(x) \approx \frac{1}{2\pi^{\frac{1}{2}}x^{\frac{1}{4}}e^{\zeta}}\,\frac{A_1(1/\zeta)}{A_2(1/\zeta)}$$

where

$$\begin{aligned}
A_1(t) = {} & 3.4653810152562903247 7\cdot 10^{-1}t^7 \\
&+1.2007595273964580554 2\cdot 10^{1}t^6 \\
&+7.6279605361523451653 8\cdot 10^{1}t^5 \\
&+1.6808922493463057626 9\cdot 10^{2}t^4 \\
&+1.5975639135016441363 9\cdot 10^{2}t^3 \\
&+7.0536090684044418311 3\cdot 10^{1}t^2 \\
&+1.4026469116338966886 4\cdot 10^{1}t \\
&+9.9999999999999999530 5\cdot 10^{-1} \,,
\end{aligned}
\qquad
\begin{aligned}
A_2(t) = {} & 5.6759453263877021284 6\cdot 10^{-1}t^7 \\
&+1.4756256258484720317 3\cdot 10^{1}t^6 \\
&+8.4513897014147462656 2\cdot 10^{1}t^5 \\
&+1.7731808814540045952 2\cdot 10^{2}t^4 \\
&+1.6423469287152970183 1\cdot 10^{2}t^3 \\
&+7.1477840082557569527 4\cdot 10^{1}t^2 \\
&+1.4095913560783402959 8\cdot 10^{1}t \\
&+1.0000000000000000047 0\cdot 10^{0} \,.
\end{aligned}$$

Since ζ is usually inexact, the accuracy of this approximation is limited by error amplification in the exponential function $\exp(\zeta)$. The rational approximation has a theoretical relative error of $2.9 \cdot 10^{-18}$. All of the minimax approximations in this section were calculated using 100 decimal arithmetic and 43 decimal function accuracy.

The corresponding expansion for the derivative of $\mathrm{Ai}(x)$ is

$$
\begin{aligned}
d_0 &= 1 \\
d_k &= -\frac{(2k+1)(2k+3)\cdots(6k-3)(6k+1)}{216^k\ k!} \\
\mathrm{Ai}'(x) &\approx -\frac{x^{\frac{1}{4}}}{2\pi^{\frac{1}{2}}e^{\zeta}}\sum_{k=0}^{\infty}\frac{(-1)^k d_k}{\zeta^k}\ .
\end{aligned}
$$

When $2 \le \zeta \le 1024$,

$$\mathrm{Ai}'(x) \approx -\frac{x^{\frac{1}{4}}}{2\pi^{\frac{1}{2}}e^{\zeta}}\,\frac{A_3(1/\zeta)}{A_4(1/\zeta)}$$

where

$$
\begin{aligned}
A_3(t) = {} & 6.1375918481403575922 5\cdot 10^{-1}t^7 \\
&+1.4745467078775532388 1\cdot 10^{1}t^6 \\
&+8.2058412347606098243 0\cdot 10^{1}t^5 \\
&+1.7118478136097638554 0\cdot 10^{2}t^4 \\
&+1.5931784713714178352 3\cdot 10^{2}t^3 \\
&+6.9977859933010301617 0\cdot 10^{1}t^2 \\
&+1.3947085698048156695 8\cdot 10^{1}t \\
&+1.0000000000000000055 0\ ,
\end{aligned}
$$

$$
\begin{aligned}
A_4(t) = {} & 3.3420367774973695304 9\cdot 10^{-1}t^7 \\
&+1.1181029730615815670 5\cdot 10^{1}t^6 \\
&+7.1172735214785996528 3\cdot 10^{1}t^5 \\
&+1.5877808437283831364 0\cdot 10^{2}t^4 \\
&+1.5320642747580922083 4\cdot 10^{2}t^3 \\
&+6.8675230459278033794 4\cdot 10^{1}t^2 \\
&+1.3849863475825944247 7\cdot 10^{1}t \\
&+0.9999999999999999450 2\ .
\end{aligned}
$$

For $\mathrm{Bi}(x)$ the asymptotic expansion is

$$\mathrm{Bi}(x) \approx \frac{e^{\zeta}}{\pi^{\frac{1}{2}}x^{\frac{1}{4}}}\sum_{k=0}^{\infty}\frac{b_k}{\zeta^k}\ .$$

A rational approximation for $\zeta > 16$ (i.e., $x > 8.3203353$) is

$$\mathrm{Bi}(x) \approx \frac{e^{\zeta}}{\pi^{\frac{1}{2}}x^{\frac{1}{4}}}\left\{1+\frac{A_5(1/\zeta)}{\zeta\ A_6(1/\zeta)}\right\}$$

where

$$
\begin{aligned}
A_5(t) = {} & -2.5324079586936415268 9\cdot 10^{-1}t^4 \\
&+5.7528516733246738422 8\cdot 10^{-1}t^3 \\
&-3.2990703687322537165 0\cdot 10^{-1}t^2 \\
&+6.4440406894819995172 7\cdot 10^{-2}t \\
&-3.8251954664133673439 4\cdot 10^{-3}\ ,
\end{aligned}
$$

$$
\begin{aligned}
A_6(t) = {} & 1.0t^5 \\
&-7.1568509505403523790 2\cdot 10^{0}t^4 \\
&+1.0603958071566469429 1\cdot 10^{1}t^3 \\
&-5.2324663647125150087 4\cdot 10^{0}t^2 \\
&+9.5739586437838383315 2\cdot 10^{-1}t \\
&-5.5082814716354961110 7\cdot 10^{-2}\ .
\end{aligned}
$$

For the derivative of $\mathrm{Bi}(x)$,

$$\mathrm{Bi}'(x) \approx \frac{x^{\frac{1}{4}}e^{\zeta}}{\pi^{\frac{1}{2}}}\sum_{k=0}^{\infty}\frac{d_k}{\zeta^k}\ .$$

If x is large and positive, the following expressions converge rapidly. The formulas containing ${}_2F_0$ (for hypergeometric functions see Section 7.1) are the same as the asymptotic expansions above. For the others, cancellation error may be severe, so they are of computational value mainly as consistency tests.

$$\begin{aligned}
\zeta &= \tfrac{2}{3}x^{\frac{3}{2}} \\
\mathrm{Ai}(x) &= \tfrac{1}{3}x^{\frac{1}{2}}\,[I_{-1/3}(\zeta) - I_{1/3}(\zeta)] \\
&= \frac{x^{\frac{1}{2}}}{\pi\sqrt{3}}\,K_{\frac{1}{3}}(\zeta) \\
&\approx \frac{1}{2\sqrt{\pi}\;e^{\zeta}x^{\frac{1}{4}}}\;{}_2F_0(\frac{1}{6},\frac{5}{6};;-\frac{1}{2\zeta}) \\
\mathrm{Bi}(x) &= \sqrt{\frac{x}{3}}\,[I_{-1/3}(\zeta) + I_{1/3}(\zeta)] \\
&\approx \frac{e^{\zeta}}{\sqrt{\pi}\;x^{\frac{1}{4}}}\;{}_2F_0(\frac{1}{6},\frac{5}{6};;\frac{1}{2\zeta}) \\
\mathrm{Ai}'(x) &= -\tfrac{1}{3}x\,[I_{-2/3}(\zeta) - I_{2/3}(\zeta)] \\
&= \frac{x}{\pi\sqrt{3}}\,K_{\frac{2}{3}}(\zeta) \\
&\approx -\frac{x^{\frac{1}{4}}}{2\sqrt{\pi}\;e^{\zeta}}\;{}_2F_0(-\frac{1}{6},\frac{7}{6};;-\frac{1}{2\zeta}) \\
\mathrm{Bi}'(x) &= \frac{x}{\sqrt{3}}[I_{-2/3}(\zeta) + I_{2/3}(\zeta)] \\
&\approx \frac{e^{\zeta}\;x^{\frac{1}{4}}}{\sqrt{\pi}}\;{}_2F_0(-\frac{1}{6},\frac{7}{6};;\frac{1}{2\zeta}) \\
\frac{1}{\sqrt{\pi}} &= 0.56418958354775628694\mathrm{8} \\
\sqrt{3} &= 1.73205080756887729352\mathrm{7}\;.
\end{aligned}$$

If x is large and negative, an asymptotic form similar to Hankel's expansion of J_n, Y_n may be used for *a priori* computation.

$$\begin{aligned}
\theta &= \zeta + \frac{\pi}{4} \\
\mathrm{Ai}(-x) &\approx \frac{1}{\pi^{\frac{1}{2}}x^{\frac{1}{4}}}\left[\sin\theta\sum_{k=0}^{\infty}\frac{(-1)^k b_{2k}}{\zeta^{2k}} - \cos\theta\sum_{k=0}^{\infty}\frac{(-1)^k b_{2k+1}}{\zeta^{2k+1}}\right]
\end{aligned}$$

where the coefficients are the same as in the asymptotic expansions given above for $x > 0$.

For computation, auxiliary functions F and G are found that satisfy

$$\mathrm{Ai}(-x) = \frac{1}{\pi^{\frac{1}{2}} x^{\frac{1}{4}}} \left[\sin\theta\ F(1/\zeta) - \cos\theta\ G(1/\zeta)\right]$$

while simultaneously satisfying a complementary expression for $\mathrm{Bi}(x)$. A double precision rational approximation for F is

$$F(1/\zeta) = 1 + \frac{A_7(1/\zeta^2)}{\zeta^2 A_8(1/\zeta^2)}$$

where

$$\begin{aligned}
A_7(t) = &\\
&-1.31696323418331795333 \cdot 10^{-1} t^8\\
&-6.26456544431912369773 \cdot 10^{-1} t^7\\
&-6.93158036036933542233 \cdot 10^{-1} t^6\\
&-2.79779981545119124951 \cdot 10^{-1} t^5\\
&-4.91900132609500318020 \cdot 10^{-2} t^4\\
&-4.06265923594885404393 \cdot 10^{-3} t^3\\
&-1.59276496239262096340 \cdot 10^{-4} t^2\\
&-2.77649108155232920844 \cdot 10^{-6} t\\
&-1.67787698489114633780 \cdot 10^{-8}\ ,
\end{aligned}$$

$$\begin{aligned}
A_8(t) = &\\
&1.0 t^9\\
&+1.33560420706553243746 \cdot 10^{1} t^8\\
&+3.26825032795224613948 \cdot 10^{1} t^7\\
&+2.67367040941499554804 \cdot 10^{1} t^6\\
&+9.18707402907259625840 \cdot 10^{0} t^5\\
&+1.47529146771666414581 \cdot 10^{0} t^4\\
&+1.15687173795188044134 \cdot 10^{-1} t^3\\
&+4.40291641615211203805 \cdot 10^{-3} t^2\\
&+7.54720348287414296618 \cdot 10^{-5} t\\
&+4.51850092970580378464 \cdot 10^{-7}\ .
\end{aligned}$$

These coefficients were computed from the following formal decomposition.

$$\begin{aligned}
F(1/\zeta) &= \pi^{\frac{1}{2}} x^{\frac{1}{4}} \left[\mathrm{Ai}(-x)\sin\theta + \mathrm{Bi}(-x)\cos\theta\right]\\
G(1/\zeta) &= \pi^{\frac{1}{2}} x^{\frac{1}{4}} \left[-\mathrm{Ai}(-x)\cos\theta + \mathrm{Bi}(-x)\sin\theta\right].
\end{aligned}$$

The rational approximation coefficients are valid for $\zeta > 2$. For the auxiliary function G,

$$G(1/\zeta) \approx \frac{A_9(1/\zeta^2)}{\zeta A_{10}(1/\zeta^2)}$$

where

$$\begin{aligned}
A_9(t) = &\\
&1.97339932091685679179 \cdot 10^{-2} t^{10}\\
&+3.91103029615688277255 \cdot 10^{-1} t^9\\
&+1.06579897599595591108 \cdot 10^{0} t^8\\
&+9.39169229816650230044 \cdot 10^{-1} t^7\\
&+3.51465656105547619242 \cdot 10^{-1} t^6\\
&+6.33888919628925490927 \cdot 10^{-2} t^5\\
&+5.85804113048388458567 \cdot 10^{-3} t^4\\
&+2.82851600836737019778 \cdot 10^{-4} t^3\\
&+6.98793669997260967291 \cdot 10^{-6} t^2\\
&+8.11789239554389293311 \cdot 10^{-8} t\\
&+3.41551784765923618484 \cdot 10^{-10},
\end{aligned}$$

$$\begin{aligned}
A_{10}(t) = &\\
&1.0 t^{10}\\
&+9.30892908077441974853 \cdot 10^{0} t^9\\
&+1.98352928718312140417 \cdot 10^{1} t^8\\
&+1.55646628932864612953 \cdot 10^{1} t^7\\
&+5.47686069422975497931 \cdot 10^{0} t^6\\
&+9.54293611618961883998 \cdot 10^{-1} t^5\\
&+8.64580826352392193095 \cdot 10^{-2} t^4\\
&+4.12656523824222607191 \cdot 10^{-3} t^3\\
&+1.01259085116509135510 \cdot 10^{-4} t^2\\
&+1.17166733214413521882 \cdot 10^{-6} t\\
&+4.91834570062930015649 \cdot 10^{-9}\ .
\end{aligned}$$

Using the same auxiliary functions F and G, $\mathrm{Bi}(x)$ may be computed from

$$\mathrm{Bi}(-x) = \frac{1}{\pi^{\frac{1}{2}} x^{\frac{1}{4}}} \left[\cos\theta \; F(1/\zeta) + \sin\theta \; G(1/\zeta)\right].$$

Note that the absolute error of the sine and cosine functions grows in direct proportion to the absolute error in ζ which is in turn proportional to $x^{3/2}$.

A similar decomposition holds for the derivatives. In terms of auxiliary functions F_1 and G_1,

$$\mathrm{Ai}'(-x) = \frac{x^{\frac{1}{4}}}{\pi^{\frac{1}{2}}} \left[-\cos\theta \; F_1(1/\zeta) - \sin\theta \; G_1(1/\zeta)\right] .$$

The definitions of F_1 and G_1 are

$$\begin{aligned} F_1(1/\zeta) &= \frac{x^{\frac{1}{4}}}{\pi^{\frac{1}{2}}} \left[-\mathrm{Ai}'(-x)\cos\theta + \mathrm{Bi}'(-x)\sin\theta\right] \\ G_1(1/\zeta) &= \frac{x^{\frac{1}{4}}}{\pi^{\frac{1}{2}}} \left[-\mathrm{Ai}'(-x)\sin\theta - \mathrm{Bi}'(-x)\cos\theta\right] . \end{aligned}$$

The rational approximation for F_1 is

$$F_1(1/\zeta) = 1 + \frac{A_{11}(1/\zeta^2)}{\zeta^2 A_{12}(1/\zeta^2)}$$

where

$$\begin{aligned} A_{11}(t) = {} & 1.8536562402253556614 2 \cdot 10^{-1} t^8 \\ & +8.8671218805258409563 7 \cdot 10^{-1} t^7 \\ & +9.8739198174739854727 2 \cdot 10^{-1} t^6 \\ & +4.0124108231800373409 2 \cdot 10^{-1} t^5 \\ & +7.1030492628963117457 9 \cdot 10^{-2} t^4 \\ & +5.9061865799566181007 1 \cdot 10^{-3} t^3 \\ & +2.3305140940177679956 9 \cdot 10^{-4} t^2 \\ & +4.0871877828903545459 8 \cdot 10^{-6} t \\ & +2.4837993290044245785 3 \cdot 10^{-8} , \end{aligned}$$

$$\begin{aligned} A_{12}(t) = {} & 1.0 t^9 \\ & +1.4734585468750254255 2 \cdot 10^{1} t^8 \\ & +3.7542393343548959446 6 \cdot 10^{1} t^7 \\ & +3.1465775120304642433 0 \cdot 10^{1} t^6 \\ & +1.0996912520729877853 6 \cdot 10^{1} t^5 \\ & +1.7888505476699941781 7 \cdot 10^{0} t^4 \\ & +1.4173327575366263687 3 \cdot 10^{-1} t^3 \\ & +5.4406606701722600362 7 \cdot 10^{-3} t^2 \\ & +9.3942129065451117166 3 \cdot 10^{-5} t \\ & +5.6597871303602700924 3 \cdot 10^{-7} . \end{aligned}$$

These rational approximation coefficients are valid for $\zeta > 2$. For the auxiliary function G_1,

$$G_1(1/\zeta) \approx \frac{A_{13}(1/\zeta^2)}{\zeta A_{14}(1/\zeta^2)}$$

where

$A_{13}(t) =$
$-3.55615429033082288335 \cdot 10^{-2}t^{10}$
$-6.37311518129435504426 \cdot 10^{-1}t^{9}$
$-1.70856738884312371053 \cdot 10^{0}t^{8}$
$-1.50221872117316635393 \cdot 10^{0}t^{7}$
$-5.63606665822102676611 \cdot 10^{-1}t^{6}$
$-1.02101031120216891789 \cdot 10^{-1}t^{5}$
$-9.48396695961445269093 \cdot 10^{-3}t^{4}$
$-4.60325307486780994357 \cdot 10^{-4}t^{3}$
$-1.14300836484517375919 \cdot 10^{-5}t^{2}$
$-1.33415518685547420648 \cdot 10^{-7}t$
$-5.63803833958893494476 \cdot 10^{-10}$,

$A_{14}(t) =$
$1.0t^{10}$
$+9.85865801696130355144 \cdot 10^{0}t^{9}$
$+2.16401867356585941885 \cdot 10^{1}t^{8}$
$+1.73130776389749389525 \cdot 10^{1}t^{7}$
$+6.17872175280828766327 \cdot 10^{0}t^{6}$
$+1.08848694396321495475 \cdot 10^{0}t^{5}$
$+9.95005543440888479402 \cdot 10^{-2}t^{4}$
$+4.78468199683886610842 \cdot 10^{-3}t^{3}$
$+1.18159633322838625562 \cdot 10^{-4}t^{2}$
$+1.37480673554219441465 \cdot 10^{-6}t$
$+5.79912514929147598821 \cdot 10^{-9}$.

For the derivative of $\mathrm{Bi}(x)$ the asymptotic expansion is

$$\mathrm{Bi}'(x) \approx \frac{e^{\zeta} x^{\frac{1}{4}}}{\pi^{\frac{1}{2}}} \sum_{k=0}^{\infty} \frac{d_k}{\zeta^k} \, .$$

When $\zeta > 16$,

$$\mathrm{Bi}'(x) \approx \frac{e^{\zeta} x^{\frac{1}{4}}}{\pi^{\frac{1}{2}}} \left\{ 1 + \frac{A_{15}(1/\zeta)}{\zeta \; A_{16}(1/\zeta)} \right\}$$

where

$A_{15}(t) =$
$4.65461162774651610328 \cdot 10^{-1}t^{4}$
$-1.08992173800493920734 \cdot 10^{0}t^{3}$
$+6.38800117371827987759 \cdot 10^{-1}t^{2}$
$-1.26844349553102907034 \cdot 10^{-1}t$
$+7.62487844342109852105 \cdot 10^{-3}$,

$A_{16}(t) =$
$1.0t^{5}$
$-8.70622787633159124240 \cdot 10^{0}t^{4}$
$+1.38993162704553213172 \cdot 10^{1}t^{3}$
$-7.14116144616431159572 \cdot 10^{0}t^{2}$
$+1.34008595960680518666 \cdot 10^{0}t$
$-7.84273211323341930448 \cdot 10^{-2}$.

Consistency tests for $x < 0$ may be constructed from

$$\begin{aligned} \zeta &= \tfrac{2}{3}(-x)^{\frac{3}{2}} \\ \mathrm{Ai}(x) &= \frac{\sqrt{-x}}{3} \left[J_{-\frac{1}{3}}(\zeta) + J_{\frac{1}{3}}(\zeta) \right] \\ \mathrm{Bi}(x) &= \sqrt{\frac{-x}{3}} \left[I_{-\frac{1}{3}}(\zeta) - I_{\frac{1}{3}}(\zeta) \right] \\ \mathrm{Ai}'(x) &= \frac{x}{3} \left[J_{-\frac{2}{3}}(\zeta) - J_{\frac{2}{3}}(\zeta) \right] \\ \mathrm{Bi}'(x) &= -\frac{x}{\sqrt{3}} \left[I_{-\frac{2}{3}}(\zeta) + I_{\frac{2}{3}}(\zeta) \right] . \end{aligned}$$

For small x the functions may be computed directly from the power series expansions. The crossover between the power series and the expansions preferred for large x occurs at $|x| = 5.5$ in double precision arithmetic.

6.14.1 airy.c

Airy function program

```
#include "mconf.h"

static double c1 = 0.35502805388781723926;
static double c2 = 0.258819403792806798405;
static double sqrt3 = 1.732050807568877293527;
static double sqpii = 5.64189583547756286948E-1;
extern double PI;

extern double MAXNUM, MACHEP;
#ifdef DEC
#define MAXAIRY 25.77
#endif
#ifdef IIEEE
#define MAXAIRY 103.892
#endif

#if DEC
static short AN[32] = {
0037661,0066561,0024675,0131301,
0041100,0017434,0034324,0101466,
0041630,0107450,0067427,0007430,
0042050,0013327,0071000,0034737,
0042037,0140642,0156417,0167366,
0041615,0011172,0075147,0051165,
0041140,0066152,0160520,0075146,
0040200,0000000,0000000,0000000,
};
static short AD[32] = {
0040021,0046740,0011422,0064606,
0041154,0014640,0024631,0062450,
0041651,0003435,0101152,0106401,
0042061,0050556,0034605,0136602,
0042044,0036024,0152377,0151414,
0041616,0172247,0072216,0115374,
0041141,0104334,0124154,0166007,
0040200,0000000,0000000,0000000,
};
static short APN[32] = {
0040035,0017522,0065145,0054755,
0041153,0166556,0161471,0057174,
0041644,0016750,0034445,0046462,
0042053,0027515,0152316,0046717,
0042037,0050536,0067023,0023264,
0041613,0172252,0007240,0131055,
0041137,0023503,0052472,0002305,
```

```
#if IIEEE
static short AN[32] = {
0xb658,0x2537,0x2dae,0x3fd6,
0x9067,0x871a,0x03e3,0x4028,
0xe1e3,0x0de2,0x11e5,0x4053,
0x073c,0xee40,0x02da,0x4065,
0xfddf,0x5ba1,0xf834,0x4063,
0xea4f,0x4f4c,0xa24f,0x4051,
0x0f4d,0x5c2a,0x0d8d,0x402c,
0x0000,0x0000,0x0000,0x3ff0,
};
static short AD[32] = {
0x4d31,0x0262,0x29bc,0x3fe2,
0x2ca5,0x0533,0x8334,0x402d,
0x51a0,0xb04d,0x20e3,0x4055,
0xb7b0,0xc730,0x2a2d,0x4066,
0xfa61,0x9a9f,0x8782,0x4064,
0xd35f,0xee91,0xde94,0x4051,
0x9d81,0x950d,0x311b,0x402c,
0x0000,0x0000,0x0000,0x3ff0,
};
static short APN[32] = {
0xab3e,0x4d4c,0xa3ea,0x3fe3,
0x2bcf,0xdc67,0x7dad,0x402d,
0xa9a6,0x0724,0x83bd,0x4054,
0xc9ba,0xba99,0x65e9,0x4065,
0x64d7,0xcdc2,0xea2b,0x4063,
0x1646,0x41d4,0x7e95,0x4051,
0x4099,0x6aa7,0xe4e8,0x402b,
```

```
0040200,0000000,0000000,0000000,
};
static short APD[32] = {
0037653,0016276,0112106,0126625,
0041062,0162577,0067111,0111761,
0041616,0054160,0140004,0137455,
0042036,0143460,0104626,0157206,
0042031,0032330,0067131,0114260,
0041611,0054667,0147207,0134564,
0041135,0114412,0070653,0146015,
0040200,0000000,0000000,0000000,
};
static short BN16[20] = {
0137601,0124307,0010213,0035210,
0040023,0042743,0101621,0016031,
0137650,0164623,0036056,0074511,
0037203,0174525,0000473,0142474,
0136172,0130041,0066726,0064324,
};
```

Leading 1.0 omitted from BD.

```
static short BD16[20] = {
0140745,0002354,0044335,0055276,
0041051,0124717,0170130,0104013,
0140647,0070135,0046473,0103501,
0040165,0013745,0033324,0127766,
0137141,0117204,0076164,0033107,
};
static short BPPN[20] = {
0037756,0050354,0167531,0135731,
0140213,0101216,0032767,0020375,
0040043,0104147,0106312,0177632,
0137401,0161574,0032015,0043714,
0036371,0155035,0143165,0142262,
};
```

Leading 1.0 omitted from BPPD.

```
static short BPPD[20] = {
0141013,0046265,0115005,0161053,
0041136,0061631,0072445,0156131,
0140744,0102145,0001127,0065304,
0040253,0103757,0146453,0102513,
0137240,0117200,0155402,0113500,
};
static short AFN[36] = {
0137406,0155546,0124127,0033732,
0140040,0057564,0141263,0041222,
0140061,0071316,0013674,0175754,
0137617,0037522,0056637,0120130,
0137111,0075567,0121755,0166122,
0136205,0020016,0043317,0002201,
```

```
0x0000,0x0000,0x0000,0x3ff0,
};
static short APD[32] = {
0xd5b3,0xd288,0x6397,0x3fd5,
0x327e,0xedc9,0x5caf,0x4026,
0x97e6,0x1800,0xcb0e,0x4051,
0xdbd1,0x1132,0xd8e6,0x4063,
0x3316,0x0dcb,0x269b,0x4063,
0xf72f,0xf9d0,0x2b36,0x4051,
0x7982,0x4e35,0xb321,0x402b,
0x0000,0x0000,0x0000,0x3ff0,
};
static short BN16[20] = {
0x6751,0xe211,0x3518,0xbfd0,
0x2383,0x7072,0x68bc,0x3fe2,
0xcf29,0x6785,0x1d32,0xbfd5,
0x78a8,0xa027,0x7f2a,0x3fb0,
0xcd1b,0x2dba,0x5604,0xbf6f,
};
static short BD16[20] = {
0xab58,0x891b,0xa09d,0xc01c,
0x1101,0xfe0b,0x3539,0x4025,
0x70e8,0xa9a7,0xee0b,0xc014,
0x95ff,0xa6da,0xa2fc,0x3fee,
0x86c9,0x8f8e,0x33d0,0xbfac,
};
static short BPPN[20] = {
0x377b,0x9deb,0xca1d,0x3fdd,
0xe420,0xc6be,0x7051,0xbff1,
0x5ff3,0xf199,0x710c,0x3fe4,
0xa8fa,0x8681,0x3c6f,0xbfc0,
0xb896,0xb8ce,0x3b43,0x3f7f,
};
static short BPPD[20] = {
0xbc45,0xb340,0x6996,0xc021,
0xbb8b,0x2ea4,0xcc73,0x402b,
0xed59,0xa04a,0x908c,0xc01c,
0x70a9,0xf9a5,0x70fd,0x3ff5,
0x52e8,0x1b60,0x13d0,0xbfb4,
};
static short AFN[36] = {
0xe6fb,0xd50a,0xdb6c,0xbfc0,
0x6852,0x9856,0x0bee,0xbfe4,
0x9f7d,0xc2f7,0x2e59,0xbfe6,
0xf40b,0x4bb3,0xe7ea,0xbfd1,
0xbd8a,0xf47d,0x2f6e,0xbfa9,
0xe090,0xc8d9,0xa401,0xbf70,
```

```
0135047,0001565,0075130,0002334,
0133472,0051700,0165021,0131551,
0131620,0020347,0132165,0013215,
};
```

Leading 1.0 omitted from AFD.

```
static short AFD[36] = {
0041125,0131131,0025627,0067623,
0041402,0135342,0021703,0154315,
0041325,0162305,0016671,0120175,
0041022,0177101,0053114,0141632,
0040274,0153131,0147364,0114306,
0037354,0166545,0120042,0150530,
0036220,0043127,0000727,0130273,
0034636,0043275,0075667,0034733,
0032762,0112715,0146250,0142474,
};
static short AGN[44] = {
0036641,0124456,0167175,0157354,
0037710,0037250,0001441,0136671,
0040210,0066031,0150401,0123532,
0040160,0066545,0003570,0153133,
0037663,0171516,0072507,0170345,
0037201,0151011,0007510,0045702,
0036277,0172317,0104572,0101030,
0035224,0045663,0000160,0136422,
0033752,0074753,0047702,0135160,
0032256,0052225,0156550,0107103,
0030273,0142443,0166277,0071720,
};
```

Leading 1.0 omitted from AGD.

```
static short AGD[40] = {
0041024,0170537,0117253,0055003,
0041236,0127256,0003570,0143240,
0041171,0004333,0172476,0160645,
0040657,0041161,0055716,0157161,
0040164,0046226,0006257,0063431,
0037261,0010357,0065445,0047563,
0036207,0034043,0057434,0116732,
0034724,0055416,0130035,0026377,
0033235,0041056,0154071,0023502,
0031250,0177071,0167254,0047242,
};
static short APFN[36] = {
0037475,0150174,0071752,0166651,
0040142,0177621,0164246,0101757,
0040174,0142670,0106760,0006573,
0037715,0067570,0116274,0022404,
0037221,0074157,0053341,0117207,
0036301,0104257,0015075,0004777,
```

```
0x009c,0xaf4b,0xe06e,0xbf24,
0x366d,0x1d42,0x4a78,0xbec7,
0xa2d2,0xf68e,0x041c,0xbe52,
};

static short AFD[36] = {
0xedf2,0x2572,0xb64b,0x402a,
0x7b1a,0x4478,0x575c,0x4040,
0x3410,0xa3b7,0xbc98,0x403a,
0x9873,0x2ac9,0x5fc8,0x4022,
0x9319,0x39de,0x9acb,0x3ff7,
0x5a2b,0xb404,0x9dac,0x3fbd,
0xf617,0xe03a,0x08ca,0x3f72,
0xe73b,0xaf76,0xc8d7,0x3f13,
0x18a7,0xb995,0x52b9,0x3e9e,
};
static short AGN[44] = {
0xbbde,0xddcf,0x3525,0x3f94,
0x37b7,0x0064,0x07d5,0x3fd9,
0x34eb,0x3a20,0x0d83,0x3ff1,
0x1acb,0xa0ef,0x0dac,0x3fee,
0xfe1d,0xcea8,0x7e69,0x3fd6,
0x0978,0x21e9,0x3a41,0x3fb0,
0x5043,0xf12f,0xfe99,0x3f77,
0x17a2,0x600e,0x8976,0x3f32,
0x574e,0x69f8,0x4f3d,0x3edd,
0x11c8,0xbbad,0xca92,0x3e75,
0xee7a,0x7d97,0x78a4,0x3df7,
};

static short AGD[40] = {
0x6b40,0xf3d5,0x9e2b,0x4022,
0x18d4,0xc0ef,0xd5d5,0x4033,
0xdc35,0x7ea7,0x211b,0x402f,
0xdbce,0x2b79,0xe84e,0x4015,
0xece3,0xc195,0x8992,0x3fee,
0xa9ee,0xed64,0x221d,0x3fb6,
0x93bb,0x6be3,0xe704,0x3f70,
0xa5a0,0xd603,0x8b61,0x3f1a,
0x24e8,0xdb07,0xa845,0x3eb3,
0x89d4,0x3dd5,0x1fc7,0x3e35,
};
static short APFN[36] = {
0x5db5,0x8e7d,0xba0f,0x3fc7,
0xd07e,0x3d14,0x5ff2,0x3fec,
0x01af,0x11be,0x98b7,0x3fef,
0x84a1,0x1397,0xadef,0x3fd9,
0x33d1,0xeadc,0x2f0d,0x3fb2,
0xa140,0xe347,0x3115,0x3f78,
```

```
0035164,0057502,0164034,0001313,
0033611,0022254,0176000,0112565,
0031725,0055523,0025153,0166057,
};
```

Leading 1.0 omitted from APFD.

```
static short APFD[36] = {
0041153,0140334,0130506,0061402,
0041426,0025551,0024440,0070611,
0041373,0134750,0047147,0176702,
0041057,0171532,0105430,0017674,
0040344,0174416,0001726,0047754,
0037421,0021207,0020167,0136264,
0036262,0043621,0151321,0124324,
0034705,0001313,0163733,0016407,
0033027,0166702,0150440,0170561,
};
static short APGN[44] = {
0137021,0124372,0176075,0075331,
0140043,0023330,0177672,0161655,
0140332,0131126,0010413,0171112,
0140300,0044263,0175560,0054070,
0140020,0044206,0142603,0073324,
0137321,0015130,0066144,0144033,
0136433,0061243,0175542,0103373,
0135361,0053721,0020441,0053203,
0134077,0141725,0160277,0130612,
0132417,0040372,0100363,0060200,
0130432,0175052,0171064,0034147,
};
```

Leading 1.0 omitted from APGD.

```
static short APGD[40] = {
0041035,0136420,0030124,0140220,
0041255,0017432,0034447,0162256,
0041212,0100456,0154544,0006321,
0040705,0134026,0127154,0123414,
0040213,0051612,0044470,0172607,
0037313,0143362,0053273,0157051,
0036234,0144322,0054536,0007264,
0034767,0146170,0054265,0170342,
0033270,0102777,0167362,0073631,
0031307,0040644,0167103,0021763,
};
#endif
```

```
0x8059,0x5d03,0x8be8,0x3f2e,
0x12af,0x9f80,0x2495,0x3ed1,
0x7d86,0x654d,0xab6a,0x3e5a,
};

static short APFD[36] = {
0xcc60,0x9628,0x781b,0x402d,
0x0e31,0x2524,0xc56d,0x4042,
0xffb8,0x09cc,0x773d,0x403f,
0x03f7,0x5163,0xfe6b,0x4025,
0xc9fd,0xc07a,0x9f21,0x3ffc,
0xf796,0xe40e,0x2450,0x3fc2,
0x351a,0x3a5a,0x48f2,0x3f76,
0x63a1,0x7cfb,0xa059,0x3f18,
0x1e2e,0x5a24,0xfdb8,0x3ea2,
};
static short APGN[44] = {
0xaf5b,0x5f87,0x351f,0xbfa2,
0x5c76,0x1ff7,0x64db,0xbfe4,
0x7e49,0xc221,0x564a,0xbffb,
0x0b07,0x7f6e,0x0916,0xbff8,
0x6edb,0xd8b0,0x0910,0xbfe2,
0x9903,0x0d8c,0x234b,0xbfba,
0x50df,0x7f6c,0x6c54,0xbf83,
0x2ad0,0x2424,0x2afa,0xbf3e,
0xf631,0xbc17,0xf87a,0xbee7,
0x6c10,0x501e,0xe81f,0xbe81,
0x870d,0x5e46,0x5f45,0xbe03,
};

static short APGD[40] = {
0x9812,0x060a,0xb7a2,0x4023,
0xfc96,0x4724,0xa3e3,0x4035,
0x819a,0xdb2c,0x5025,0x4031,
0x94e2,0xd5cd,0xb702,0x4018,
0x1eb1,0x4927,0x6a71,0x3ff1,
0x7bc5,0x4ad7,0x78de,0x3fb9,
0xc1d7,0x4b2b,0x991a,0x3f73,
0xbe1c,0x0b16,0xf98f,0x3f1e,
0x4ef3,0xfdde,0x10bf,0x3eb7,
0x647e,0x9dc8,0xe834,0x3e38,
};
#endif
```

```
double airy( x, ai, aip, bi, bip )
double x, *ai, *aip, *bi, *bip;
	{
	double z, zz, t, f, g, uf, ug, k, zeta, theta;
```

```
int domflg;
double fabs(), exp(), sqrt();
double polevl(), p1evl(), sin(), cos();

domflg = 0;
if( x > MAXAIRY )
      {
      *ai = 0;
      *aip = 0;
      *bi = MAXNUM;
      *bip = MAXNUM;
      goto aidone;
      }
if( x < -2.09 )
      {
      domflg = 15;
      t = sqrt(-x);
      zeta = -2.0 * x * t / 3.0;
      t = sqrt(t);
      k = sqpii / t;
      z = 1.0/zeta;
      zz = z * z;
      uf = 1.0 + zz * polevl( zz, AFN, 8 )
           / p1evl( zz, AFD, 9 );
      ug = z * polevl( zz, AGN, 10 )
           / p1evl( zz, AGD, 10 );
      theta = zeta + 0.25 * PI;
      f = sin( theta );
      g = cos( theta );
      *ai = k * (f * uf - g * ug);
      *bi = k * (g * uf + f * ug);
      uf = 1.0 + zz * polevl( zz, APFN, 8 )
           / p1evl( zz, APFD, 9 );
      ug = z * polevl( zz, APGN, 10 )
           / p1evl( zz, APGD, 10 );
      k = sqpii * t;
      *aip = -k * (g * uf + f * ug);
      *bip = k * (f * uf - g * ug);
      goto aidone;
      }
if( x >= 2.09 ) cbrt(9)
      {
      domflg = 5;
      t = sqrt(x);
      zeta = 2.0 * x * t / 3.0;
```

```
        g = exp( zeta );
        t = sqrt(t);
        k = 2.0 * t * g;
        z = 1.0/zeta;
        f = polevl( z, AN, 7 ) / polevl( z, AD, 7 );
        *ai = sqpii * f / k;
        k = -0.5 * sqpii * t / g;
        f = polevl( z, APN, 7 ) / polevl( z, APD, 7 );
        *aip = f * k;
        if( x > 8.3203353 ) ζ > 16
            {
            f = z * polevl( z, BN16, 4 )
                / p1evl( z, BD16, 5 );
            k = sqpii * g;
            *bi = k * (1.0 + f) / t;
            f = z * polevl( z, BPPN, 4 )
                / p1evl( z, BPPD, 5 );
            *bip = k * t * (1.0 + f);
            goto aidone;
            }
        }
    f = 1.0;
    g = x;
    t = 1.0;
    uf = 1.0;
    ug = x;
    k = 1.0;
    z = x * x * x;
    while( t > MACHEP )
        {
        uf *= z;
        k += 1.0;
        uf /=k;
        ug *= z;
        k += 1.0;
        ug /=k;
        uf /=k;
        f += uf;
        k += 1.0;
        ug /=k;
        g += ug;
        t = fabs(uf/f);
        }
    uf = c1 * f;
    ug = c2 * g;
```

```
	if( (domflg & 1) == 0 )
		*ai = uf - ug;
	if( (domflg & 2) == 0 )
		*bi = sqrt3 * (uf + ug);
```

The deriviative of Ai

```
	k = 4.0;
	uf = x * x/2.0;
	ug = z/3.0;
	f = uf;
	g = 1.0 + ug;
	uf /= 3.0;
	t = 1.0;
	while( t > MACHEP )
		{
		uf *= z;
		ug /=k;
		k += 1.0;
		ug *= z;
		uf /=k;
		f += uf;
		k += 1.0;
		ug /=k;
		uf /=k;
		g += ug;
		k += 1.0;
		t = fabs(ug/g);
		}
	uf = c1 * f;
	ug = c2 * g;
	if( (domflg & 4) == 0 )
		*aip = uf - ug;
	if( (domflg & 8) == 0 )
		*bip = sqrt3 * (uf + ug);
aidone:
	return(0.0);
	}
```

6.15 $Y_n(x)$

When n is a positive integer, the forward recurrence

$$Y_{\nu+1}(x) = \frac{2\nu}{x} Y_\nu(x) - Y_{\nu-1}(x)$$

can be used to find Y_n starting with values for Y_0 and Y_1 computed by the methods given in the previous sections. If n is negative, use the reflection formula

$$Y_{-n}(x) = (-1)^n Y_n(x) \; .$$

For *a priori* computation the following power series is recommended. This provides a substantially different calculation method as a check.

$$\begin{aligned} Y_n(x) &= -\pi^{-1}(\tfrac{1}{2}x)^{-n}\sum_{k=0}^{n-1}\frac{(n-k-1)!}{k!}(x^2/4)^k \\ &\quad +2\pi^{-1}\ln(x/2)J_n(x) \\ &\quad -\pi^{-1}(x/2)^n\sum_{k=0}^{\infty}[\psi(k+1)+\psi(n+k+1)]\frac{(-x^2/4)^k}{k!\,(n+k)!} \end{aligned}$$

where $\psi(x)$ is as defined in Section 7.3.

When x and n are large, special asymptotic expansions for the case $x \approx n$ are needed. The procedure is very similar to the corresponding cases of $J_\nu(x)$; see Watson's treatise[3] and also AMS55 #9.3.

When ν is not an integer, the formula

$$Y_\nu(x) = \frac{J_\nu(x)\cos(\nu\pi) - J_{-\nu}(x)}{\sin(\nu\pi)}$$

is useful unless ν is very nearly integer valued. See Section 7.2 for an application to Struve functions.

6.15.1 yn.c

```
extern double MAXNUM, MAXLOG;

double yn( n, x )
int n;
double x;
    {
    double an, anm1, anm2, r;
    double y0(), y1(), log();
    int k, sign;

    if( n < 0 )
        {
        n = -n;
        if( (n & 1) == 0 )
            sign = 1;
```

[3]Watson, G. N. *A Treatise on the Theory of Bessel Functions*, 2nd ed. Cambridge University Press, 1944.

```
        else
                sign = -1;
        }
else
        sign = 1;
if( n == 0 )
        return( sign * y0(x) );
if( n == 1 )
        return( sign * y1(x) );
if( x <= 0.0 )
        {
        mtherr( "yn", SING );
        return( -MAXNUM );
        }
Forward recurrence on n
anm2 = y0(x);
anm1 = y1(x);
k = 1;
r = 2 * k;
do
        {
        an = r * anm1 / x - anm2;
        anm2 = anm1;
        anm1 = an;
        r += 2.0;
        ++k;
        }
while( k < n );
return( sign * an );
}
```

6.16 Testing

Table 6.5 gives typical test results for the Bessel function subroutines described in this chapter. The error criterion is absolute error for $J_0(x)$, $J_1(x)$, and $J_\nu(x)$. The criterion was relative error for those lines in the table having a star at the right. Figures for Y_0 and Y_1 indicate absolute error, except relative when $Y > 1$. The test interval for `iv.c` was $0 \le x \le 28, 0 \le v \le 30$. For $K_n(x)$, n and x ranged from 0 to 30; the error was high only near the crossover point $x = 9.55$ between the two expansions used.

$J_\nu(x)$ was subjected to several tests. Results for integer ν are indicated by "i", where x and ν both vary from -125 or -33 to +125. Otherwise, x ranged from 0 to 125, ν varied as indicated by "Domain." Error criterion

Table 6.5: Accuracies of Bessel Function Subroutines

Function	Arithmetic	Domain	Trials	Peak	RMS
J_0	DEC	$0, 30$	10000	$4.4 \cdot 10^{-17}$	$6.3 \cdot 10^{-18}$
J_0	IEEE	$0, 30$	60000	$4.2 \cdot 10^{-16}$	$1.1 \cdot 10^{-16}$
Y_0	DEC	$0, 30$	9400	$7.0 \cdot 10^{-17}$	$7.9 \cdot 10^{-18}$
Y_0	IEEE	$0, 30$	30000	$1.3 \cdot 10^{-15}$	$1.6 \cdot 10^{-16}$
J_1	DEC	$0, 30$	10000	$4.0 \cdot 10^{-17}$	$1.1 \cdot 10^{-17}$
J_1	IEEE	$0, 30$	30000	$2.6 \cdot 10^{-16}$	$1.1 \cdot 10^{-16}$
Y_1	DEC	$0, 30$	10000	$8.6 \cdot 10^{-17}$	$1.3 \cdot 10^{-17}$
Y_1	IEEE	$0, 30$	30000	$1.0 \cdot 10^{-15}$	$1.3 \cdot 10^{-16}$
Y_n	DEC	$0, 30$	2200	$2.9 \cdot 10^{-16}$	$5.3 \cdot 10^{-17}$
Y_n	IEEE	$0, 30$	30000	$3.4 \cdot 10^{-15}$	$4.3 \cdot 10^{-16}$
J_ν	DEC	$-33, 125$	5000	$4.0 \cdot 10^{-13}$	$6.8 \cdot 10^{-15}$
J_ν	DEC	$-33, 125$	6000	$2.2 \cdot 10^{-16}$i	$2.7 \cdot 10^{-17}$i
J_ν	IEEE	$0, 125$	90000	$2.3 \cdot 10^{-12}$	$1.7 \cdot 10^{-14}$
J_ν	IEEE	$-125, 0$	30000	$6.7 \cdot 10^{-11}$	$4.4 \cdot 10^{-13}$
J_ν	IEEE	$-+125$	30000	$2.4 \cdot 10^{-15}$i	$1.9 \cdot 10^{-16}$i
I_0	DEC	$0, 30$	6000	$8.2 \cdot 10^{-17}$*	$1.9 \cdot 10^{-17}$*
I_0	IEEE	$0, 30$	30000	$5.8 \cdot 10^{-16}$*	$1.4 \cdot 10^{-16}$*
I_1	DEC	$0, 30$	3400	$1.2 \cdot 10^{-16}$*	$2.3 \cdot 10^{-17}$*
I_1	IEEE	$0, 30$	30000	$1.9 \cdot 10^{-15}$*	$2.1 \cdot 10^{-16}$*
I_ν	DEC	$0, 30$	2000	$3.1 \cdot 10^{-15}$*	$5.4 \cdot 10^{-16}$*
I_ν	IEEE	$0, 30$	20000	$1.7 \cdot 10^{-14}$*	$2.7 \cdot 10^{-15}$*
K_0	DEC	$0, 30$	3100	$1.3 \cdot 10^{-16}$*	$2.1 \cdot 10^{-17}$*
K_0	IEEE	$0, 30$	30000	$1.2 \cdot 10^{-15}$*	$1.6 \cdot 10^{-16}$*
K_1	DEC	$0, 30$	3300	$8.9 \cdot 10^{-17}$*	$2.2 \cdot 10^{-17}$*
K_1	IEEE	$0, 30$	30000	$1.2 \cdot 10^{-16}$*	$1.6 \cdot 10^{-16}$*
K_n	DEC	$0, 30$	3000	$1.3 \cdot 10^{-9}$*	$5.8 \cdot 10^{-11}$*
K_n	IEEE	$0, 30$	90000	$1.8 \cdot 10^{-8}$*	$3.0 \cdot 10^{-10}$*
Ai	IEEE	$-10, 0$	10000	$1.6 \cdot 10^{-15}$	$2.7 \cdot 10^{-16}$
Ai	IEEE	$0, 10$	10000	$2.3 \cdot 10^{-14}$*	$1.8 \cdot 10^{-15}$*
Ai′	IEEE	$-10, 0$	10000	$4.6 \cdot 10^{-15}$	$7.6 \cdot 10^{-16}$
Ai′	IEEE	$0, 10$	10000	$1.8 \cdot 10^{-14}$*	$1.5 \cdot 10^{-15}$*
Bi	IEEE	$-10, 10$	30000	$4.2 \cdot 10^{-15}$	$5.3 \cdot 10^{-16}$
Bi′	IEEE	$-10, 10$	30000	$4.9 \cdot 10^{-15}$	$7.3 \cdot 10^{-16}$
Ai	DEC	$-10, 0$	5000	$1.7 \cdot 10^{-16}$	$2.8 \cdot 10^{-17}$
Ai	DEC	$0, 10$	5000	$2.1 \cdot 10^{-15}$*	$1.7 \cdot 10^{-16}$*
Ai′	DEC	$-10, 0$	5000	$4.7 \cdot 10^{-16}$	$7.8 \cdot 10^{-17}$
Ai′	DEC	$0, 10$	12000	$1.8 \cdot 10^{-15}$*	$1.5 \cdot 10^{-16}$*
Bi	DEC	$-10, 10$	10000	$5.5 \cdot 10^{-16}$	$6.8 \cdot 10^{-17}$
Bi′	DEC	$-10, 10$	7000	$5.3 \cdot 10^{-15}$	$8.7 \cdot 10^{-17}$

was absolute, except relative when $|J_\nu(x)| > 1$.

7

Other Special Functions

7.1 Hypergeometric Functions

A general hypergeometric function[1] may be written

$$ {}_pF_q(a_1,\ldots,a_p;b_1,\ldots,b_q;x) = \sum_{k=0}^{\infty} \frac{(a_1)_k\cdots(a_p)_k}{(b_1)_k\cdots(b_q)_k\; k!}\, x^k $$

where Pochhammer's symbol $(a)_k$ is defined by

$$ \begin{aligned} (a)_0 &= 1 \\ (a)_k &= a(a+1)\cdots(a+k-1) = \frac{\Gamma(a+k)}{\Gamma(a)}, \quad k > 0\,. \end{aligned} $$

There are numerous special cases such as the exponential function,

$$ \begin{aligned} e^x &= {}_0F_0(\,;\,;x) \\ &= \sum_{k=0}^{\infty} \frac{1}{k!}\, x^k \end{aligned} $$

Bessel function,

$$ J_\nu(x) = \frac{(\frac{1}{2}x)^\nu}{\Gamma(\nu+1)}\ {}_0F_1(\,;\nu+1;-\tfrac{1}{4}x^2) $$

incomplete gamma integral,

$$ P(a,x) = \frac{x^a}{\Gamma(a+1)}\,{}_1F_1(a;a+1;-x) $$

and incomplete beta integral,

$$ I_x(a,b) = \frac{x^a\Gamma(a+b)}{\Gamma(a+1)\Gamma(b)}\,{}_2F_1(a,1-b;a+1;x)\,. $$

[1] For a large collection of such functions, see Luke, Yudell L., *Integrals of Bessel Functions*, McGraw-Hill, 1962.

7.1.1 $_2F_1$

The Gauss hypergeometric series is defined by

$$
\begin{aligned}
&{}_2F_1(a,b;c;x) \\
&= 1+\sum_{k=1}^{\infty} \frac{a(a+1)\cdots(a+k-1)\; b(b+1)\cdots(b+k-1)}{c(c+1)\cdots(c+k-1)\; k!}\; x^k \\
&= \sum_{k=0}^{\infty} \frac{(a)_k (b)_k}{(c)_k\; k!}\; x^k \;.
\end{aligned}
$$

A program to compute the Gauss series should check for at least the following special cases. If c is a negative integer, the series will diverge unless either a or b is a negative integer $-m$ such that $c < -m < 0$. If a or b is a negative integer $-m$, the series reduces to a polynomial of degree m. There is divergence if $x = 1$ and $\Re(c-a-b) \le 0$. There is divergence if $|x| = 1$ and $\Re(c-a-b) \le -1$. If either $c-a$ or $c-b$ is a negative integer then the transformation

$$ {}_2F_1(a,b;c;x) = (1-x)^{c-a-b}{}_2F_1(c-a,c-b;c;x) $$

will reduce the series to a polynomial. By the same transformation, if $c = a$ or $c = b$, then

$$ {}_2F_1(a,b;c;x) = (1-x)^{c-a-b} \;. $$

For $x < -0.5$ and $b > 0$, use

$$ {}_2F_1(a,b;c;x) = (1-x)^{-a}{}_2F_1(a,c-b;c;-x/(1-x)) \;. $$

For $x < -0.5$ and $b <= 0$, use

$$ {}_2F_1(a,b;c;x) = (1-x)^{-b}{}_2F_1(c-a,b;c;-x/(1-x)) \;. $$

If $x > 0.75$ and $c-a-b$ is not an integer, use the transformation

$$
\begin{aligned}
&{}_2F_1(a,b;c;x) = \\
&\quad \frac{\Gamma(c)\,\Gamma(d)}{\Gamma(c-a)\,\Gamma(c-b)}\;{}_2F_1(a,b;1-d;1-x) \\
&\quad +(1-x)^d\,\frac{\Gamma(c)\Gamma(-d)}{\Gamma(a)\,\Gamma(b)}\;{}_2F_1(c-a,c-b;1+d;1-x)
\end{aligned}
$$

where $d = c-a-b$. If d is an integer and $x > 0.75$, the following expansion is useful. Set

$$ d_1 = \begin{cases} d, & d \ge 0 \\ 0, & d < 0 \;, \end{cases} $$

$$ d_2 = \begin{cases} 0, & d \ge 0 \\ d, & d < 0 \;. \end{cases} $$

Arithmetic	Domain	Trials	Peak	RMS
DEC	0, 7	500	$4.0 \cdot 10^{-11}$	$2.2 \cdot 10^{-12}$
IEEE	0, 7	30000	$9.3 \cdot 10^{-8}$	$1.1 \cdot 10^{-9}$

Table 7.1: `hyp2f1.c`, relative error

Then compute

$$F_1 = \frac{\Gamma(|d|)\,\Gamma(c)\,(1-x)^{d_2}}{\Gamma(a+d_1)\,\Gamma(b+d_1)} \sum_{k=0}^{|d|-1} \frac{(a+d_2)_k\,(b+d_2)_k}{(1+d_2)_k\,k!}(1-x)^k$$

$$F_2 = \frac{\Gamma(c)\,(1-x)^{d_1}}{\Gamma(a+d_2)\,\Gamma(b+d_2)} \sum_{k=0}^{\infty} \frac{(a+d_1)_k\,(b+d_1)_k\,(1-x)^k}{(|d|+k)!\,k!}\Psi_k$$

where

$$\Psi_k = \psi(1+k) + \psi(1+|d|+k) - \psi(a+d_1+k) - \psi(b+d_1+k) - \ln(1-x)$$

and $F_1 = 0$ if $d = 0$. The terms are combined to obtain

$${}_2F_1(a,b;c;x) = F_1 + (-1)^{|d|} F_2$$

(compare AMS55 §15.3.10-12). Table 7.1 shows the accuracy of the Gauss series computed with the above special cases taken into account. Relative errors are presented for all four arguments a, b, c, x ranging randomly from 0 to 7. Several special cases were also tested at a total of 17,000 random points with parameters in the range -7 to 7. Peak error in DEC arithmetic was $3.1 \cdot 10^{-12}$, rms error $1.6 \cdot 10^{-15}$. The program rejects $|x| > 1$ except in the cited special cases.

7.1.2 hyp2f1.c

Gauss hypergeometric function program

```
#include "mconf.h"
#ifdef DEC
#define EPS 1.0e-14
#define EPS2 1.0e-11
#endif
```

```
#ifdef IBMPC
#define EPS 1.0e-13
#define EPS2 1.0e-10
#endif

#ifdef MIEEE
#define EPS 1.0e-13
#define EPS2 1.0e-10
#endif

#ifdef UNK
#define EPS 1.0e-13
#define EPS2 1.0e-10
#endif

extern double MAXNUM;
double fabs(), pow(), round(), hys2f1(), gamma();
double psi(), log(), exp();

double hyp2f1( a, b, c, x )
double a, b, c, x;
    {
    double d, d1, d2, e;
    double p, q, r, s, t, y, y1, ax;
    double ia, ib, ic;
    int flag, i, aic;

    ax = fabs(x);
    s = 1.0 - x;
    flag = 0;
    if( a <= 0 )
        {
        ia = round(a); nearest integer to a
        if( fabs(a-ia) < EPS ) a is a negative integer
            flag |= 1;
        }
    if( b <= 0 )
        {
        ib = round(b);
        if( fabs(b-ib) < EPS ) b is a negative integer
            flag |= 2;
        }
    if( ax < 1.0 )
        {
        if( fabs(b-c) < EPS ) b = c
```

```
            {
            y = pow( s, -a ); s to the -a power
            goto hypdon;
            }
        if( fabs(a-c) < EPS ) a = c
            {
            y = pow( s, -b ); s to the -b power
            goto hypdon;
            }
        }
    if( c <= 0.0 )
        {
        ic = round(c); nearest integer to c
        if( fabs(c-ic) < EPS ) c is a negative integer
            {
check if termination before overflow
            if( (flag & 1) && (ia > ic) )
                goto hypok;
            if( (flag & 2) && (ib > ic) )
                goto hypok;
            goto hypdiv;
            }
        }
    if( flag ) function is a polynomial
        goto hypok;
    if( ax > 1.0 ) series diverges
        goto hypdiv;
    d = c - a - b;
    if( (fabs(ax)-1.0) < EPS ) |x| = 1.0
        {
        if( d <= -1.0 )
            goto hypdiv;
        if( (d <= 0.0) && (x > 0.0) )
            goto hypdiv;
        }
    p = c - a;
    ia = round(p);
    if( (ia <= 0.0) && (fabs(p-ia) < EPS) )
        goto hypf; c - a is a negative integer
    r = c - b;
    ib = round(r); nearest integer to r
    if( (ib <= 0.0) && (fabs(r-ib) < EPS) )
        goto hypf; c - b is a negative integer
    if( x < -0.5 )
        {
```

```
        if( r < c )
            y = pow( s, -a )
                * hys2f1( a, r, c, -x/s );
        else
            y = pow( s, -b )
                * hys2f1( p, b, c, -x/s );
        goto hypdon;
        }
    ic = round(d); nearest integer to ic
    if( x > 0.75 )
        {
        if( fabs(d-ic) > EPS2 )
            {
            q = hys2f1( a, b, 1.0-d, s );
Note, use signed log gamma function if available
            q *= gamma(d) /(gamma(c-a)
                * gamma(c-b));
            r = hys2f1( c-a, c-b, d+1.0, s );
            r *= gamma(-d)/(gamma(a)
                * gamma(b));
            r *= pow( s, d );
            y = (q + r) * gamma(c);
            goto hypdon;
            }
        else
            {
Psi function expansion
            if( ic >= 0.0 )
                {
                e = d;
                d1 = d;
                d2 = 0.0;
                aic = ic;
                }
            else
                {
                e = -d;
                d1 = 0.0;
                d2 = d;
                aic = -ic;
                }
            ax = log(s);
sum for t = 0
            y = psi(1.0) + psi(1.0+e) - psi(a+d1)
                - psi(b+d1) - ax;
```

```
		y /= gamma(e+1.0);
Pochhammer symbol for t=1
		p = (a+d1) * (b+d1) * s / gamma(e+2.0);
		t = 1.0;
		do
			{
			r = psi(1.0+t) + psi(1.0+t+e)
			- psi(a+t+d1) - psi(b+t+d1) - ax;
			q = p * r;
			y += q;
			p *= s * (a+t+d1) / (t+1.0);
			p *= (b+t+d1) / (t+1.0+e);
			t += 1.0;
			}
		while( fabs(q/y) > EPS );
		if( ic == 0.0 )
			{
			y *= gamma(c)
				/(gamma(a)*gamma(b));
			goto psidon;
			}
		y1 = 1.0;
		if( aic == 1 )
			goto nosum;
		t = 0.0;
		p = 1.0;
		for( i=1; i<aic; i++ )
			{
			r = 1.0-e+t;
			p *= s * (a+t+d2) * (b+t+d2) / r;
			t += 1.0;
			p /= t;
			y1 += p;
			}
nosum:
		p = gamma(c);
		y1 *= gamma(e) * p
			/ (gamma(a+d1) * gamma(b+d1));
		y *= p / (gamma(a+d2) * gamma(b+d2));
		if( (aic & 1) != 0 )
			y = -y;
		q = pow( s, ic ); s to the ic power
		if( ic > 0.0 )
			y *= q;
		else
```

```
                    y1 *= q;
                y += y1;
psidon:
                goto hypdon;
                }
            }
```

Defining power series if no special cases and $x \leq 0.5$

```
hypok:
    y = hys2f1( a, b, c, x );

hypdon:
    return(y);
```

The transformation for $c - a$ or $c - b$ a negative integer

```
hypf:
    y = pow( s, d ) * hys2f1( c-a, c-b, c, x );
    goto hypdon;
```

Alarm exit

```
hypdiv:
    mtherr( "hyp2f1", OVERFLOW );
    return( MAXNUM );
    }
```

Defining power series expansion of Gauss hypergeometric function

```
static double hys2f1( a, b, c, x )
double a, b, c, x;
    {
    double f, g, h, k, s, u;

    f = a;
    g = b;
    h = c;
    k = 1.0;
    s = 1.0;
    u = 1.0;
    do
        {
        if( fabs(h) < EPS )
            return( MAXNUM );
        u = u * (f * g * x / (h * k));
```

```
        s += u;
        f += 1.0;
        g += 1.0;
        h += 1.0;
        k += 1.0;
        }
    while( fabs(u/s) > EPS );
    return(s);
    }
```

7.1.3 $_1F_1$

The fundamental ascending series for the confluent hypergeometric function is

$$
\begin{aligned}
{}_1F_1(a;b;x) &= \sum_{k=0}^{\infty} \frac{(a)_k}{(b)_k\ k!} x^k \\
&= 1 + \sum_{k=1}^{\infty} \frac{a(a+1)\cdots(a+k-1)}{b(b+1)\cdots(b+k-1)\ k!} x^k \ .
\end{aligned}
$$

As is evident from the formula, b must not be a negative integer or zero unless a is a smaller nonpositive integer, $0 \geq a > b$. A number of transcendental functions are special cases of this power series. A computer routine for the confluent hypergeometric function may attempt both a direct summation of the series and an asymptotic expansion. In each case error due to roundoff, cancellation, and nonconvergence is estimated. The program returns the result with smaller estimated error. Cancellation error can be extremely severe. The program may estimate the error of its own output by keeping track of the largest value summed. A lower bound on the absolute error of the result is proportional to this largest value. Additional errors can arise in computing the terms themselves, as when $a+k$ or $b+k$ passes near zero.

Even though the series may be convergent, the intermediate terms may become quite large. Therefore it is advisable to check continually for overflow conditions. In accumulating the sum for any hypergeometric function, each term t_k is equal to the previous term multiplied by a function u_k of k. For $_1F_1$ this would be

$$
\begin{aligned}
u_k &= \frac{(a+k-1)\ x}{(b+k-1)\ k} \\
t_k &= t_{k-1}\ u_k \ .
\end{aligned}
$$

To avoid overflow, the magnitude of u_k may be compared with that of `MAXNUM`$/t_{k-1}$. If $|u_k|$ is greater, then the product $t_{k-1}u_k$ will overflow and the procedure cannot continue.

Arithmetic	Domain	Trials	Peak	RMS
DEC	0, 30	2000	$1.2 \cdot 10^{-15}$	$1.3 \cdot 10^{-16}$
IEEE	0, 30	30000	$1.8 \cdot 10^{-14}$	$1.1 \cdot 10^{-15}$

Table 7.2: Accuracy of Confluent Hypergeometric Function

The asymptotic expansion is

$$
\begin{aligned}
{}_1F_1(a;b;x) \quad &\approx \quad \frac{\Gamma(b)(-x)^{-a}}{\Gamma(b-a)}\ {}_2F_0(a, 1+a-b;;-1/x) \\
&\quad + \frac{\Gamma(b)e^x x^{a-b}}{\Gamma(a)}\ {}_2F_0(b-a, 1-a;;1/x)
\end{aligned}
$$

If x is large and negative, the first term tends to dominate because of the factor e^x in the second term. The second term dominates if x is large and positive. Hence the program takes only the dominant term. The fact that the other term may have an imaginary component thus becomes irrelevant. The ${}_2F_0$ series is divergent except in special cases. Summation stops when its terms reach a minimum absolute magnitude.

Table 7.2 gives relative error for random test points (a, b, x), all three variables ranging from 0 to 30. Larger errors can be observed when b is near a negative integer or zero. Certain combinations of arguments yield serious cancellation error in the power series summation and also are not in the region of near convergence of the asymptotic series. A warning message is printed if the self-estimated relative error is greater than $1.0 \cdot 10^{-12}$.

Under some circumstances a Kummer transformation such as

$$
{}_1F_1(a;b;x) = e^x\,{}_1F_1(b-a;b;-x)
$$

will help; but this has not been included in the program.

7.1.4 hyp1f1.c

Confluent hypergeometric function program

```
extern double MAXNUM, MACHEP;

double hyp1f1( a, b, x)
double a, b, x;
    {
```

```
	double asum, psum, temp, acanc, pcanc;
	double exp(), fabs(), hy1f1a(), hy1f1p();

	psum = hy1f1p( a, b, x, &pcanc );
	if( pcanc < 1.0e-15 )
		goto done;
```

Try the asymptotic series

```
	asum = hy1f1a( a, b, x, &acanc );
```

Pick the result with less estimated error

```
	if( acanc < pcanc )
		{
		pcanc = acanc;
		psum = asum;
		}
done:
	if( pcanc > 1.0e-12 )
		mtherr( "hyp1f1", PLOSS );
	return( psum );
	}
```

Power series summation for confluent hypergeometric function

```
static double hy1f1p( a, b, x, err )
double a, b, x;
double *err;
	{
	double n, a0, sum, t, u, temp;
	double fabs();
	double an, bn, maxt, pcanc;

	an = a;
	bn = b;
	a0 = 1.0;
	sum = 1.0;
	n = 1.0;
	t = 1.0;
	maxt = 0.0;
	while( t > MACHEP )
		{
```

check bn first since if both
an and *bn* are zero it is a singularity.

```
		if( bn == 0 )
			{
			mtherr( "hyp1f1", SING );
```

```
                return( MAXNUM );
                }
        if( an == 0 )
                return( sum );
        if( n > 200 )
                goto pdone;
        u = x * ( an / (bn * n) );
Check for overflow
        temp = fabs(u);
        if( (temp > 1.0 ) && (maxt > (MAXNUM/temp)) )
                {
                pcanc = 1.0; estimate 100% error
                goto overf;
                }
        a0 *= u;
        sum += a0;
        t = fabs(a0);
        if( t > maxt )
                maxt = t;
        an += 1.0;
        bn += 1.0;
        n += 1.0;
        }
pdone:
Rough estimate of error due to roundoff and cancellation
    if( sum != 0.0 )
        maxt /= fabs(sum);
    maxt *= MACHEP; this way avoids multiply overflow
    pcanc = fabs( MACHEP * n + maxt );
overf:
    *err = pcanc;
    return( sum );
    }


Asymptotic expansion of 1F1
static double hy1f1a( a, b, x, err )
double a, b, x;
double *err;
    {
    double h1, h2, t, u, temp, acanc, asum, err1, err2;
    double exp(), log(), gamma();
    double lgam(), fabs(), hyp2f0();

    if( x == 0 )
```

```
        {
        acanc = 1.0;
        asum = MAXNUM;
        goto adone;
        }
    temp = log( fabs(x) );
    t = x + temp * (a-b);
    u = -temp * a;
    if( b > 0 )
        {
        temp = lgam(b);
        t += temp;
        u += temp;
        }
    h1 = hyp2f0( a, a-b+1, -1.0/x, 1, &err1 );
    temp = exp(u) / gamma(b-a);
    h1 *= temp;
    err1 *= temp;
    h2 = hyp2f0( b-a, 1.0-a, 1.0/x, 2, &err2 );
    if( a < 0 )
        temp = exp(t) / gamma(a);
    else
        temp = exp( t - lgam(a) );
    h2 *= temp;
    err2 *= temp;
    if( x < 0 )
        asum = h1;
    else
        asum = h2;
    acanc = fabs(err1) + fabs(err2);
    if( b < 0 )
        {
        temp = gamma(b);
        asum *= temp;
        acanc *= fabs(temp);
        }
    if( asum != 0.0 )
        acanc /= fabs(asum);
Bias decision toward power series expansion
    acanc *= 30.0;
adone:
    *err = acanc;
    return( asum );
    }
```

7.1.5 $_2F_0$

This program sums the divergent series $_2F_0(a, b; ; x)$ up to the smallest term in the expansion.

7.1.6 hyp2f0.c

```
double hyp2f0( a, b, x, type, err )
double a, b, x;
int type; determines what converging factor to use
double *err;
      {
      double fabs();
      double a0, alast, t, tlast, maxt;
      double n, an, bn, u, sum, temp;

      an = a;
      bn = b;
      a0 = 1.0e0;
      alast = 1.0e0;
      sum = 0.0;
      n = 1.0e0;
      t = 1.0e0;
      tlast = 1.0e9;
      maxt = 0.0;
      do
            {
            if( an == 0 )
                  goto pdone;
            if( bn == 0 )
                  goto pdone;
            u = an * (bn * x / n);
check for overflow
            temp = fabs(u);
            if( (temp > 1.0 ) && (maxt > (MAXNUM/temp)) )
                  goto error;
            a0 *= u;
            t = fabs(a0);
terminating condition for asymptotic series
            if( t > tlast )
                  goto ndone;
            tlast = t;
            sum += alast; the sum is one term behind
            alast = a0;
            if( n > 200 )
```

```
                goto ndone;
            an += 1.0e0;
            bn += 1.0e0;
            n += 1.0e0;
            if( t > maxt )
                maxt = t;
            }
        while( t > MACHEP );
pdone:  Series converged to machine accuracy.
Estimate error due to roundoff and cancellation.
        *err = fabs( MACHEP * (n + maxt) );
        alast = a0;
        goto done;
ndone:  Series did not converge.
The following "converging factors" are supposed to improve accuracy.
        n -= 1.0;
        x = 1.0/x;
        switch( type )  type given as subroutine argument
            {
            case 1:
            alast *= 0.5 +
                (0.125 + 0.25*b - 0.5*a + 0.25*x - 0.25*n)/x;
            break;
            case 2:
            alast *= 2.0/3.0 - b + 2.0*a + x - n;
            break;
            default:
            }

Estimate error due to roundoff, cancellation, and nonconvergence.
        *err = MACHEP * (n + maxt) + fabs ( a0 );
done:
        sum += alast;
        return( sum );
error:
        *err = MAXNUM;
        mtherr( "hyp1f1", OVERFLOW );
        return( sum );
        }
```

7.2 Struve Functions

Struve functions have the general expansion

$$\begin{aligned}\mathbf{H}_\nu(x) &= (x/2)^{\nu+1}\sum_{k=0}^{\infty}\frac{(-1)^k(x/2)^{2k}}{\Gamma(k+\frac{3}{2})\Gamma(k+\nu+\frac{3}{2})}\\ &= \frac{(x/2)^{\nu+1}}{\Gamma(\frac{3}{2})\Gamma(\nu+\frac{3}{2})}{}_1F_2(1;\tfrac{3}{2},\nu+\tfrac{3}{2};-x^2/4)\ .\end{aligned}$$

A special case is

$$\mathbf{H}_{-n-\frac{1}{2}}(x) = (-1)^n J_{n+\frac{1}{2}}(x)\ .$$

The asymptotic expansion for large x is

$$\mathbf{H}_\nu(x) \approx Y_\nu(x) + \frac{(x/2)^{\nu-1}}{\Gamma(\frac{1}{2})\Gamma(\nu+\frac{1}{2})}{}_3F_0(1,\tfrac{1}{2},\tfrac{1}{2}-\nu;;-4/x^2)\ .$$

It is convenient to organize the calculation using routines for the hypergeometric functions (though the series are actually simpler special cases) and a routine to find $Y_\nu(x)$ for noninteger ν. The program rejects negative x unless ν is an integer. Also, the asymptotic formula is not very useful when $x < 0$.

7.2.1 hyp1f2.c

```
extern double PI, MACHEP;
static double stop = 1.37e-17;

double hyp1f2( a, b, c, x, err )
double a, b, c, x;
double *err;
    {
    double n, a0, sum, t;
    double an, bn, cn, max, z;
    double fabs();

    an = a;
    bn = b;
    cn = c;
    a0 = 1.0;
    sum = 1.0;
    n = 1.0;
    t = 1.0;
    max = 0.0;
    do
        {
```

```
        if( an == 0 )
            goto done;
        if( bn == 0 )
            goto error;
        if( cn == 0 )
            goto error;
        if( (a0 > 1.0e34) || (n > 200) )
            goto error;
        a0 *= (an * x) / (bn * cn * n);
        sum += a0;
        an += 1.0;
        bn += 1.0;
        cn += 1.0;
        n += 1.0;
        z = fabs( a0 );
        if( z > max )
            max = z;
        if( sum != 0 )
            t = fabs( a0 / sum );
        else
            t = z;
        }
    while( t > stop );

    done:
    *err = fabs( MACHEP*max /sum );
    goto xit;

    error:
    *err = 1.0e38;

    xit:
    return(sum);
    }
```

7.2.2 hyp3f0.c

```
double hyp3f0( a, b, c, x, err )
double a, b, c, x;
double *err;
    {
    double n, a0, sum, t, conv, conv1;
    double an, bn, cn, max, z;
    double fabs();
```

```
an = a;
bn = b;
cn = c;
a0 = 1.0;
sum = 1.0;
n = 1.0;
t = 1.0;
max = 0.0;
conv = 1.0e38;
conv1 = conv;
do
      {
      if( an == 0.0 )
            goto done;
      if( bn == 0.0 )
            goto done;
      if( cn == 0.0 )
            goto done;
      if( (a0 > 1.0e34) || (n > 200) )
            goto error;
      a0 *= (an * bn * cn * x) / n;
      an += 1.0;
      bn += 1.0;
      cn += 1.0;
      n += 1.0;
      z = fabs( a0 );
      if( z > max )
            max = z;
      if( z >= conv )
            {
            if( (z < max) && (z > conv1) )
                  goto done;
            }
      conv1 = conv;
      conv = z;
      sum += a0;
      if( sum != 0 )
            t = fabs( a0 / sum );
      else
            t = z;
      }
while( t > stop );

done:
t = fabs( MACHEP*max/sum );
```

```
    max = fabs( conv/sum );
    if( max > t )
        t = max;
    goto xit;

    error:
    t = 1.0e38;

    xit:
    *err = t;
    return(sum);
    }
```

7.2.3 yv.c

Bessel function of second kind of noninteger order.

```
static double yv( v, x )
double v, x;
    {
    double y, t;
    int n;
    double floor(), yn(), jv(), sin(), cos();

    y = floor( v );
    if( y == v )
        {
        n = v;
        y = yn( n, x );
        return( y );
        }
    t = PI * v;
    y = (cos(t) * jv( v, x ) - jv( -v, x ))/sin(t);
    return( y );
    }
```

7.2.4 struve.c

```
double struve( v, x )
double v, x;
    {
    double y, ya, f, g, h, t;
    double onef2err, threef0err;
    double gamma(), pow(), sqrt();
    double yv(), jv(), fabs(), floor();
```

```
f = floor(v);
if( (v < 0) && ( v-f == 0.5 ) )
      {
      y = jv( -v, x );
      f = 1.0 - f;
      g = 2.0 * floor(f/2.0);
      if( g != f )
            y = -y;
      return(y);
      }
t = 0.25*x*x;
f = fabs(x);
g = 1.5 * fabs(v);
if( (f > 30.0) && (f > g) )
      onef2err = 1.0e38;
else
      y = hyp1f2( 1.0, 1.5, 1.5+v, -t, &onef2err );
if( (f < 18.0) || (x < 0.0) )
      threef0err = 1.0e38;
else
      ya = hyp3f0( 1.0, 0.5, 0.5-v, -1.0/t, &threef0err );
f = sqrt( PI );
h = pow( 0.5*x, v-1.0 );

if( onef2err <= threef0err )
      {
      g = gamma( v + 1.5 );
      y = y * h * t / ( 0.5 * f * g );
      return(y);
      }
else
      {
      g = gamma( v + 0.5 );
      ya = ya * h / ( f * g );
      ya = ya + yv( v, x );
      return(ya);
      }
}
```

7.3 $\psi(x)$

Closely related to $\Gamma(x)$ is the psi function

$$\psi(x) = \frac{d}{dx} \ln \Gamma(x) \; .$$

If x is an integer, then

$$\begin{aligned}\psi(1) &= -\gamma \\ \psi(n) &= -\gamma + \sum_{k=1}^{n-1} \frac{1}{k} \quad , \; n > 1\end{aligned}$$

where $\gamma = 0.57721566490153286061$ is Euler's constant. This formula is suitable for computation when $1 \leq n \leq 10$. If x is negative, it may be transformed to a positive argument by the reflection formula

$$\psi(x) = \psi(1-x) - \frac{\pi}{\tan \pi x} \; .$$

Error conditions occur at the singularites for x equal to a negative integer or zero. For general positive x, the argument is made greater than 10 using the recurrence

$$\psi(x+1) = \psi(x) + \frac{1}{x}$$

so that an asymptotic expansion can be applied. This can be done by the following computer statements. Set $s = x$, $w = 0$. While $s < 10$,

$$\begin{aligned}w &:= w + \frac{1}{s} \\ s &:= s + 1 \; .\end{aligned}$$

Then the asymptotic expansion

$$\psi(x) \approx \ln x - \frac{1}{2x} - \sum_{k=1}^{\infty} \frac{B_{2k}}{2kx^{2k}} \; ,$$

where B_{2k} are Bernoulli numbers, is applied in the following form:

$$\psi(x) \approx \ln s - \frac{1}{2s} - s^{-2} P(s^{-2}) - w \; .$$

The coefficients are

$P(t) =$
$+8.33333333333333333333 \cdot 10^{-2} t^6$
$-2.10927960927960927961 \cdot 10^{-2} t^5$
$+7.57575757575757575758 \cdot 10^{-3} t^4$
$-4.16666666666666666667 \cdot 10^{-3} t^3$
$+3.96825396825396825397 \cdot 10^{-3} t^2$
$-8.33333333333333333333 \cdot 10^{-3} t$
$+8.33333333333333333333 \cdot 10^{-2}$.

If $s > 10^9$, then

$$\psi(x) \approx \ln s - \frac{1}{2s} - w \; .$$

Table 7.3 gives experimental test results for the psi function.

Arithmetic	Domain	Trials	Peak	RMS
DEC	0,30	2500	$1.7 \cdot 10^{-16}$	$2.0 \cdot 10^{-17}$
DEC	$-2, 0$	1000	$4.2 \cdot 10^{-15}$	$3.0 \cdot 10^{-16}$
IEEE	0,30	30000	$1.3 \cdot 10^{-15}$	$1.4 \cdot 10^{-16}$
IEEE	$-2, 0$	20000	$3.1 \cdot 10^{-12}$	$2.4 \cdot 10^{-14}$

Table 7.3: psi.c, relative error (absolute error, if < 1)

7.3.1 psi.c

```
#include "mconf.h"

#ifdef DEC
static short A[] = {
0037252,0125252,0125252,0125253,
0136654,0145314,0126312,0146255,
0036370,0037017,0101740,0174076,
0136210,0104210,0104210,0104211,
0036202,0004040,0101010,0020202,
0136410,0104210,0104210,0104211,
0037252,0125252,0125252,0125253
};
#endif

#ifdef IIEEE
static short A[] = {
0x5555,0x5555,0x5555,0x3fb5,
0x5996,0x9599,0x9959,0xbf95,
0x1f08,0xf07c,0x07c1,0x3f7f,
0x1111,0x1111,0x1111,0xbf71,
0x0410,0x1041,0x4104,0x3f70,
0x1111,0x1111,0x1111,0xbf81,
0x5555,0x5555,0x5555,0x3fb5
};
#endif


#define EUL 0.57721566490153286061
extern double PI, MAXNUM;

double psi(x)
double x;
    {
    double s, w, y, z;
    int i, n, flag;
    double floor(), log(), tan(), polevl();

    if( x <= 0.0 )
        {
```

Check for negative integer x.

```
        if( x == floor(x) )
            {
            mtherr( "psi", SING );
```

```
                return( MAXNUM );
                }
```

Use reflection formula if x is negative and not an integer.

```
            z = psi(1.0-x) - PI/tan(PI*x);
            return(z);
            }
```

Check for x a positive integer up to 10.

```
        if( (x <= 10.0) && (x == floor(x)) )
            {
            y = 0.0;
            n = x;
            for( i=1; i<n; i++ )
                {
                w = i;
                y += 1.0/w;
                }
            return( y - EUL );
            }
```

Transform x to be greater than 10.

```
        s = x;
        w = 0.0;
        while( s < 10.0 )
            {
            w += 1.0/s;
            s += 1.0;
            }
        if( s < 1.0e9 )
            {
```

Asymptotic expansion

```
            z = 1.0/(s * s);
            y = z * polevl( z, A, 6 );
            }
        else
            y = 0.0;
        y = log(s) - (0.5/s) - y - w;
        return(y);
        }
```

7.4 Exponential Integral

If $x > 0$ and $n \geq 0$ is an integer,

$$E_n(x) = \int_1^\infty t^{-n} e^{-xt} dt \; .$$

The power series expansion uses the ψ function

$$\begin{aligned}\psi(1) &= -\gamma\\ \psi(n) &= -\gamma + \sum_{k=1}^{n-1} \frac{1}{k}\end{aligned}$$

in terms of which

$$E_n(x) = \frac{(-x)^{n-1}[\psi(n) - \ln x]}{(n-1)!} - \sum_{k=0,k\neq n-1}^{\infty} \frac{(-x)^k}{(k-n+1)\; k!}\;.$$

The summation omits the term for $k = n - 1$. Singularities occur at $x = 0$ if $n = 0$ or $n = 1$. If $n > 1$,

$$E_n(0) = \frac{1}{n-1}\;.$$

If $n = 0$ the integral has the elementary solution

$$E_0(x) = \frac{e^{-x}}{x}\;.$$

When $x > 1$ the following continued fraction is preferable to the power series.

$$e^x\; E_n(x) = \frac{1}{x+}\;\frac{n}{1+}\;\frac{1}{x+}\;\frac{n+1}{1+}\;\frac{2}{x+}\cdots$$

If n is large ($n > 5000$ with double precision arithmetic) use the asymptotic expansion

$$E_n(x) \approx \frac{e^{-x}}{x+n}\left(1 + \frac{n}{(x+n)^2} + \frac{n(n-2x)}{(x+n)^4} + \frac{n(6x^2 - 8nx + n^2)}{(x+n)^6}\right)\;.$$

If x is large compared to n, the asymptotic expansion is

$$E_n(x) \approx \frac{e^{-x}}{x}\left(1 - \frac{n}{x} + \frac{n(n+1)}{x^2} - \frac{n(n+1)(n+2)}{x^3} + \ldots\right)\;.$$

The continued fraction seems to handle large x adequately, so the program does not use this expansion. Table 7.4 shows typical test results when n and x range from 0 to 30.

7.4.1 en.c

Exponential integral program

```
#define EUL 0.57721566490153286060G5
#define BIG 1.44115188075855872E+17
extern double MAXNUM, MACHEP, MAXLOG;
```

Arithmetic	Domain	Trials	Peak	RMS
DEC	0, 30	5000	$2.0 \cdot 10^{-16}$	$4.6 \cdot 10^{-17}$
IEEE	0, 30	10000	$1.7 \cdot 10^{-15}$	$3.6 \cdot 10^{-16}$

Table 7.4: en(n, x), relative error

```
double en( n, x )
int n;
double x;
    {
    double ans, r, t, yk, xk;
    double pk, pkm1, pkm2, qk, qkm1, qkm2;
    double psi, z;
    int i, k;
    static double big = BIG;
    double pow(), gamma(), log(), exp(), fabs();

    if( (n < 0) || (x < 0) )
        {
        mtherr( "expn", DOMAIN );
        return( MAXNUM );
        }
    if( x > MAXLOG )
        return( 0.0 );
    if( x == 0.0 )
        {
        if( n < 2 )
            {
            mtherr( "expn", SING );
            return( MAXNUM );
            }
        else
            return( 1.0/(n-1.0) );
        }
    if( n == 0 )
        return( exp(-x)/x );
```
Expansion for large n

```
if( n > 5000 )
	{
	xk = x + n;
	yk = 1.0 / (xk * xk);
	t = n;
	ans = yk * t * (6.0 * x * x - 8.0 * t * x + t * t);
	ans = yk * (ans + t * (t - 2.0 * x));
	ans = yk * (ans + t);
	ans = (ans + 1.0) * exp( -x ) / xk;
	goto done;
	}
if( x > 1.0 )
	goto cfrac;
```

Power series expansion

```
psi = -EUL - log(x);
for( i=1; i<n; i++ )
	psi = psi + 1.0/i;
z = -x;
xk = 0.0;
yk = 1.0;
pk = 1.0 - n;
if( n == 1 )
	ans = 0.0;
else
	ans = 1.0/pk;
do
	{
	xk += 1.0;
	yk *= z/xk;
	pk += 1.0;
	if( pk != 0.0 )
		ans += yk/pk;
	if( ans != 0.0 )
		t = fabs(yk/ans);
	else
		t = 1.0;
	}
while( t > MACHEP );
t = n;
r = n - 1;
ans = (pow(z, r) * psi / gamma(t)) - ans;
goto done;
```

Continued fraction

```
cfrac:
	k = 1;
```

```
    pkm2 = 1.0;
    qkm2 = x;
    pkm1 = 1.0;
    qkm1 = x + n;
    ans = pkm1/qkm1;
    do
        {
        k += 1;
        if( k & 1 )
            {
            yk = 1.0;
            xk = n + (k-1)/2;
            }
        else
            {
            yk = x;
            xk = k/2;
            }
        pk = pkm1 * yk + pkm2 * xk;
        qk = qkm1 * yk + qkm2 * xk;
        if( qk != 0 )
            {
            r = pk/qk;
            t = fabs( (ans - r)/r );
            ans = r;
            }
        else
            t = 1.0;
        pkm2 = pkm1;
        pkm1 = pk;
        qkm2 = qkm1;
        qkm1 = qk;
        if( fabs(pk) > big )
            {
            pkm2 /= big;
            pkm1 /= big;
            qkm2 /= big;
            qkm1 /= big;
            }
        }
    while( t > MACHEP );
    ans *= exp( -x );
done:
    return( ans );
    }
```

7.5 Sine and Cosine Integrals

The sine integral $\mathrm{Si}(x)$ has the ascending power series

$$\begin{aligned} \mathrm{Si}(x) &= \int_0^x \frac{\sin t}{t} dt \\ &= \sum_{k=0}^{\infty} \frac{x(-x^2)^k}{(2k+1)(2k+1)!} . \end{aligned}$$

For the cosine integral $\mathrm{Ci}(x)$ the expansion is

$$\begin{aligned} \mathrm{Ci}(x) &= \gamma + \ln x + \int_0^x \frac{\cos t \ - 1}{t} dt \\ &= \gamma + \ln x + \sum_{k=1}^{\infty} \frac{(-x^2)^k}{2k(2k)!} \end{aligned}$$

where $\gamma = 0.57721566490153286061$ is Euler's constant. If x is relatively small, the integrals are approximated directly by rational functions.

$$\mathrm{Si}(x) \approx x \, \frac{SN(x^2)}{SD(x^2)}$$

where

$$\begin{aligned} SN(t) = \; & -8.39167827910303881427 \cdot 10^{-11} t^5 \\ & +4.62591714427012837309 \cdot 10^{-8} t^4 \\ & -9.75759303843632795789 \cdot 10^{-6} t^3 \\ & +9.76945438170435310816 \cdot 10^{-4} t^2 \\ & -4.13470316229406538752 \cdot 10^{-2} t \\ & +1.00000000000000000302 \ , \end{aligned}$$

$$\begin{aligned} SD(t) = \; & +2.03269266195951942049 \cdot 10^{-12} t^5 \\ & +1.27997891179943299903 \cdot 10^{-9} t^4 \\ & +4.41827842801218905784 \cdot 10^{-7} t^3 \\ & +9.96412122043875552487 \cdot 10^{-5} t^2 \\ & +1.42085239326149893930 \cdot 10^{-2} t \\ & +0.99999999999999996984 \ . \end{aligned}$$

This holds for $0 \le x \le 4$. In the same interval, the approximation for Ci is

$$\mathrm{Ci}(x) \approx \gamma + \ln x + \frac{x^2 CN(x^2)}{CD(x^2)}$$

where

$$\begin{aligned} CN(t) = \; & 2.02524002389102268789 \cdot 10^{-11} t^5 \\ & -1.35249504915790756375 \cdot 10^{-8} t^4 \\ & +3.59325051419993077021 \cdot 10^{-6} t^3 \\ & -4.74007206873407909465 \cdot 10^{-4} t^2 \\ & +2.89159652607555242092 \cdot 10^{-2} t \\ & -1.00000000000000000080 \ , \end{aligned}$$

$$\begin{aligned} CD(t) = \; & +4.07746040061880559506 \cdot 10^{-12} t^5 \\ & +3.06780997581887812692 \cdot 10^{-9} t^4 \\ & +1.23210355685883423679 \cdot 10^{-6} t^3 \\ & +3.17442024775032769882 \cdot 10^{-4} t^2 \\ & +5.10028056236446052392 \cdot 10^{-2} t \\ & +4.00000000000000000080 \ . \end{aligned}$$

For moderate to large x auxiliary functions $f(x)$ and $g(x)$ are employed such that

$$\begin{aligned}\mathrm{Si}(x) &= \pi/2 - f(x)\cos x - g(x)\sin x\\ \mathrm{Ci}(x) &= f(x)\sin x - g(x)\cos x\ .\end{aligned}$$

For large x these functions have the asymptotic expansions

$$f(x) \approx \frac{1}{x}\left(1 - \frac{2!}{x^2} + \frac{4!}{x^4} - \frac{6!}{x^6} + \ldots\right)$$

$$g(x) \approx \frac{1}{x^2}\left(1 - \frac{3!}{x^2} + \frac{5!}{x^4} - \frac{7!}{x^6} + \ldots\right)\ .$$

The minimax rational approximations of this form have all positive coefficients, so are numerically better behaved than the approximations to the ascending series. For $4 \le x \le 8$,

$$\mathrm{Si}(x) \approx \pi/2 - (1/x)\frac{FN_4(1/x^2)}{FD_4(1/x^2)}\cos x - (1/x^2)\frac{GN_4(1/x^2)}{GD_4(1/x^2)}\sin x$$

where

$FN_4(t) =$
$+4.23612862892216586994 \cdot 10^{0}t^{6}$
$+5.45937717161812843388 \cdot 10^{0}t^{5}$
$+1.62083287701538329132 \cdot 10^{0}t^{4}$
$+1.67006611831323023771 \cdot 10^{-1}t^{3}$
$+6.81020132472518137426 \cdot 10^{-3}t^{2}$
$+1.08936580650328664411 \cdot 10^{-4}t$
$+5.48900223421373614008 \cdot 10^{-7}$,

$FD_4(t) =$
$+1.0t^{7}$
$+8.16496634205391016773 \cdot 10^{0}t^{6}$
$+7.30828822505564552187 \cdot 10^{0}t^{5}$
$+1.86792257950184183883 \cdot 10^{0}t^{4}$
$+1.78792052963149907262 \cdot 10^{-1}t^{3}$
$+7.01710668322789753610 \cdot 10^{-3}t^{2}$
$+1.10034357153915731354 \cdot 10^{-4}t$
$+5.48900252756255700982 \cdot 10^{-7}$;

$$\mathrm{Ci}(x) \approx (1/x)\frac{FN_4(1/x^2)}{FD_4(1/x^2)}\sin x - (1/x^2)\frac{GN_4(1/x^2)}{GD_4(1/x^2)}\cos x$$

$GN_4(t) =$
$+8.71001698973114191777 \cdot 10^{-2}t^{7}$
$+6.11379109952219284151 \cdot 10^{-1}t^{6}$
$+3.97180296392337498885 \cdot 10^{-1}t^{5}$
$+7.48527737628469092119 \cdot 10^{-2}t^{4}$
$+5.38868681462177273157 \cdot 10^{-3}t^{3}$
$+1.61999794598934024525 \cdot 10^{-4}t^{2}$
$+1.97963874140963632189 \cdot 10^{-6}t$
$+7.82579040744090311069 \cdot 10^{-9}$,

$GD_4(t) =$
$+1.0t^{7}$
$+1.64402202413355338886 \cdot 10^{0}t^{6}$
$+6.66296701268987968381 \cdot 10^{-1}t^{5}$
$+9.88771761277688796203 \cdot 10^{-2}t^{4}$
$+6.22396345441768420760 \cdot 10^{-3}t^{3}$
$+1.73221081474177119497 \cdot 10^{-4}t^{2}$
$+2.02659182086343991969 \cdot 10^{-6}t$
$+7.82579218933534490868 \cdot 10^{-9}$.

For $8 \le x \le \infty$, the approximations are

$$\mathrm{Si}(x) \approx \pi/2 - (1/x)\frac{FN_8(1/x^2)}{FD_8(1/x^2)}\cos x - (1/x^2)\frac{GN_8(1/x^2)}{GD_8(1/x^2)}\sin x$$

where

$$\begin{array}{ll}
FN_8(t) = & FD_8(t) = \\
+4.5588087347046531520 6\cdot 10^{-1}t^8 & +1.0t^8 \\
+7.1371527410014671137 4\cdot 10^{-1}t^7 & +9.1746361187368405370 3\cdot 10^{-1}t^7 \\
+1.6030015822231945632 0\cdot 10^{-1}t^6 & +1.7868554533207453632 1\cdot 10^{-1}t^6 \\
+1.1606422940812440791 5\cdot 10^{-2}t^5 & +1.2225359477197129303 2\cdot 10^{-2}t^5 \\
+3.4955644244785905560 5\cdot 10^{-4}t^4 & +3.5869648188185158029 7\cdot 10^{-4}t^4 \\
+4.8621543082645474948 2\cdot 10^{-6}t^3 & +4.9243506431788146439 3\cdot 10^{-6}t^3 \\
+3.2009279009100490280 6\cdot 10^{-8}t^2 & +3.2195693910104601837 7\cdot 10^{-8}t^2 \\
+9.4177957612851293659 2\cdot 10^{-11}t & +9.4372059035027673237 6\cdot 10^{-11}t \\
+9.7050711088195202463 1\cdot 10^{-14}\ , & +9.7050711088195202572 5\cdot 10^{-14}\ ;
\end{array}$$

finally,

$$\mathrm{Ci}(x) \approx (1/x)\frac{FN_8(1/x^2)}{FD_8(1/x^2)}\sin x - (1/x^2)\frac{GN_8(1/x^2)}{GD_8(1/x^2)}\cos x$$

$$\begin{array}{ll}
GN_8(t) = & GD_8(t) = \\
& 1.0t^9 \\
+6.9735995344327621493 4\cdot 10^{-1}t^8 & +1.6854889881101164001 7\cdot 10^{0}t^8 \\
+3.3041097930563206322 5\cdot 10^{-1}t^7 & +4.8785225869530496748 6\cdot 10^{-1}t^7 \\
+3.8487876764997429592 0\cdot 10^{-2}t^6 & +4.6791319425962580632 0\cdot 10^{-2}t^6 \\
+1.7171823905234790355 8\cdot 10^{-3}t^5 & +1.9028442667439952363 8\cdot 10^{-3}t^5 \\
+3.4894116550227943677 7\cdot 10^{-5}t^4 & +3.6847550444256110816 2\cdot 10^{-5}t^4 \\
+3.4713116708411667380 0\cdot 10^{-7}t^3 & +3.5704322344374083877 1\cdot 10^{-7}t^3 \\
+1.7040445278204452618 9\cdot 10^{-9}t^2 & +1.7269374896631614673 6\cdot 10^{-9}t^2 \\
+3.8594592543027660045 3\cdot 10^{-12}t & +3.8783016602395470675 2\cdot 10^{-12}t \\
+3.1404009894636333464 0\cdot 10^{-15}\ , & +3.1404009894636333524 2\cdot 10^{-15}\ .
\end{array}$$

If x is very large, the asymptotic expansions simplify to

$$\begin{aligned}
\mathrm{Si}(x) &\approx \frac{\pi}{2} - \frac{\cos x}{x} \\
\mathrm{Ci}(x) &\approx \frac{\sin x}{x}\ .
\end{aligned}$$

If $x < 0$,

$$\begin{aligned}
\mathrm{Si}(x) &= -\mathrm{Si}(-x) \\
\mathrm{Ci}(x) &= \mathrm{Ci}(-x) + i\pi\ .
\end{aligned}$$

Test results are shown in Table 7.5.

7.5.1 sici.c

Sine and cosine integral program

Arithmetic	Function	Trials	Peak	RMS
IEEE	Si	30000	$4.4 \cdot 10^{-16}$	$7.3 \cdot 10^{-17}$
IEEE	Ci	30000	$6.9 \cdot 10^{-16}$	$5.1 \cdot 10^{-17}$
DEC	Si	5000	$4.4 \cdot 10^{-17}$	$9.0 \cdot 10^{-18}$
DEC	Ci	5300	$7.9 \cdot 10^{-17}$	$5.2 \cdot 10^{-18}$

Table 7.5: `sici.c`, absolute error except relative when the function is greater than 1. Argument ranges from 0 to 50.

```
#include "mconf.h"

#if DEC
static short SN[] = {
0127670,0104362,0167505,0035161,
0032106,0127177,0032131,0056461,
0134043,0132213,0000476,0172351,
0035600,0006331,0064761,0032665,
0137051,0055601,0044667,0017645,
0040200,0000000,0000000,0000000,
};
static short SD[] = {
0026417,0004674,0052064,0001573,
0030657,0165501,0014666,0131526,
0032755,0032133,0034147,0024124,
0034720,0173167,0166624,0154477,
0036550,0145336,0063534,0063220,
0040200,0000000,0000000,0000000,
};
static short CN[] = {
0027262,0022131,0160257,0020166,
0131550,0055534,0077637,0000557,
0033561,0021622,0161463,0026575,
0135370,0102053,0116333,0000466,
0036754,0160454,0122022,0024622,
0140200,0000000,0000000,0000000,
};
static short CD[] = {
0026617,0073177,0107543,0104425,
0031122,0150573,0156453,0041517,
0033245,0057301,0077706,0110510,
0035246,0067130,0165424,0044543,

#if IIEEE
static short SN[] = {
0xa74e,0x5de8,0x111e,0xbdd7,
0x2ba6,0xe68b,0xd5cf,0x3e68,
0xde9d,0x6027,0x7691,0xbee4,
0x26b7,0x2d3e,0x019b,0x3f50,
0xe3f5,0x2936,0x2b70,0xbfa5,
0x0000,0x0000,0x0000,0x3ff0,
};
static short SD[] = {
0x806f,0x8a86,0xe137,0x3d81,
0xd66b,0x2336,0xfd68,0x3e15,
0xe50a,0x670c,0xa68b,0x3e9d,
0x9b28,0xfdb2,0x1ece,0x3f1a,
0x8cd2,0xcceb,0x195b,0x3f8d,
0x0000,0x0000,0x0000,0x3ff0,
};
static short CN[] = {
0xe40f,0x3c15,0x448b,0x3db6,
0xe02e,0x8ff3,0x0b6b,0xbe4d,
0x65b0,0x5c66,0x2472,0x3ece,
0x6027,0x739b,0x1085,0xbf3f,
0x4532,0x9482,0x9c25,0x3f9d,
0x0000,0x0000,0x0000,0xbff0,
};
static short CD[] = {
0x7123,0xf1ec,0xeecf,0x3d91,
0x686a,0x7ba5,0x5a2f,0x3e2a,
0xd229,0x2ff8,0xabd8,0x3eb4,
0x892c,0x1d62,0xcdcb,0x3f34,
```

```
0037120,0164121,0061206,0053657,
0040600,0000000,0000000,0000000,
};
static short FN4[] = {
0040607,0107135,0120133,0153471,
0040656,0131467,0140424,0017567,
0040317,0073563,0121610,0002511,
0037453,0001710,0000040,0006334,
0036337,0024033,0176003,0171425,
0034744,0072341,0121657,0126035,
0033023,0054042,0154652,0000451,
};
```

Leading 1.0 omitted from FD4.

```
static short FD4[] = {
0041002,0121663,0137500,0177450,
0040751,0156577,0042213,0061552,
0040357,0014026,0045465,0147265,
0037467,0012503,0110413,0131772,
0036345,0167701,0155706,0160551,
0034746,0141076,0162250,0123547,
0033023,0054043,0056706,0151050,
};
static short FN8[] = {
0037751,0064467,0142332,0164573,
0040066,0133013,0050352,0071102,
0037444,0022671,0102157,0013535,
0036476,0024335,0136423,0146444,
0035267,0042253,0164110,0110460,
0033643,0022626,0062535,0060320,
0032011,0075223,0010110,0153413,
0027717,0014572,0011360,0014034,
0025332,0104755,0004563,0152354,
};
```

Leading 1.0 omitted from FD8.

```
static short FD8[] = {
0040152,0157345,0030104,0075616,
0037466,0174527,0172740,0071060,
0036510,0046337,0144272,0156552,
0035274,0007555,0042537,0015572,
0033645,0035731,0112465,0026474,
0032012,0043612,0030613,0030123,
0027717,0103277,0004564,0151000,
0025332,0104755,0004563,0152354,
};
static short GN4[] = {
0037262,0060622,0164572,0157515,
0040034,0101527,0061263,0147204,
0037713,0055467,0037475,0144512,
0037231,0046151,0035234,0045261,
```

```
0xcaf6,0x2c50,0x1d0a,0x3faa,
0x0000,0x0000,0x0000,0x4010,
};
static short FN4[] = {
0x7ae7,0xb40b,0xf1cb,0x4010,
0x83ef,0xf822,0xd666,0x4015,
0x00a9,0x7471,0xeeee,0x3ff9,
0x019c,0x0004,0x6079,0x3fc5,
0x7e63,0x7f80,0xe503,0x3f7b,
0xf584,0x3475,0x8e9c,0x3f1c,
0x4025,0x5b35,0x6b04,0x3ea2,
};
static short FD4[] = {
0x1fe5,0x77e8,0x5476,0x4020,
0x6c6d,0xe891,0x3baf,0x401d,
0xb9d7,0xc966,0xe302,0x3ffd,
0x767f,0x7221,0xe2a8,0x3fc6,
0xdc2d,0x3b78,0xbdf8,0x3f7c,
0x14ed,0xdc95,0xd847,0x3f1c,
0xda45,0x6bb8,0x6b04,0x3ea2,
};
static short FN8[] = {
0x5d2f,0xf89b,0x2d26,0x3fdd,
0x4e48,0x6a1d,0xd6c1,0x3fe6,
0xe2ec,0x308d,0x84b7,0x3fc4,
0x79a4,0xb7a2,0xc51b,0x3f87,
0x1226,0x7d09,0xe895,0x3f36,
0xac1a,0xccab,0x64b2,0x3ed4,
0x1ae1,0x6209,0x2f52,0x3e61,
0x0304,0x425e,0xe32f,0x3dd9,
0x7a9d,0xa12e,0x513d,0x3d3b,
};
static short FD8[] = {
0x8f72,0xa608,0x5bdc,0x3fed,
0x0e46,0xfebc,0xdf2a,0x3fc6,
0x5bad,0xf917,0x099b,0x3f89,
0xe36f,0xa8ab,0x81ed,0x3f37,
0xa5a8,0x32a6,0xa77b,0x3ed4,
0x660a,0x4631,0x48f1,0x3e61,
0x9a40,0xe12e,0xf0d7,0x3dd9,
0x7a9d,0xa12e,0x513d,0x3d3b,
};
static short GN4[] = {
0x5bea,0x5d2f,0x4c32,0x3fb6,
0x79d1,0xec56,0x906a,0x3fe3,
0xb929,0xe7e7,0x6b66,0x3fd9,
0x8956,0x2753,0x298d,0x3fb3,
```

```
0036260,0111624,0150617,0053536,
0035051,0157175,0016675,0155456,
0033404,0154757,0041211,0000055,
0031406,0071060,0130322,0033322,
};
```
Leading 1.0 omitted from GD4.
```
static short GD4[] = {
0040322,0067520,0046707,0053275,
0040052,0111153,0126542,0005516,
0037312,0100035,0167121,0014552,
0036313,0171143,0137176,0014213,
0035065,0121256,0012033,0150603,
0033410,0000225,0013121,0071643,
0031406,0071062,0131152,0150454,
};
static short GN8[] = {
0040062,0103056,0110624,0033123,
0037651,0025640,0136266,0145647,
0037035,0122566,0137770,0061777,
0035741,0011424,0065311,0013370,
0034422,0055505,0134324,0016755,
0032672,0056530,0022565,0014747,
0030752,0031674,0114735,0013162,
0026607,0145353,0022020,0123625,
0024142,0045054,0060033,0016505,
};
```
Leading 1.0 omitted from GD8.
```
static short GD8[] = {
0040327,0137032,0064331,0136425,
0037771,0143705,0070300,0105711,
0037077,0124101,0025275,0035356,
0035771,0064333,0145103,0105357,
0034432,0106301,0105311,0010713,
0032677,0127645,0120034,0157551,
0030755,0054466,0010743,0105566,
0026610,0072242,0142530,0135744,
0024142,0045054,0060033,0016505,
};
#endif
```

```
0xeaec,0x9a31,0x1272,0x3f76,
0xbb66,0xa3b7,0x3bcf,0x3f25,
0x2006,0xe851,0x9b3d,0x3ec0,
0x46da,0x161a,0xce46,0x3e40,
};

static short GD4[] = {
0xead8,0x09b8,0x4dea,0x3ffa,
0x416a,0x75ac,0x524d,0x3fe5,
0x232d,0xbdca,0x5003,0x3fb9,
0xc311,0x77cf,0x7e4c,0x3f79,
0x7a30,0xc283,0xb455,0x3f26,
0x2e74,0xa2ca,0x0012,0x3ec1,
0x5a26,0x564d,0xce46,0x3e40,
};
static short GN8[] = {
0x86ca,0xd232,0x50c5,0x3fe6,
0xd975,0x1796,0x2574,0x3fd5,
0x0c80,0xd7ff,0xb4ae,0x3fa3,
0x22df,0x8d59,0x2262,0x3f5c,
0x83be,0xb71a,0x4b68,0x3f02,
0xa33d,0x04ae,0x4bab,0x3e97,
0xa2ce,0x933b,0x4677,0x3e1d,
0x14f3,0x6482,0xf95d,0x3d90,
0x63a9,0x8c03,0x4945,0x3cec,
};

static short GD8[] = {
0x37a3,0x4d1b,0xf7c3,0x3ffa,
0x1179,0xae18,0x38f8,0x3fdf,
0xa75e,0x2557,0xf508,0x3fa7,
0x715e,0x7948,0x2d1b,0x3f5f,
0x2239,0x3159,0x5198,0x3f03,
0x9bed,0xb403,0xf5f4,0x3e97,
0x716f,0xc23c,0xab26,0x3e1d,
0x177c,0x58ab,0x0e94,0x3d91,
0x63a9,0x8c03,0x4945,0x3cec,
};
#endif
```

```
#define EUL 0.57721566490153286061
extern double MAXNUM, PIO2, MACHEP;

int sici( x, si, ci )
double x;
double *si, *ci;
    {
```

```
double z, c, s, f, g;
short sign;
double log(), sin(), cos(), polevl(), p1evl();

if( x < 0.0 )
        {
        sign = -1;
        x = -x;
        }
else
        sign = 0;
if( x == 0.0 )
        {
        *si = 0.0;
        *ci = -MAXNUM;
        return( 0 );
        }
if( x > 1.0e9 )
        {
        *si = PIO2 - cos(x)/x;
        *ci = sin(x)/x;
        }
if( x > 4.0 )
        goto asympt;
z = x * x;
s = x * polevl( z, SN, 5 ) / polevl( z, SD, 5 );
c = z * polevl( z, CN, 5 ) / polevl( z, CD, 5 );
if( sign )
        s = -s;
*si = s;
```

Return real part of Ci if $x < 0$

```
*ci = EUL + log(x) + c;
return(0);
asympt:
s = sin(x);
c = cos(x);
z = 1.0/(x*x);
if( x < 8.0 )
        {
        f = polevl( z, FN4, 6 ) / (x * p1evl( z, FD4, 7 ));
        g = z * polevl( z, GN4, 7 ) / p1evl( z, GD4, 7 );
        }
else
        {
        f = polevl( z, FN8, 8 ) / (x * p1evl( z, FD8, 8 ));
```

```
        g = z * polevl( z, GN8, 8 ) / p1evl( z, GD8, 9 );
        }
    *si = PIO2 - f * c - g * s;
    if( sign )
        *si = -( *si );
    *ci = f * s - g * c;
    return(0);
    }
```

7.5.2 Hyperbolic Sine and Cosine Integrals

The definitions are

$$\mathrm{Chi}(x) = \gamma + \ln x + \int_0^x \frac{\cosh t\ - 1}{t} dt$$

and

$$\mathrm{Shi}(x) = \int_0^x \frac{\sinh t}{t} dt \ .$$

For *a priori* computation the power series expansions are

$$\mathrm{Shi}(x) = \sum_{k=0}^{\infty} \frac{x^{2k+1}}{(2k+1)(2k+1)!}$$

and

$$\mathrm{Chi}(x) = \gamma + \ln x + \sum_{k=1}^{\infty} \frac{x^{2k}}{2k(2k)!} \ .$$

For large x both functions have the asymptotic expansion

$$\mathrm{Chi}(x) \approx \mathrm{Shi}(x) \approx \frac{e^x}{2x} \sum_{k=0}^{\infty} \frac{k!}{x^k}$$

which converges to double precision accuracy if $x > 42$.

Since the terms are all positive, there is no cancellation error and both power series expansions are suitable for computation. For moderate x use the following Chebyshev expansions. If $8 \le x < 18$,

$$\begin{aligned} w &= \frac{144/x - 13}{5} \\ \mathrm{Shi}(x) &= \frac{e^x}{x} \sum_{k=0}^{\infty}{}' S1_k T_k(w) \\ \mathrm{Chi}(x) &= \gamma + \ln x + \frac{e^x}{x} \sum_{k=0}^{\infty}{}' C1_k T_k(w) \ . \end{aligned}$$

Table 7.6: Chebyshev Coefficients for Shi(x)

k	$S1_k$	$S2_k$
22		$-1.053 \cdot 10^{-17}$
21	$+1.839 \cdot 10^{-17}$	$+2.624 \cdot 10^{-17}$
20	$-9.555 \cdot 10^{-17}$	$+8.821 \cdot 10^{-17}$
19	$+2.0433 \cdot 10^{-16}$	$-3.3846 \cdot 10^{-16}$
18	$+1.09897 \cdot 10^{-15}$	$-8.3061 \cdot 10^{-16}$
17	$-1.313135 \cdot 10^{-14}$	$+3.93398 \cdot 10^{-15}$
16	$+5.939762 \cdot 10^{-14}$	$+1.017656 \cdot 10^{-14}$
15	$-3.471970 \cdot 10^{-14}$	$-4.211282 \cdot 10^{-14}$
14	$-1.40059765 \cdot 10^{-12}$	$-1.6081820 \cdot 10^{-13}$
13	$+9.49044626 \cdot 10^{-12}$	$+3.3471495 \cdot 10^{-13}$
12	$-1.615961811 \cdot 10^{-11}$	$+2.72600352 \cdot 10^{-12}$
11	$-1.7789978444 \cdot 10^{-10}$	$+1.66894955 \cdot 10^{-12}$
10	$+1.35455469767 \cdot 10^{-9}$	$-3.492781410 \cdot 10^{-11}$
9	$-1.03257121793 \cdot 10^{-9}$	$-1.5858066167 \cdot 10^{-10}$
8	$-3.566996111150 \cdot 10^{-8}$	$-1.7928943718 \cdot 10^{-10}$
7	$+1.4481887738427 \cdot 10^{-7}$	$+1.76281629144 \cdot 10^{-9}$
6	$+7.8201821518405 \cdot 10^{-7}$	$+1.690502288794 \cdot 10^{-8}$
5	$-5.39919118403805 \cdot 10^{-6}$	$+1.2539177122849 \cdot 10^{-7}$
4	$-3.124582021689598 \cdot 10^{-5}$	$+1.16229947068677 \cdot 10^{-6}$
3	$+8.901367419507275 \cdot 10^{-5}$	$+1.610382601173763 \cdot 10^{-5}$
2	$+2.02558474743846862 \cdot 10^{-3}$	$+3.4981037560105397 \cdot 10^{-4}$
1	$+2.960644408556332570 \cdot 10^{-2}$	$+1.284780652596476108 \cdot 10^{-2}$
0	$+1.118477510472570366 25$	$+1.036657225887983267 12$

For $18 \leq x \leq 88$,

$$
\begin{aligned}
w &= \frac{1584/x - 53}{35} \\
\operatorname{Shi}(x) &= \frac{e^x}{x} \sum_{k=0}^{\infty}{}' S2_k T_k(w) \\
\operatorname{Chi}(x) &= \gamma + \ln x + \frac{e^x}{x} \sum_{k=0}^{\infty}{}' C2_k T_k(w) \;.
\end{aligned}
$$

The Chebyshev coefficients are given in Table 7.6 and Table 7.7. Overflow in the exponential function must be noted for large x.

Table 7.7: Chebyshev Coefficients for Chi(x)

k	$C1_k$	$C2_k$
23		$+8.07 \cdot 10^{-18}$
22	$-8.12 \cdot 10^{-18}$	$-2.081 \cdot 10^{-17}$
21	$+2.176 \cdot 10^{-17}$	$-5.981 \cdot 10^{-17}$
20	$+5.226 \cdot 10^{-17}$	$+2.6853 \cdot 10^{-16}$
19	$-9.4881 \cdot 10^{-16}$	$+4.5231 \cdot 10^{-16}$
18	$+5.35546 \cdot 10^{-15}$	$-3.10735 \cdot 10^{-15}$
17	$-1.210100 \cdot 10^{-14}$	$-4.42823 \cdot 10^{-15}$
16	$-6.008652 \cdot 10^{-14}$	$+3.496397 \cdot 10^{-14}$
15	$+7.1633965 \cdot 10^{-13}$	$+6.634067 \cdot 10^{-14}$
14	$-2.93496073 \cdot 10^{-12}$	$-3.7190245 \cdot 10^{-13}$
13	$-1.40359438 \cdot 10^{-12}$	$-1.27135418 \cdot 10^{-12}$
12	$+8.763022886 \cdot 10^{-11}$	$+2.74851142 \cdot 10^{-12}$
11	$-4.4009247621 \cdot 10^{-10}$	$+2.337818440 \cdot 10^{-11}$
10	$-1.8799207564 \cdot 10^{-10}$	$+2.714360064 \cdot 10^{-11}$
9	$+1.314581509895 \cdot 10^{-8}$	$-2.5660018000 \cdot 10^{-10}$
8	$-4.755139309248 \cdot 10^{-8}$	$-1.61021375164 \cdot 10^{-9}$
7	$-2.2177501880185 \cdot 10^{-7}$	$-4.72543064876 \cdot 10^{-9}$
6	$+1.94635531373272 \cdot 10^{-6}$	$-3.00095178029 \cdot 10^{-9}$
5	$+4.33505889257316 \cdot 10^{-6}$	$+7.793874743909 \cdot 10^{-8}$
4	$-6.133870010764943 \cdot 10^{-5}$	$+1.06942765566402 \cdot 10^{-6}$
3	$-3.1308547749299747 \cdot 10^{-4}$	$+1.595031648023132 \cdot 10^{-5}$
2	$+4.9716478982311606 \cdot 10^{-4}$	$+3.4959257515377800 \cdot 10^{-4}$
1	$+2.643474960313745266 \cdot 10^{-2}$	$+1.284753875300652474 \cdot 10^{-2}$
0	$+1.11446150876699213025$	$+1.03665693917934275131$

Arithmetic	Function	Trials	Peak	RMS
DEC	Shi	3000	$9.1 \cdot 10^{-17}$	
IEEE	Shi	30000	$6.9 \cdot 10^{-16}$	$1.6 \cdot 10^{-16}$
DEC	Chi	2500	$9.3 \cdot 10^{-17}$	
IEEE	Chi	30000	$8.4 \cdot 10^{-16}$	$1.4 \cdot 10^{-16}$

Table 7.8: `shichi.c`, relative error, except absolute error when magnitude of Chi is less than 1. Argument ranges from 0 to 88.

Chi(x) has a logarithmic singularity at $x = 0$. If x is negative,

$$\mathrm{Shi}(-x) = -\mathrm{Shi}(x)$$

but Chi(x) is complex if $x < 0$. Table 7.8 gives the test results.

7.5.3 shichi.c

Hyperbolic sine and cosine integral program

```
#include "mconf.h"

#ifdef DEC
static short S1[] = {
0022251,0115635,0165120,0006574,
0122734,0050751,0020305,0101356,
0023153,0111154,0011103,0177462,
0023636,0060321,0060253,0124246,
0124554,0106655,0152525,0166400,
0025205,0140145,0171006,0106556,
0125034,0056427,0004205,0176022,
0126305,0016731,0025011,0134453,
0027046,0172453,0112604,0116235,
0127216,0022071,0116600,0137667,
0130103,0115126,0071104,0052535,
0030672,0025450,0010071,0141414,
0130615,0165136,0132137,0177737,
0132031,0031611,0074436,0175407,
0032433,0077602,0104345,0060076,
0033121,0165741,0167177,0172433,
0133665,0025262,0174621,0022612,
0134403,0006761,0124566,0145405,

#ifdef IIEEE
static short S1[] = {
0x01b0,0xbd4a,0x3373,0x3c75,
0xb05e,0x2418,0x8a3d,0xbc9b,
0x7fe6,0x8248,0x724d,0x3cad,
0x7515,0x2c15,0xcc1a,0x3cd3,
0xbda0,0xbaaa,0x91b5,0xbd0d,
0xd1ae,0xbe40,0xb80c,0x3d30,
0xbf82,0xe110,0x8ba2,0xbd23,
0x3725,0x2541,0xa3bb,0xbd78,
0x9394,0x72b0,0xdea5,0x3da4,
0x17f7,0x33b0,0xc487,0xbdb1,
0x8aac,0xce48,0x734a,0xbde8,
0x3862,0x0207,0x4565,0x3e17,
0xfffc,0xd68b,0xbd4b,0xbe11,
0xdf61,0x2f23,0x2671,0xbe63,
0xac08,0x511c,0x6ff0,0x3e83,
0xfea3,0x3dcf,0x3d7c,0x3eaa,
0x24b1,0x5f32,0xa556,0xbed6,
0xd961,0x352e,0x61be,0xbf00,
```

```
0034672,0126332,0034737,0116744,
0036004,0137654,0037332,0131766,
0036762,0104466,0121445,0124326,
0040217,0025105,0062145,0042640
};
static short S2[] = {
0122102,0041774,0016051,0055137,
0022362,0010125,0007651,0015773,
0022713,0062551,0040227,0071645,
0123303,0015732,0025731,0146570,
0123557,0064016,0002067,0067711,
0024215,0136214,0132374,0124234,
0024467,0051425,0071066,0064210,
0125075,0124305,0135123,0024170,
0125465,0010261,0005560,0034232,
0025674,0066602,0030724,0174557,
0026477,0151520,0051510,0067250,
0026352,0161076,0113154,0116271,
0127431,0116470,0177465,0127274,
0130056,0056174,0170315,0013321,
0130105,0020575,0075327,0036710,
0030762,0043625,0113046,0125035,
0031621,0033211,0154354,0022077,
0032406,0121555,0074270,0041141,
0033234,0000116,0041611,0173743,
0034207,0013263,0174715,0115563,
0035267,0063300,0175753,0117266,
0036522,0077633,0033255,0136200,
0040204,0130457,0014454,0166254
};
static short C1[] = {
0122025,0157055,0021702,0021427,
0022310,0130043,0123265,0022340,
0022561,0002231,0017746,0013043,
0123610,0136375,0002352,0024467,
0024300,0171555,0141300,0000446,
0124531,0176777,0126210,0035616,
0125207,0046604,0167760,0077132,
0026111,0120666,0026606,0064143,
0126516,0103615,0054127,0005436,
0126305,0104721,0025415,0004134,
0027700,0131556,0164725,0157553,
0130361,0170602,0077274,0055406,
0130116,0131420,0125472,0017231,
0031541,0153747,0177312,0056304,
0132114,0035517,0041545,0043151,
0132556,0020415,0110044,0172442,
0033402,0117041,0031152,0010364,
0033621,0072737,0050647,0013720,
```

```
0xf3bd,0x473b,0x559b,0x3f17,
0x567f,0x87db,0x97f5,0x3f60,
0xb51b,0xd464,0x5126,0x3f9e,
0xa8b4,0xac8c,0xe548,0x3ff1
};
static short S2[] = {
0x2b4c,0x8385,0x487f,0xbc68,
0x237f,0xa1f5,0x420a,0x3c7e,
0xee75,0x2812,0x6cad,0x3c99,
0x39af,0x457b,0x637b,0xbcb8,
0xedf9,0xc086,0xed01,0xbccd,
0x9513,0x969f,0xb791,0x3cf1,
0xcd11,0xae46,0xea62,0x3d06,
0x650f,0xb74a,0xb518,0xbd27,
0x0713,0x216e,0xa216,0xbd46,
0x9f2e,0x463a,0x8db0,0x3d57,
0x0dd5,0x0a69,0xfa6a,0x3d87,
0x9397,0xd2cd,0x5c47,0x3d7d,
0xb5d8,0x1fe6,0x33a7,0xbdc3,
0xa2da,0x9e19,0xcb8f,0xbde5,
0xe7b9,0xaf5a,0xa42f,0xbde8,
0xd544,0xb2c4,0x48f2,0x3e1e,
0x8488,0x3b1d,0x26d1,0x3e52,
0x084c,0xaf17,0xd46d,0x3e80,
0x3efc,0xc871,0x8009,0x3eb3,
0xb36e,0x7f39,0xe2d6,0x3ef0,
0x73d7,0x1f7d,0xecd8,0x3f36,
0xb790,0x66d5,0x4ff3,0x3f8a,
0x9d96,0xe325,0x9625,0x3ff0
};
static short C1[] = {
0x4463,0xa478,0xbbc5,0xbc62,
0xa49c,0x74d6,0x1604,0x3c79,
0xc2c4,0x23fc,0x2093,0x3c8e,
0x4527,0xa09d,0x179f,0xbcd1,
0x0025,0xb858,0x1e6d,0x3cf8,
0x0772,0xf591,0x3fbf,0xbd0b,
0x0fcb,0x9dfe,0xe9b0,0xbd30,
0xcd0c,0xc5b0,0x3436,0x3d69,
0xe164,0xab0a,0xd0f1,0xbd89,
0xa10c,0x2561,0xb13a,0xbd78,
0xbbed,0xdd3a,0x166d,0x3dd8,
0x8b61,0x4fd7,0x3e30,0xbdfe,
0x43d3,0x1567,0xd662,0xbde9,
0x4b98,0xffd9,0x3afc,0x3e4c,
0xa8cd,0xe86c,0x8769,0xbe69,
0x9ea4,0xb204,0xc421,0xbe8d,
0x421f,0x264d,0x53c4,0x3ec0,
0xe2fa,0xea34,0x2ebb,0x3ed2,
```

```
0134600,0121366,0140010,0063265,
0135244,0022637,0013756,0044742,
0035402,0052052,0006523,0043564,
0036730,0106660,0020277,0162146,
0040216,0123254,0135147,0005724
};
static short C2[] = {
0022024,0154550,0104311,0144257,
0122277,0165037,0133443,0155601,
0122611,0165102,0157053,0055252,
0023232,0146235,0153511,0113222,
0023402,0057340,0145304,0010471,
0124137,0164171,0113071,0100002,
0124237,0105473,0056130,0022022,
0025035,0073266,0056746,0164433,
0025225,0061313,0055600,0165407,
0125721,0056312,0107613,0051215,
0126262,0166534,0115336,0066653,
0026501,0064307,0127442,0065573,
0027315,0121375,0142020,0045356,
0027356,0140764,0070641,0046570,
0130215,0010503,0146335,0177737,
0130735,0047134,0015215,0163665,
0131242,0056523,0155276,0050053,
0131116,0034515,0050707,0163512,
0032247,0057507,0107545,0032007,
0033217,0104501,0021706,0025047,
0034205,0146413,0033746,0076562,
0035267,0044605,0065355,0002772,
0036522,0077173,0130716,0170304,
0040204,0130454,0130571,0027270
};
#endif
```

```
0x0cd7,0xd801,0x145e,0xbf10,
0xc93c,0xe2fd,0x84b3,0xbf34,
0x68ef,0x41aa,0x4a85,0x3f40,
0xfc8d,0x0417,0x11b6,0x3f9b,
0xe17b,0x974c,0xd4d5,0x3ff1
};
static short C2[] = {
0x3916,0x1119,0x9b2d,0x3c62,
0x7b70,0xf6e4,0xfd43,0xbc77,
0x6b55,0x5bc5,0x3d48,0xbc91,
0x32d2,0xbae9,0x5993,0x3cb3,
0x8227,0x1958,0x4bdc,0x3cc0,
0x3000,0x32c7,0xfd0f,0xbceb,
0x0482,0x6b8b,0xf167,0xbcf3,
0xdd23,0xcbbc,0xaed6,0x3d23,
0x1d61,0x6b70,0xac59,0x3d32,
0x6a52,0x51f1,0x2b99,0xbd5a,
0xcdb5,0x935b,0x5dab,0xbd76,
0x4d6f,0xf5e4,0x2d18,0x3d88,
0x095e,0xb882,0xb45f,0x3db9,
0x29af,0x8e34,0xd83e,0x3dbd,
0xbffc,0x799b,0xa228,0xbdf1,
0xbcf7,0x8351,0xa9cb,0xbe1b,
0xca05,0x7b57,0x4baa,0xbe34,
0xfce9,0xaa38,0xc729,0xbe29,
0xa681,0xf1ec,0xebe8,0x3e74,
0xc545,0x2478,0xf128,0x3eb1,
0xcfae,0x66fc,0xb9a1,0x3ef0,
0xa0bf,0xad5d,0xe930,0x3f36,
0xde19,0x7639,0x4fcf,0x3f8a,
0x25d7,0x962f,0x9625,0x3ff0
};
#endif
```

```
#define EUL 0.57721566490153286061
extern double MACHEP, MAXNUM, PIO2;

int shichi( x, si, ci )
double x;
double *si, *ci;
	{
	double k, z, t, c, s, a;
	short sign;
	double log(), exp(), fabs(), chbevl();

	if( x < 0.0 )
		{
```

```
		sign = -1;
		x = -x;
		}
	else
		sign = 0;
	if( x == 0.0 )
		{
		*si = 0.0;
		*ci = -MAXNUM;
		return( 0 );
		}
	if( x >= 8.0 )
		goto chb;
	z = x * x;
Power series expansion
	a = 1.0;
	s = 1.0;
	c = 0.0;
	k = 2.0;
	do
		{
		a *= z/k;
		c += a/k;
		k += 1.0;
		a /= k;
		s += a/k;
		k += 1.0;
		}
	while( fabs(a/s) > MACHEP );
	s *= x;
	goto done;
chb:
	if( x < 18.0 )
		{
		a = (576.0/x - 52.0)/10.0;
		k = exp(x) / x;
		s = k * chbevl( a, S1, 22 );
		c = k * chbevl( a, C1, 23 );
		goto done;
		}
	if( x <= 88.0 )
		{
		a = (6336.0/x - 212.0)/70.0;
		k = exp(x) / x;
		s = k * chbevl( a, S2, 23 );
```

```
		c = k * chbevl( a, C2, 24 );
		goto done;
		}
	else
		{
		if( sign )
			*si = -MAXNUM;
		else
			*si = MAXNUM;
		*ci = MAXNUM;
		return(0);
		}
done:
	if( sign )
		s = -s;
	*si = s;
	*ci = EUL + log(x) + c;
	return(0);
	}
```

7.6 Dilogarithm

Also called Spence's integral, the definition is

$$f(x) = -\int_1^x \frac{\ln t}{t-1} dt \ .$$

The series expansion

$$f(x) = \sum_{k=1}^{\infty} \frac{(1-x)^k}{k^2}$$

converges if $0 \le x \le 2$. Special cases are

$$\begin{aligned} f(1) &= 0 \\ f(0) &= \pi^2/6 \ . \end{aligned}$$

Approximation based on the power series converges best in the interval $0.5 \le x \le 1.5$. If $x > 1.5$ use the transformation

$$f(1/x) = -f(x) - \tfrac{1}{2}(\ln x)^2 \ .$$

If $0 < x < 0.5$, use

$$f(x) = -f(1-x) - \ln x \ \ln(1-x) + \frac{\pi^2}{6}.$$

Arithmetic	Domain	Trials	Peak	RMS
IEEE	0, 4	30000	$3.9 \cdot 10^{-15}$	$5.4 \cdot 10^{-16}$
DEC	0, 4	3000	$2.5 \cdot 10^{-16}$	$4.5 \cdot 10^{-17}$

Table 7.9: `spence.c`, relative error

Having transformed x into the interval from 0.5 to 1.5, compute

$$\begin{aligned} w &= x-1 \\ f(x) &= -w\,\frac{A(w)}{B(w)} \end{aligned}$$

where

$A(t) =$
$4.65128586073990045278 \cdot 10^{-5}t^7$
$+7.31589045238094711071 \cdot 10^{-3}t^6$
$+1.33847639578309018650 \cdot 10^{-1}t^5$
$+8.79691311754530315341 \cdot 10^{-1}t^4$
$+2.71149851196553469920t^3$
$+4.25697156008121755724t^2$
$+3.29771340985225106936t$
$+1.00000000000000000126$,

$B(t) =$
$6.90990488912553276999 \cdot 10^{-4}t^7$
$+2.54043763932544379113 \cdot 10^{-2}t^6$
$+2.82974860602568089943 \cdot 10^{-1}t^5$
$+1.41172597751831069617t^4$
$+3.63800533345137075418t^3$
$+5.03278880143316990390t^2$
$+3.54771340985225096217t$
$+0.99999999999999998740$.

Test results are given in Table 7.9.

7.6.1 spence.c

Dilogarithm program

```
#include "mconf.h"

#if DEC
static short A[32] = {
0034503,0013315,0034120,0157771,
0036357,0135043,0016766,0150637,
0037411,0007533,0005212,0161475,
0040141,0031563,0023217,0120331,
0040455,0104461,0007002,0155522,
0040610,0034434,0065721,0120465,
0040523,0006674,0105671,0054427,
0040200,0000000,0000000,0000000,
```

```
#if IIEEE
static short A[32] = {
0x1bff,0xa70a,0x62d9,0x3f08,
0xda34,0x63be,0xf744,0x3f7d,
0x5c68,0x6151,0x21eb,0x3fc1,
0xf41b,0x64d1,0x266e,0x3fec,
0x5b6a,0x21c0,0xb126,0x4005,
0x3427,0x8d7a,0x0723,0x4011,
0x2b23,0x9177,0x61b7,0x400a,
0x0000,0x0000,0x0000,0x3ff0,
```

```
};
static short B[32] = {
0035465,0021626,0032367,0144157,
0036720,0016326,0134431,0000406,
0037620,0161024,0133701,0120766,
0040264,0131557,0152055,0064512,
0040550,0152424,0051166,0034272,
0040641,0006233,0014672,0111572,
0040543,0006674,0105671,0054425,
0040200,0000000,0000000,0000000,
};
#endif
```

```
};
static short B[32] = {
0xf90e,0xc69e,0xa472,0x3f46,
0x2021,0xd723,0x039a,0x3f9a,
0x343f,0x96f8,0x1c42,0x3fd2,
0xad29,0xfa85,0x966d,0x3ff6,
0xc717,0x8a4e,0x1aa2,0x400d,
0x526f,0x6337,0x2193,0x4014,
0x2b23,0x9177,0x61b7,0x400c,
0x0000,0x0000,0x0000,0x3ff0,
};
#endif
```

```
extern double PI, MACHEP;

double spence(x)
double x;
    {
    double a, w, y, k, z;
    double fabs(), log(), polevl();
    int flag;

    if( x < 0.0 )
        {
        mtherr( "spence", DOMAIN );
        return(0.0);
        }
    if( x == 1.0 )
        return( 0.0 );
    if( x == 0.0 )
        return( PI*PI/6.0 );
    flag = 0;
    if( x > 2.0 )
        {
        x = 1.0/x;
        flag |= 2;
        }
    if( x > 1.5 )
        {
        w = (1.0/x) - 1.0;
        flag |= 2;
        }
    else if( x < 0.5 )
            {
            w = -x;
```

```
            flag |= 1;
            }
        else
            w = x - 1.0;
    y = -w * polevl( w, A, 7) / polevl( w, B, 7 );
    if( flag & 1 )
        y = (PI * PI)/6.0 - log(x) * log(1.0-x) - y;
    if( flag & 2 )
        {
        z = log(x);
        y = -0.5 * z * z - y;
        }
    return( y );
    }
```

7.7 Dawson's Integral

This integral has the power series expansion

$$\begin{aligned} F(x) &= e^{-x^2} \int_0^x e^{t^2} dt \\ &= x\left(1+\sum_{k=1}^{\infty} \frac{(-x^2)^k}{\frac{3}{2}\cdot\frac{5}{2}\cdots\frac{2k+1}{2}}\right) . \end{aligned}$$

The approximation in the interval $0 \le x \le 3.25$ takes the form of the power series:

$$F(x) \approx x\frac{AN(x^2)}{AD(x^2)}$$

where

$$\begin{aligned} AN(t) = {}& 1.13681498971755972054 \cdot 10^{-11} t^9 \\ &+8.49262267667473811108 \cdot 10^{-10} t^8 \\ &+1.94434204175553054283 \cdot 10^{-8} t^7 \\ &+9.53151741254484363489 \cdot 10^{-7} t^6 \\ &+3.07828309874913200438 \cdot 10^{-6} t^5 \\ &+3.52513368520288738649 \cdot 10^{-4} t^4 \\ &-8.50149846724410912031 \cdot 10^{-4} t^3 \\ &+4.22618223005546594270 \cdot 10^{-2} t^2 \\ &-9.17480371773452345351 \cdot 10^{-2} t \\ &+0.999999999999999994612 , \end{aligned}$$

$$\begin{aligned} AD(t) = {}& 2.40372073066762605484 \cdot 10^{-11} t^{10} \\ &+1.48864681368493396752 \cdot 10^{-9} t^9 \\ &+5.21265281010541664570 \cdot 10^{-8} t^8 \\ &+1.27258478273186970203 \cdot 10^{-6} t^7 \\ &+2.32490249820789513991 \cdot 10^{-5} t^6 \\ &+3.25524741826057911661 \cdot 10^{-4} t^5 \\ &+3.48805814657162590916 \cdot 10^{-3} t^4 \\ &+2.79448531198828973716 \cdot 10^{-2} t^3 \\ &+1.58874241960120565368 \cdot 10^{-1} t^2 \\ &+5.74918629489320327824 \cdot 10^{-1} t \\ &+1.00000000000000000539 . \end{aligned}$$

An asymptotic expansion for large x is

$$F(x) = \frac{1}{2x}\left(1 + \frac{\frac{1}{2}}{x^2} - \frac{\frac{1}{2}\cdot\frac{3}{2}}{x^4} + \ldots\right).$$

For $3.25 \le x \le 6.25$ an expansion in $1/x$ is used, with the Cody and Waite form

$$\begin{aligned} w &= 1/x \\ 2F(x) &= w + w^3 \frac{BN(w^2)}{BD(w^2)} \end{aligned}$$

where

$BN(t) =$	$BD(t) =$
$5.08955156417900903354 \cdot 10^{-1} t^{10}$	$1.0 t^{10}$
$-2.44754418142697847934 \cdot 10^{-1} t^{9}$	$-6.31839869873368190192 \cdot 10^{-1} t^{9}$
$+9.41512335303534411857 \cdot 10^{-2} t^{8}$	$+2.36706788228248691528 \cdot 10^{-1} t^{8}$
$-2.18711255142039025206 \cdot 10^{-2} t^{7}$	$-5.31806367003223277662 \cdot 10^{-2} t^{7}$
$+3.66207612329569181322 \cdot 10^{-3} t^{6}$	$+8.48041718586295374409 \cdot 10^{-3} t^{6}$
$-4.23209114460388756528 \cdot 10^{-4} t^{5}$	$-9.47996768486665330168 \cdot 10^{-4} t^{5}$
$+3.59641304793896631888 \cdot 10^{-5} t^{4}$	$+7.81025592944552338085 \cdot 10^{-5} t^{4}$
$-2.14640351719968974225 \cdot 10^{-6} t^{3}$	$-4.55875153252442634831 \cdot 10^{-6} t^{3}$
$+9.10010780076391431042 \cdot 10^{-8} t^{2}$	$+1.89100358111421846170 \cdot 10^{-7} t^{2}$
$-2.40274520828250956942 \cdot 10^{-9} t$	$-4.91324691331920606875 \cdot 10^{-9} t$
$+3.59233385440928410398 \cdot 10^{-11}$,	$+7.18466403235734541950 \cdot 10^{-11}$.

The asymptotic form is used also for $6.25 \le x \le \infty$:

$$\begin{aligned} w &= 1/x \\ 2F(x) &= w + w^3 \frac{CN(w^2)}{CD(w^2)} \end{aligned}$$

where

$CN(t) =$	$CD(t) =$
	$1.0 t^{5}$
$-5.90592860534773254987 \cdot 10^{-1} t^{4}$	$-2.69820057197544900361 \cdot 10^{0} t^{4}$
$+6.29235242724368800674 \cdot 10^{-1} t^{3}$	$+1.73270799045947845857 \cdot 10^{0} t^{3}$
$-1.72858975380388136411 \cdot 10^{-1} t^{2}$	$-3.93708582281939493482 \cdot 10^{-1} t^{2}$
$+1.64837047825189632310 \cdot 10^{-2} t$	$+3.44278924041233391079 \cdot 10^{-2} t$
$-4.86827613020462700845 \cdot 10^{-4}$,	$-9.73655226040941223894 \cdot 10^{-4}$.

Table 7.10 shows the test results.

7.7.1 dawsn.c

Dawson's integral program

Arithmetic	Domain	Trials	Peak	RMS
IEEE	0, 10	10000	$6.9 \cdot 10^{-16}$	$1.0 \cdot 10^{-16}$
DEC	0, 10	6000	$7.4 \cdot 10^{-17}$	$1.4 \cdot 10^{-17}$

Table 7.10: `dawsn.c`, relative error

```
#include "mconf.h"

#if DEC
static short AN[40] = {
0027107,0176630,0075752,0107612,
0030551,0070604,0166707,0127727,
0031647,0002210,0117120,0056376,
0033177,0156026,0141275,0140627,
0033516,0112200,0037035,0165515,
0035270,0150613,0016423,0105634,
0135536,0156227,0023515,0044413,
0037055,0015273,0105147,0064025,
0137273,0163145,0014460,0166465,
0040200,0000000,0000000,0000000,
};
static short AD[44] = {
0027323,0067372,0115566,0131320,
0030714,0114432,0074206,0006637,
0032137,0160671,0044203,0026344,
0033252,0146656,0020247,0100231,
0034303,0003346,0123260,0022433,
0035252,0125460,0173041,0155415,
0036144,0113747,0125203,0124617,
0036744,0166232,0143671,0133670,
0037442,0127755,0162625,0000100,
0040023,0026736,0003604,0106265,
0040200,0000000,0000000,0000000,
};
static short BN[44] = {
0040002,0045342,0113762,0004360,
0137572,0120346,0172745,0144046,
0037300,0151134,0123440,0117047,
0136663,0025423,0014755,0046026,
0036157,0177561,0027535,0046744,
0135335,0161052,0071243,0146535,
```

```
#if IIEEE
static short AN[40] = {
0x51f1,0x0f7d,0xffb3,0x3da8,
0xf5fb,0x9db8,0x2e30,0x3e0d,
0x0ba0,0x13ca,0xe091,0x3e54,
0xb833,0xd857,0xfb82,0x3eaf,
0xbd6a,0x07c3,0xd290,0x3ec9,
0x7174,0x63a2,0x1a31,0x3f37,
0xa921,0xe4e9,0xdb92,0xbf4b,
0xed03,0x714c,0xa357,0x3fa5,
0x1da7,0xa326,0x7ccc,0xbfb7,
0x0000,0x0000,0x0000,0x3ff0,
};
static short AD[44] = {
0xd65a,0x536e,0x6ddf,0x3dba,
0xc1b4,0x4f10,0x9323,0x3e19,
0x659c,0x2910,0xfc37,0x3e6b,
0xf013,0xc414,0x59b5,0x3eb5,
0x04a3,0xd4d6,0x60dc,0x3ef8,
0x3b62,0x1ec4,0x5566,0x3f35,
0x7532,0xf550,0x92fc,0x3f6c,
0x36f7,0x58f7,0x9d93,0x3f9c,
0xa008,0xbcb2,0x55fd,0x3fc4,
0x9197,0xc0f0,0x65bb,0x3fe2,
0x0000,0x0000,0x0000,0x3ff0,
};
static short BN[44] = {
0x411e,0x52fe,0x495c,0x3fe0,
0xb905,0xdebc,0x541c,0xbfcf,
0x13c5,0x94e4,0x1a4b,0x3fb8,
0xa983,0x633d,0x6562,0xbf96,
0xa9bd,0x25eb,0xffee,0x3f6d,
0x79ac,0x4e54,0xbc45,0xbf3b,
```

```
0034426,0154060,0164506,0135625,
0133420,0005356,0100017,0151334,
0032303,0066137,0024013,0046212,
0131045,0016612,0066270,0047574,
0027435,0177025,0060625,0116363,
};
Leading 1.0 omitted from BD.
static short BD[40] = {
0140041,0140101,0174552,0037073,
0037562,0061503,0124271,0160756,
0137131,0151760,0073210,0110534,
0036412,0170562,0117017,0155377,
0135570,0101374,0074056,0037276,
0034643,0145376,0001516,0060636,
0133630,0173540,0121344,0155231,
0032513,0005602,0134516,0007144,
0131250,0150540,0075747,0105341,
0027635,0177020,0012465,0125402,
};
static short CN[20] = {
0140027,0030427,0176477,0074402,
0040041,0012617,0112375,0162657,
0137461,0000761,0074120,0135160,
0036607,0004325,0117246,0115525,
0135377,0036345,0064750,0047732,
};
Leading 1.0 omitted from CD.
static short CD[20] = {
0140454,0127521,0071653,0133415,
0040335,0144540,0016105,0045241,
0137711,0112053,0155034,0062237,
0037015,0002102,0177442,0074546,
0135577,0036345,0064750,0052152,
};
#endif
```

```
0xd773,0x1d28,0xdb06,0x3f02,
0xfa5b,0xd001,0x015d,0xbec2,
0x6991,0xe501,0x6d8b,0x3e78,
0x09f0,0x4d97,0xa3b1,0xbe24,
0xb39e,0xac32,0xbfc2,0x3dc3,
};
static short BD[40] = {
0x47c7,0x3f2d,0x3808,0xbfe4,
0x3c3e,0x7517,0x4c68,0x3fce,
0x122b,0x0ed1,0x3a7e,0xbfab,
0xfb60,0x53c1,0x5e2e,0x3f81,
0xc7d8,0x8f05,0x105f,0xbf4f,
0xcc34,0xc069,0x795f,0x3f14,
0x9b53,0x145c,0x1eec,0xbed3,
0xc1cd,0x5729,0x6170,0x3e89,
0xf15c,0x0f7c,0x1a2c,0xbe35,
0xb560,0x02a6,0xbfc2,0x3dd3,
};
static short CN[20] = {
0xef20,0xffa7,0xe622,0xbfe2,
0xbcb6,0xf29f,0x22b1,0x3fe4,
0x174e,0x2f0a,0x203e,0xbfc6,
0xd36b,0xb3d4,0xe11a,0x3f90,
0x09fb,0xad3d,0xe79c,0xbf3f,
};
static short CD[20] = {
0x76e2,0x2e75,0x95ea,0xc005,
0xa954,0x0388,0xb92c,0x3ffb,
0x8c94,0x7b43,0x3285,0xbfd9,
0x4f2d,0x5fe4,0xa088,0x3fa1,
0x0a8d,0xad3d,0xe79c,0xbf4f,
};
#endif
```

```
extern double PI, MACHEP;

double dawsn( xx )
double xx;
    {
    double ans, q, r, s, t, x, y;
    double spi;
    int i, sign, flag;
    double chbevl(), sqrt(), fabs(), polevl(), p1evl();

    sign = 1;
```

```
if( xx < 0.0 )
    {
    sign = -1;
    xx = -xx;
    }
if( xx < 3.25 )
    {
    x = xx*xx;
    y = xx * polevl( x, AN, 9 )/polevl( x, AD, 10 );
    return( sign * y );
    }
x = 1.0/(xx*xx);
if( xx < 6.25 )
    {
    y = 1.0/xx + x * polevl( x, BN, 10)
        / (p1evl( x, BD, 10) * xx);
    return( sign * 0.5*y );
    }
if( xx > 1.0e9 )
    return( (sign * 0.5)/xx );
y = 1.0/xx + x * polevl( x, CN, 4)
    / (p1evl( x, CD, 5) * xx);
return( sign * 0.5 * y );
}
```

7.8 Fresnel Integrals

$$\begin{aligned} C(x) &= \int_0^x \cos \tfrac{1}{2}\pi t^2 \, dt \\ S(x) &= \int_0^x \sin \tfrac{1}{2}\pi t^2 \, dt \, . \end{aligned}$$

These integrals have the power series expansions

$$C(x) = \sum_{k=0}^{\infty} \frac{(-1)^k (\frac{1}{2}\pi)^{2k}}{(2k)!\,(4k+1)}\, x^{4k+1}$$

$$S(x) = \sum_{k=0}^{\infty} \frac{(-1)^k (\frac{1}{2}\pi)^{2k+1}}{(2k+1)!\,(4k+3)}\, x^{4k+3} \, .$$

Both are odd functions:

$$\begin{aligned} C(-x) &= -C(x) \\ S(-x) &= -S(x) \, . \end{aligned}$$

The following are double precision approximations on the interval $0 \leq x^2 \leq 2.5625$.

$$S(x) = x^3 \frac{SN(x^4)}{SD(x^4)}$$

where

$SN(t) =$
$-2.99181919401019853726 \cdot 10^3 t^5$
$+7.08840045257738576863 \cdot 10^5 t^4$
$-6.29741486205862506537 \cdot 10^7 t^3$
$+2.54890880573376359104 \cdot 10^9 t^2$
$-4.42979518059697779103 \cdot 10^{10} t$
$+3.18016297876567817986 \cdot 10^{11}$,

$SD(t) =$
$1.0t^6$
$+2.81376268889994315696 \cdot 10^2 t^5$
$+4.55847810806532581675 \cdot 10^4 t^4$
$+5.17343888770096400730 \cdot 10^6 t^3$
$+4.19320245898111231129 \cdot 10^8 t^2$
$+2.24411795645340920940 \cdot 10^{10} t$
$+6.07366389490084639049 \cdot 10^{11}$.

$$C(x) = x \frac{CN(x^4)}{CD(x^4)}$$

where

$CN(t) =$
$-4.98843114573573548651 \cdot 10^{-8} t^5$
$+9.50428062829859605134 \cdot 10^{-6} t^4$
$-6.45191435683965050962 \cdot 10^{-4} t^3$
$+1.88843319396703850064 \cdot 10^{-2} t^2$
$-2.05525900955013891793 \cdot 10^{-1} t,$
$+0.99999999999999998822$,

$CD(t) =$
$3.99982968972495980367 \cdot 10^{-12} t^6$
$+9.15439215774657478799 \cdot 10^{-10} t^5$
$+1.25001862479598821474 \cdot 10^{-7} t^4$
$+1.22262789024179030997 \cdot 10^{-5} t^3$
$+8.68029542941784300606 \cdot 10^{-4} t^2$
$+4.12142090722199792936 \cdot 10^{-2} t$
$+1.00000000000000000118$.

S and C can be expressed in terms of auxiliary functions f and g such that

$$\begin{aligned} C(x) &= \tfrac{1}{2} + f(x)\sin(\tfrac{1}{2}\pi x^2) - g(x)\cos(\tfrac{1}{2}\pi x^2) \\ S(x) &= \tfrac{1}{2} - f(x)\cos(\tfrac{1}{2}\pi x^2) - g(x)\sin(\tfrac{1}{2}\pi x^2) \ . \end{aligned}$$

For large x the auxiliary functions have the asymptotic expansions

$$\begin{aligned} \pi x f(x) &\approx 1 + \sum_{k=1}^{\infty} (-1)^k \frac{1 \cdot 3 \cdots (4k-1)}{(\pi x^2)^{2k}} \\ \pi x g(x) &\approx \sum_{k=1}^{\infty} (-1)^k \frac{1 \cdot 3 \cdots (4k+1)}{(\pi x^2)^{2k+1}} \ . \end{aligned}$$

For sufficiently large x the sine and cosine of x^2 will suffer loss of precision. With typical trigonometry routines limited to $x < 10^9$, set $S(x) = C(x) = 0.5$ if $x > 36974.0$.

The following are double precision approximations for f and g that use the form of the asymptotic expansions. They are valid for $\sqrt{8/\pi} \leq x \leq \infty$.

$$u = \frac{1}{\pi x^2}$$

$$\pi x f(x) \approx 1 - u^2 \frac{FN(u^2)}{FD(u^2)}$$

where

$$\begin{aligned} FN(t) = {} & 4.21543555043677546506 \cdot 10^{-1} t^9 \\ & +1.43407919780758885261 \cdot 10^{-1} t^8 \\ & +1.15220955073585758835 \cdot 10^{-2} t^7 \\ & +3.45017939782574027900 \cdot 10^{-4} t^6 \\ & +4.63613749287867322088 \cdot 10^{-6} t^5 \\ & +3.05568983790257605827 \cdot 10^{-8} t^4 \\ & +1.02304514164907233465 \cdot 10^{-10} t^3 \\ & +1.72010743268161828879 \cdot 10^{-13} t^2 \\ & +1.34283276233062758925 \cdot 10^{-16} t \\ & +3.76329711269987889006 \cdot 10^{-20} , \end{aligned}$$

$$\begin{aligned} FD(t) = {} & 1.0 t^{10} \\ & +7.51586398353378947175 \cdot 10^{-1} t^9 \\ & +1.16888925859191382142 \cdot 10^{-1} t^8 \\ & +6.44051526508858611005 \cdot 10^{-3} t^7 \\ & +1.55934409164153020873 \cdot 10^{-4} t^6 \\ & +1.84627567348930545870 \cdot 10^{-6} t^5 \\ & +1.12699224763999035261 \cdot 10^{-8} t^4 \\ & +3.60140029589371370404 \cdot 10^{-11} t^3 \\ & +5.88754533621578410010 \cdot 10^{-14} t^2 \\ & +4.52001434074129701496 \cdot 10^{-17} t \\ & +1.25443237090011264384 \cdot 10^{-20} . \end{aligned}$$

The approximation for $g(x)$ is

$$\begin{aligned} u &= \frac{1}{\pi x^2} \\ \pi x g(x) &\approx u \frac{GN(u^2)}{GD(u^2)} \end{aligned}$$

where

$$\begin{aligned} GN(t) = {} & 5.04442073643383265887 \cdot 10^{-1} t^{10} \\ & +1.97102833525523411709 \cdot 10^{-1} t^9 \\ & +1.87648584092575249293 \cdot 10^{-2} t^8 \\ & +6.84079380915393090172 \cdot 10^{-4} t^7 \\ & +1.15138826111884280931 \cdot 10^{-5} t^6 \\ & +9.82852443688422223854 \cdot 10^{-8} t^5 \\ & +4.45344415861750144738 \cdot 10^{-10} t^4 \\ & +1.08268041139020870318 \cdot 10^{-12} t^3 \\ & +1.37555460633261799868 \cdot 10^{-15} t^2 \\ & +8.36354435630677421531 \cdot 10^{-19} t \\ & +1.86958710162783235106 \cdot 10^{-22} , \end{aligned}$$

$$\begin{aligned} GD(t) = {} & 1.0 t^{11} \\ & +1.47495759925128324529 \cdot 10^{0} t^{10} \\ & +3.37748989120019970451 \cdot 10^{-1} t^9 \\ & +2.53603741420338795122 \cdot 10^{-2} t^8 \\ & +8.14679107184306179049 \cdot 10^{-4} t^7 \\ & +1.27545075667729118702 \cdot 10^{-5} t^6 \\ & +1.04314589657571990585 \cdot 10^{-7} t^5 \\ & +4.60680728146520428211 \cdot 10^{-10} t^4 \\ & +1.10273215066240270757 \cdot 10^{-12} t^3 \\ & +1.38796531259578871258 \cdot 10^{-15} t^2 \\ & +8.39158816283118707363 \cdot 10^{-19} t \\ & +1.86958710162783236342 \cdot 10^{-22} . \end{aligned}$$

Accuracy figures are presented in Table 7.11.

7.8.1 fresnl.c

Fresnel integral program.

```
#include "mconf.h"
```

Table 7.11: `fresnl.c`, relative error, $0 \le x \le 10$.

Arithmetic	Function	Trials	Peak	RMS
IEEE	S(x)	10000	$2.0 \cdot 10^{-15}$	$3.2 \cdot 10^{-16}$
IEEE	C(x)	10000	$1.8 \cdot 10^{-15}$	$3.3 \cdot 10^{-16}$
DEC	S(x)	6000	$2.2 \cdot 10^{-16}$	$3.9 \cdot 10^{-17}$
DEC	C(x)	5000	$2.3 \cdot 10^{-16}$	$3.9 \cdot 10^{-17}$

$S(x)$ for small x

```
#if DEC
static short sn[24] = {
0143072,0176433,0065455,0127034,
0045055,0007200,0134540,0026661,
0146560,0035061,0023667,0127545,
0050027,0166503,0002673,0153756,
0151045,0002721,0121737,0102066,
0051624,0013177,0033451,0021271,
};
```

Leading 1.0 omitted from sd.

```
static short sd[24] = {
0042214,0130051,0112070,0101617,
0044062,0010307,0172346,0152510,
0045635,0160575,0143200,0136642,
0047307,0171215,0127457,0052361,
0050647,0031447,0032621,0013510,
0052015,0064733,0117362,0012653,
};
```

$C(x)$ for small x

```
static short cn[24] = {
0132126,0040141,0063733,0013231,
0034037,0072223,0010200,0075637,
0135451,0021020,0073264,0036057,
0036632,0131520,0101316,0060233,
0137522,0072541,0136124,0132202,
0040200,0000000,0000000,0000000,
};
static short cd[28] = {
0026614,0135503,0051776,0032631,
0030573,0121116,0154033,0126712,
0032406,0034100,0012442,0106212,
0034115,0017567,0150520,0164623,
```

```
#if IIEEE
static short sn[24] = {
0xb5c3,0x6d65,0x5fa3,0xc0a7,
0x05b6,0x172c,0xa1d0,0x4125,
0xf5ed,0x24f6,0x0746,0xc18e,
0x7afe,0x60b7,0xfda8,0x41e2,
0xf087,0x347b,0xa0ba,0xc224,
0x2457,0xe6e5,0x82cf,0x4252,
};

static short sd[24] = {
0x1072,0x3287,0x9605,0x4071,
0xdaa9,0xfe9c,0x4218,0x40e6,
0x17b4,0xb8d0,0xbc2f,0x4153,
0xea9e,0xb5e5,0xfe51,0x41b8,
0x22e9,0xe6b2,0xe664,0x4214,
0x42b5,0x73de,0xad3b,0x4261,
};

static short cn[24] = {
0x62d3,0x2cfb,0xc80c,0xbe6a,
0x0f74,0x6210,0xee92,0x3ee3,
0x8786,0x0ed6,0x2442,0xbf45,
0xcc13,0x1059,0x566a,0x3f93,
0x9690,0x378a,0x4eac,0xbfca,
0x0000,0x0000,0x0000,0x3ff0,
};
static short cd[28] = {
0xc6b3,0x6a7f,0x9768,0x3d91,
0x75b9,0xdb03,0x7449,0x3e0f,
0x5191,0x02a4,0xc708,0x3e80,
0x1d32,0xfa2a,0xa3ee,0x3ee9,
```

```
0035543,0106171,0177336,0146351,
0037050,0150073,0000607,0171635,
0040200,0000000,0000000,0000000,
};
```

Auxiliary function $f(x)$

```
static short fn[40] = {
0037727,0152216,0106601,0016214,
0037422,0154606,0112710,0071355,
0036474,0143453,0154253,0166545,
0035264,0161606,0022250,0073743,
0033633,0110036,0024653,0136246,
0032003,0036652,0041164,0036413,
0027740,0174122,0046305,0036726,
0025501,0125270,0121317,0167667,
0023032,0150555,0076175,0047443,
0020061,0133570,0070130,0027657,
};
```

Leading 1.0 omitted from fd.

```
static short fd[40] = {
0040100,0063767,0054413,0151452,
0037357,0061566,0007243,0065754,
0036323,0005365,0033552,0133625,
0035043,0101123,0000275,0165402,
0033367,0146614,0110623,0023647,
0031501,0116644,0125222,0144263,
0027436,0062051,0117235,0001411,
0025204,0111543,0056370,0036201,
0022520,0071351,0015227,0122144,
0017554,0172240,0112713,0005006,
};
```

Auxiliary function $g(x)$

```
static short gn[44] = {
0040001,0021435,0120406,0053123,
0037511,0152523,0037703,0122011,
0036631,0134302,0122721,0110235,
0035463,0051712,0043215,0114732,
0034101,0025677,0147725,0057630,
0032323,0010342,0067523,0002206,
0030364,0152247,0110007,0054107,
0026230,0057654,0035464,0047124,
0023706,0036401,0167705,0045440,
0021166,0154447,0105632,0142461,
0016142,0002353,0011175,0170530,
};
```

Leading 1.0 omitted from gd.

```
static short gd[44] = {
0040274,0145551,0016742,0127005,
0037654,0166557,0076416,0015165,
0036717,0140217,0030675,0050111,
```

```
0xd99d,0x3fdb,0x718f,0x3f4c,
0xfe74,0x6030,0x1a07,0x3fa5,
0x0000,0x0000,0x0000,0x3ff0,
};

static short fn[40] = {
0x2391,0xd1b0,0xfa91,0x3fda,
0x0e5e,0xd2b9,0x5b30,0x3fc2,
0x7dad,0x7b15,0x98e5,0x3f87,
0x0efc,0xc495,0x9c70,0x3f36,
0x7795,0xc535,0x7203,0x3ed3,
0x87a1,0x484e,0x67b5,0x3e60,
0xa7bb,0x4998,0x1f0a,0x3ddc,
0xfdf7,0x1459,0x3557,0x3d48,
0xa9e4,0xaf8f,0x5a2d,0x3ca3,
0x05f6,0x0e0b,0x36ef,0x3be6,
};

static short fd[40] = {
0x7a65,0xeb21,0x0cfe,0x3fe8,
0x6d7d,0xc1d4,0xec6e,0x3fbd,
0x56f3,0xa6ed,0x615e,0x3f7a,
0xbd60,0x6017,0x704a,0x3f24,
0x64f5,0x9232,0xf9b1,0x3ebe,
0x5916,0x9552,0x33b4,0x3e48,
0xa061,0x33d3,0xcc85,0x3dc3,
0x0790,0x6b9f,0x926c,0x3d30,
0xf48d,0x2352,0x0e5d,0x3c8a,
0x6141,0x12b9,0x9e94,0x3bcd,
};

static short gn[44] = {
0xcaca,0xb420,0x2463,0x3fe0,
0x7481,0x67f8,0x3aaa,0x3fc9,
0x3214,0x54ba,0x3718,0x3f93,
0xb33b,0x48d1,0x6a79,0x3f46,
0xabf3,0xf9fa,0x2577,0x3ee8,
0x6091,0x4dea,0x621c,0x3e7a,
0xeb09,0xf200,0x9a94,0x3dfe,
0x89cb,0x8766,0x0bf5,0x3d73,
0xa964,0x3df8,0xc7a0,0x3cd8,
0x58a6,0xf173,0xdb24,0x3c2e,
0xbe2b,0x624f,0x409d,0x3b6c,
};

static short gd[44] = {
0x55c1,0x23bc,0x996d,0x3ff7,
0xc34f,0xefa1,0x9dad,0x3fd5,
0xaa09,0xe637,0xf811,0x3f99,
```

```
0035525,0110060,0076405,0070502,   0xae28,0x0fa0,0xb206,0x3f4a,
0034125,0176061,0060120,0031730,   0x067b,0x2c0a,0xbf86,0x3eea,
0032340,0001615,0054343,0120501,   0x7428,0xab1c,0x0071,0x3e7c,
0030375,0041414,0070747,0107060,   0xf1c6,0x8e3c,0xa861,0x3dff,
0026233,0031034,0160757,0074526,   0xef2b,0x9c3d,0x6643,0x3d73,
0023710,0003341,0137100,0144664,   0x1936,0x37c8,0x00dc,0x3cd9,
0021167,0126414,0023774,0015435,   0x8364,0x84ff,0xf5a1,0x3c2e,
0016142,0002353,0011175,0170530,   0xbe2b,0x624f,0x409d,0x3b6c,
};                                 };
#endif                             #endif
```

```
extern double PI, PIO2, MACHEP;

int fresnl( xxa, ssa, cca )
double xxa, *ssa, *cca;
    {
    double f, g, cc, ss, c, s, t, u;
    double nn, x, x2, x3, x4;
    double fabs(), cos(), sin(), polevl(), p1evl();
    int sign;

    x = fabs(xxa);
    x2 = x * x;
    if( x2 < 2.5625 )
        {
        t = x2 * x2;
        ss = x * x2 * polevl( t, sn, 5)
            / p1evl( t, sd, 6 );
        cc = x * polevl( t, cn, 5)/polevl(t, cd, 6 );
        goto done;
        }
    if( x > 36974.0 )
        {
        cc = 0.5;
        ss = 0.5;
        goto done;
        }
```

Auxiliary functions for large argument

```
    x2 = x * x;
    t = PI * x2;
    u = 1.0/(t * t);
    t = 1.0/t;
    f = 1.0 - u * polevl( u, fn, 9)/p1evl(u, fd, 10);
    g = t * polevl( u, gn, 10)/p1evl(u, gd, 11);
    t = PIO2 * x2;
```

```
    c = cos(t);
    s = sin(t);
    t = PI * x;
    cc = 0.5 + (f * s - g * c)/t;
    ss = 0.5 - (f * c + g * s)/t;
done:
    if( xxa < 0.0 )
        {
        cc = -cc;
        ss = -ss;
        }
    *cca = cc;
    *ssa = ss;
    return(0);
    }
```

7.9 Elliptic Functions

7.9.1 $K(m)$

The complete elliptic integral of the first kind is the definite integral

$$K(m) = \int_0^{\pi/2} \frac{dt}{\sqrt{1 - m \, \sin^2 t}} \, .$$

Its domain of definition is $0 \le m \le 1$. There is a logarithmic singularity at $m = 1$.

Hastings' form of approximation[2] for this function,

$$\begin{aligned} m' &= 1 - m \\ K(m) &= P(m') - \ln m' \, Q(m') \end{aligned}$$

has the following coefficients for double precision accuracy.

$P(t) =$	$Q(t) =$
$1.37982864606273237150 \cdot 10^{-4} t^{10}$	$+2.94078955048598507511 \cdot 10^{-5} t^{10}$
$+2.28025724005875567385 \cdot 10^{-3} t^{9}$	$+9.14184723865917226571 \cdot 10^{-4} t^{9}$
$+7.97404013220415179367 \cdot 10^{-3} t^{8}$	$+5.94058303753167793257 \cdot 10^{-3} t^{8}$
$+9.85821379021226008714 \cdot 10^{-3} t^{7}$	$+1.54850516649762399335 \cdot 10^{-2} t^{7}$
$+6.87489687449949877925 \cdot 10^{-3} t^{6}$	$+2.39089602715924892727 \cdot 10^{-2} t^{6}$
$+6.18901033637687613229 \cdot 10^{-3} t^{5}$	$+3.01204715227604046988 \cdot 10^{-2} t^{5}$
$+8.79078273952743772254 \cdot 10^{-3} t^{4}$	$+3.73774314173823228969 \cdot 10^{-2} t^{4}$
$+1.49380448916805252718 \cdot 10^{-2} t^{3}$	$+4.88280347570998239232 \cdot 10^{-2} t^{3}$
$+3.08851465246711995998 \cdot 10^{-2} t^{2}$	$+7.03124996963957469739 \cdot 10^{-2} t^{2}$
$+9.65735902811690126535 \cdot 10^{-2} t$	$+1.24999999999870820058 \cdot 10^{-1} t$
$+1.38629436111989062502 \, ,$	$+4.99999999999999999821 \cdot 10^{-1} \, .$

[2]Cecil Hastings, Jr., *Approximations for Digital Computers*, Princeton, 1955.

When m' is very close to zero, for example less than the value of the machine roundoff error, use

$$K(m) \approx \ln 4 - \tfrac{1}{2} \ln m'$$

with the constant

$$\ln 4 = 1.3862943611198906188 \; .$$

For checking purposes,

$$K(0) = \frac{\pi}{2} \; .$$

7.9.2 ellpk.c

Complete elliptic integral of the first kind

```
#include "mconf.h"

#ifdef DEC
static short P[] =
{
0035020,0127576,0040430,0051544,
0036025,0070136,0042703,0153716,
0036402,0122614,0062555,0077777,
0036441,0102130,0072334,0025172,
0036341,0043320,0117242,0172076,
0036312,0146456,0077242,0154141,
0036420,0003467,0013727,0035407,
0036564,0137263,0110651,0020237,
0036775,0001330,0144056,0020305,
0037305,0144137,0157521,0141734,
0040261,0071027,0173721,0147572
};
static short Q[] =
{
0034366,0130371,0103453,0077633,
0035557,0122745,0173515,0113016,
0036302,0124470,0167304,0074473,
0036575,0132403,0117226,0117576,
0036703,0156271,0047124,0147733,
0036766,0137465,0002053,0157312,
0037031,0014423,0154274,0176515,
0037107,0177747,0143216,0016145,
0037217,0177777,0172621,0074000,
0037377,0177777,0177776,0156435,
0040000,0000000,0000000,0000000
};
```

ln 4

```
#ifdef IIEEE
static short P[] =
{
0x0a6d,0xc823,0x15ef,0x3f22,
0x7afa,0xc8b8,0xae0b,0x3f62,
0xb000,0x8cad,0x54b1,0x3f80,
0x854f,0x0e9b,0x308b,0x3f84,
0x5e88,0x13d4,0x28da,0x3f7c,
0x5b0c,0xcfd4,0x59a5,0x3f79,
0xe761,0xe2fa,0x00e6,0x3f82,
0x2414,0x7235,0x97d6,0x3f8e,
0xc419,0x1905,0xa05b,0x3f9f,
0x387c,0xfbea,0xb90b,0x3fb8,
0x39ef,0xfefa,0x2e42,0x3ff6
};
static short Q[] =
{
0x6ff3,0x30e5,0xd61f,0x3efe,
0xb2c2,0xbee9,0xf4bc,0x3f4d,
0x8f27,0x1dd8,0x5527,0x3f78,
0xd3f0,0x73d2,0xb6a0,0x3f8f,
0x99fb,0x29ca,0x7b97,0x3f98,
0x7bd9,0xa085,0xd7e6,0x3f9e,
0x9faa,0x7b17,0x2322,0x3fa3,
0xc38d,0xf8d1,0xfffc,0x3fa8,
0x2f00,0xfeb2,0xffff,0x3fb1,
0xdba4,0xffff,0xffff,0x3fbf,
0x0000,0x0000,0x0000,0x3fe0
};
```

```
#if UNK
static double C1 = 1.3862943611198906188E0;
#endif
static short ac1[] = {
0040261,0071027,0173721,0147572};
#define C1 (*(double *)ac1)
#endif
static short ac1[] = {
0x39ef,0xfefa,0x2e42,0x3ff6};
#define C1 (*(double *)ac1)
#endif

extern double MACHEP, MAXNUM;

double ellpk(x)
double x;
    {
    double polevl(), p1evl(), log();

    if( (x < 0.0) || (x > 1.0) )
        {
        mtherr( "ellpk", DOMAIN );
        return( 0.0 );
        }
    if( x > MACHEP )
        {
        return( polevl(x,P,10)
            - log(x) * polevl(x,Q,10) );
        }
    else
        {
        if( x == 0.0 )
            {
            mtherr( "ellpk", SING );
            return( MAXNUM );
            }
        else
            {
            return( C1 - 0.5 * log(x) );
            }
        }
    }
```

7.9.3 $F(\phi|m)$

$$F(\phi|m) = \int_0^\phi \frac{dt}{\sqrt{1 - m\ sin^2 t}}$$

is the incomplete elliptic integral of the first kind. Its arguments are the amplitude ϕ and the modulus m. It may be computed by the Gauss-Legendre *arithmetic-geometric mean algorithm* (A.G.M.) or by expansion in theta functions. The A.G.M. begins by initializing

$$
\begin{aligned}
m' &= 1 - m \\
a_0 &= 1 \\
b_0 &= \sqrt{m'} \\
c_0 &= \sqrt{m} \\
d_0 &= 1 \\
t_0 &= \tan\phi \\
\mu_0 &= \frac{\phi + \pi/2}{\pi} .
\end{aligned}
$$

The iterative step is

$$
\begin{aligned}
\phi_{i+1} &= \phi_i + \mu_i \pi + \tan^{-1}(b_i t_i / a_i) \\
\mu_{i+1} &= \frac{\phi_i + \pi/2}{\pi} \\
t_{i+1} &= t_i \, \frac{1 + b_i/a_i}{1 - t_i^2 b_i/a_i} \\
c_{i+1} &= \tfrac{1}{2}(a_i - b_i) \\
a_{i+1} &= \tfrac{1}{2} t_i (a_i + b_i) \\
b_{i+1} &= \sqrt{a_i b_i} \\
d_{i+1} &= 2 d_i .
\end{aligned}
$$

Iteration continues until $|c/a|$ is below the machine roundoff error. Then

$$F(\phi|m) \approx \frac{\mu\pi + \tan^{-1} t}{a\, d}$$

where all the values are taken from the result of the final iteration step. If $\phi < 0$, then replace ϕ by $-\phi$ and invert the sign of the answer. Other special cases are

$$F(\phi|0) = \phi$$

and

$$F(\phi|1) = \ln\left[\tan(\tfrac{1}{4}\pi + \tfrac{1}{2}\phi)\right] .$$

7.9.4 ellik.c

Incomplete elliptic integral of the first kind

```
extern double PI, PIO2, MACHEP;
```

```
double ellik( phi, m )
double phi, m;
     {
     double a, b, c, temp;
     double t, step;
     double sqrt(), fabs(), log(), tan(), atan();
     int d, mod, sign;

     if( m == 0.0 )
          return( phi );
     if( phi < 0.0 )
          {
          phi = -phi;
          sign = -1;
          }
     else
          sign = 0;
     a = 1.0;
     b = 1.0 - m;
     if( b == 0.0 )
          return( log( tan( (PIO2 + phi)/2.0 ) ) );
     b = sqrt(b);
     c = sqrt(m);
     d = 1;
     t = tan( phi );
     mod = (phi + PIO2)/PI;
     while( fabs(c/a) > MACHEP )
          {
          temp = b/a;
          phi = phi + atan(t*temp) + mod * PI;
          mod = (phi + PIO2)/PI;
          t = t * ( 1.0 + temp )/( 1.0 - temp * t * t );
          c = ( a - b )/2.0;
          temp = sqrt( a * b );
          a = ( a + b )/2.0;
          b = temp;
          d += d;
          }
     temp = (atan(t) + mod * PI)/(d * a);
     if( sign < 0 )
          temp = -temp;
     return( temp );
     }
```

7.9.5 $E(m)$

The complete elliptic integral of the second kind is

$$E(m) = \int_0^{\pi/2} \sqrt{1 - m\sin^2 t}\, dt$$

defined for $0 \le m \le 1$. There are no singularites;

$$\begin{aligned} E(1) &= 1 \\ E(0) &= \pi/2 . \end{aligned}$$

The integral is approximated by

$$\begin{aligned} m' &= 1 - m \\ E(m) &\approx P(m') - m'Q(m')\ln m' \end{aligned}$$

where

$$\begin{aligned} P(t) = {} & 1.53552577301013293365 \cdot 10^{-4} t^{10} \\ & +2.50888492163602060990 \cdot 10^{-3} t^{9} \\ & +8.68786816565889628429 \cdot 10^{-3} t^{8} \\ & +1.07350949056076193403 \cdot 10^{-2} t^{7} \\ & +7.77395492516787092951 \cdot 10^{-3} t^{6} \\ & +7.58395289413514708519 \cdot 10^{-3} t^{5} \\ & +1.15688436810574127319 \cdot 10^{-2} t^{4} \\ & +2.18317996015557253103 \cdot 10^{-2} t^{3} \\ & +5.68051945617860553470 \cdot 10^{-2} t^{2} \\ & +4.43147180560990850618 \cdot 10^{-1} t \\ & +1.00000000000000000299 , \end{aligned}$$

$$\begin{aligned} Q(t) = {} & 3.27954898576485872656 \cdot 10^{-5} t^{9} \\ & +1.00962792679356715133 \cdot 10^{-3} t^{8} \\ & +6.50609489976927491433 \cdot 10^{-3} t^{7} \\ & +1.68862163993311317300 \cdot 10^{-2} t^{6} \\ & +2.61769742454493659583 \cdot 10^{-2} t^{5} \\ & +3.34833904888224918614 \cdot 10^{-2} t^{4} \\ & +4.27180926518931511717 \cdot 10^{-2} t^{3} \\ & +5.85936634471101055642 \cdot 10^{-2} t^{2} \\ & +9.37499997197644278445 \cdot 10^{-2} t \\ & +2.49999999999888314361 \cdot 10^{-1} . \end{aligned}$$

7.9.6 ellpe.c

Complete elliptic integral of the second kind

```
#include "mconf.h"

#ifdef DEC
static short P[] = {
0035041,0001364,0141572,0117555,
0036044,0066032,0130027,0033404,
0036416,0053617,0064456,0102632,
0036457,0161100,0061177,0122612,
0036376,0136251,0012403,0124162,
0036370,0101316,0151715,0131613,
0036475,0105477,0050317,0133272,
0036662,0154232,0024645,0171552,
```

```
#ifdef IIEEE
static short P[] = {
0x53ee,0x986f,0x205e,0x3f24,
0xe6e0,0x5602,0x8d83,0x3f64,
0xd0b3,0xed25,0xcaf1,0x3f81,
0xf4b1,0x0c4f,0xfc48,0x3f85,
0x750e,0x22a0,0xd795,0x3f7f,
0xb671,0xda79,0x1059,0x3f7f,
0xf6d7,0xea19,0xb167,0x3f87,
0xbe6d,0x4534,0x5b13,0x3f96,
```

```
0037150,0126220,0047054,0030064,    0x8607,0x09c5,0x1592,0x3fad,
0037742,0162057,0167645,0165612,    0xbd71,0xfdf4,0x5c85,0x3fdc,
0040200,0000000,0000000,0000000     0x0000,0x0000,0x0000,0x3ff0
};                                  };
static short Q[] = {                static short Q[] = {
0034411,0106743,0115771,0055462,    0x2b66,0x737f,0x31bc,0x3f01,
0035604,0052575,0155171,0045540,    0x296c,0xbb4f,0x8aaf,0x3f50,
0036325,0030424,0064332,0167756,    0x5dfe,0x8d1b,0xa622,0x3f7a,
0036612,0052366,0063006,0115175,    0xd350,0xccc0,0x4a9e,0x3f91,
0036726,0070430,0004533,0124654,    0x7535,0x012b,0xce23,0x3f9a,
0037011,0022741,0030675,0030711,    0xa639,0x2637,0x24bc,0x3fa1,
0037056,0174452,0127062,0132122,    0x568a,0x55c6,0xdf25,0x3fa5,
0037157,0177750,0142041,0072523,    0x2eaa,0x1884,0xfffd,0x3fad,
0037277,0177777,0173137,0002627,    0xe0b3,0xfecb,0xffff,0x3fb7,
0037577,0177777,0177777,0101101     0xf048,0xffff,0xffff,0x3fcf
};                                  };
#endif                              #endif
```

```
double ellpe(x)
double x;
	{
	double polevl(), log();

	if( (x <= 0.0) || (x > 1.0) )
		{
		if( x == 0.0 )
			return( 1.0 );
		mtherr( "ellpe", DOMAIN );
		return( 0.0 );
		}
	return( polevl(x,P,10)
		- log(x) * (x * polevl(x,Q,9)) );
	}
```

7.9.7 $E(\phi|m)$

$$E(\phi|m) = \int_0^{\phi} \sqrt{1 - m \sin^2 t}\; dt$$

is the incomplete elliptic integral of the second kind. It is computed by the arithmetic-geometric mean algorithm. Special cases are

$$\begin{aligned} E(\phi|0) &= \phi \\ E(\phi|1) &= \sin \phi \\ E(-\phi|m) &= E(\phi|m) \,. \end{aligned}$$

Initial conditions for the iteration are

$$\begin{aligned} m' &= 1 - m \\ a_0 &= 1 \\ b_0 &= \sqrt{m'} \\ c_0 &= \sqrt{m} \\ d_0 &= 1 \\ e_0 &= 0 \\ t_0 &= \tan\phi \\ \mu_0 &= \frac{\phi + \pi/2}{\pi} . \end{aligned}$$

The iterative step is

$$\begin{aligned} \phi_{i+1} &= \phi_i + \mu_i\pi + \tan^{-1}(b_i t_i/a_i) \\ \mu_{i+1} &= \frac{\phi_i + \pi/2}{\pi} \\ t_{i+1} &= t_i \, \frac{1 + b_i/a_i}{1 - t_i^2 b_i/a_i} \\ c_{i+1} &= \tfrac{1}{2}(a_i - b_i) \\ a_{i+1} &= \tfrac{1}{2}(a_i + b_i) \\ b_{i+1} &= \sqrt{a_i b_i} \\ d_{i+1} &= 2d_i \\ e_{i+1} &= e_i + c_{i+1}\sin\phi_{i+1} . \end{aligned}$$

This iteration is identical to the procedure for computing $F(\phi|m)$ but with the addition of the term e. Iteration continues until $|c/a|$ is below the machine roundoff error. Then

$$E(\phi|m) \approx \frac{E(m)}{K(m)} \, \frac{\mu\pi + \tan^{-1} t}{a\ d} + e$$

where all the values are taken from the result of the final iteration step.

7.9.8 ellie.c

Incomplete elliptic integral of the second kind

```
extern double PI, PIO2, MACHEP;

double ellie( phi, m )
double phi, m;
```

```
{
double a, b, c, e, temp;
double lphi, t, step;
double sqrt(), fabs(), log(), sin(), tan(), atan();
double ellpe(), ellpk();
int d, mod, sign;

if( m == 0.0 )
     return( phi );
if( m == 1.0 )
     return( sin(phi) );
lphi = phi;
if( lphi < 0.0 )
     lphi = -lphi;
a = 1.0;
b = 1.0 - m;
b = sqrt(b);
c = sqrt(m);
d = 1;
e = 0.0;
t = tan( lphi );
mod = (lphi + PIO2)/PI;
while( fabs(c/a) > MACHEP )
     {
     temp = b/a;
     lphi = lphi + atan(t*temp) + mod * PI;
     mod = (lphi + PIO2)/PI;
     t = t * ( 1.0 + temp )/( 1.0 - temp * t * t );
     c = ( a - b )/2.0;
     temp = sqrt( a * b );
     a = ( a + b )/2.0;
     b = temp;
     d += d;
     e += c * sin(lphi);
     }
b = 1.0 - m;
temp = ellpe(b)/ellpk(b);
temp *= (atan(t) + mod * PI)/(d * a);
temp += e;
if( phi < 0.0 )
     temp = -temp;
return( temp );
}
```

7.9.9 Jacobian Elliptic Functions

These functions are periodic, with quarter-period on the real axis equal to the complete elliptic integral $K(1-m)$. They are denoted by functions $\mathrm{sn}(u|m)$, $\mathrm{cn}(u|m)$, and $\mathrm{dn}(u|m)$ of parameter m between 0 and 1, and real argument u. If $u = F(\phi|m)$, then $\mathrm{sn}(u|m) = \sin\phi$ and $\mathrm{cn}(u|m) = \cos\phi$. ϕ is called the amplitude of u. Computation is by means of the arithmetic-geometric mean algorithm, except when m is close to 0 or 1. In that case with m close to 1, the approximation given here applies only for $\phi < \pi/2$.

The approximations for $0 \le m < 10^{-9}$ are

$$\begin{aligned} d &= \tfrac{1}{4}m(u - \sin u \cos u) \\ \mathrm{sn}(u|m) &\approx \sin u - d\cos u \\ \mathrm{cn}(u|m) &\approx \cos u + d\sin u \\ \mathrm{dn}(u|m) &\approx 1 - \tfrac{1}{2}m\sin^2 u \\ \phi &\approx u - d\,. \end{aligned}$$

If $1 \ge m \ge 1 - 10^{-9}$, the approximations are

$$\begin{aligned} d &= \tfrac{1}{4}(1-m) \\ \mathrm{sn}(u|m) &\approx \tanh u + d\frac{\cosh u \sinh u - u}{\cosh^2 u} \\ \phi &\approx 2\tan^{-1}(e^u) - \pi/2 + d\frac{\cosh u \sinh u - u}{\cosh u} \\ e &= d\phi \tanh u \\ \mathrm{cn}(u|m) &\approx \phi - e(\cosh u \sinh u - u) \\ \mathrm{dn}(u|m) &\approx \phi + e(\cosh u \sinh u + u)\,. \end{aligned}$$

Initialization for the arithmetic-geometric mean scale is

$$\begin{aligned} a_0 &= 1 \\ b_0 &= \sqrt{1-m} \\ c_0 &= \sqrt{m} \\ w_0 &= 1\,. \end{aligned}$$

The iteration step is

$$\begin{aligned} a_{i+1} &= \tfrac{1}{2}(a_i + b_i) \\ b_{i+1} &= \sqrt{a_i b_i} \\ c_{i+1} &= \tfrac{1}{2}(a_i - b_i) \\ w_{i+1} &= 2w_i\,. \end{aligned}$$

This continues until $|c/a|$ is less than the machine roundoff error: Storage arrays must be supplied to hold the intermediate values a_i and c_i. Eight

Function	Arithmetic	Domain	Trials	Peak	RMS
ellpk	DEC	$0,1$	16000	$3.5 \cdot 10^{-17}$	$1.1 \cdot 10^{-17}$
ellpk	IEEE	$0,1$	30000	$2.5 \cdot 10^{-16}$	$6.8 \cdot 10^{-17}$
ellik	DEC	$0,1$	3700	$8.1 \cdot 10^{-17}$	$2.5 \cdot 10^{-17}$
ellik	IEEE	$0,1$	10000	$6.0 \cdot 10^{-16}$	$1.4 \cdot 10^{-16}$
ellpe	DEC	$0,1$	13000	$3.1 \cdot 10^{-17}$	$9.4 \cdot 10^{-18}$
ellpe	IEEE	$0,1$	10000	$2.1 \cdot 10^{-16}$	$7.3 \cdot 10^{-17}$
ellie	DEC	$0,1$	2000	$1.9 \cdot 10^{-16}$	$3.4 \cdot 10^{-17}$
ellie	IEEE	$0,1$	10000	$2.2 \cdot 10^{-15}$	$2.1 \cdot 10^{-16}$
sn	DEC	$0,1$	1800	$4.5 \cdot 10^{-16}$	$8.7 \cdot 10^{-17}$
sn	IEEE	$0,1$	50000	$4.1 \cdot 10^{-15}$	$4.6 \cdot 10^{-16}$
cn	IEEE	$0,1$	40000	$3.6 \cdot 10^{-15}$	$4.4 \cdot 10^{-16}$
dn	IEEE	$0,1$	10000	$1.3 \cdot 10^{-12}$	$1.8 \cdot 10^{-14}$
phi	IEEE	$0,1$	10000	$9.2 \cdot 10^{-16}$	$1.4 \cdot 10^{-16}$

Table 7.12: Elliptic functions, relative error, except absolute for sn, cn, dn, phi.

locations are sufficient for a double precision routine. Having computed a, c, and w, a reduced value of ϕ is found by backward recurrence. The initial condition is

$$\phi_n = w_n a_n u \; .$$

The recurrence is

$$\phi_{i-1} = \tfrac{1}{2}[\phi_i + \sin^{-1}(c_i \sin\phi_i / a_i)] \; .$$

This is repeated until $i = 0$. At each step the previous value of ϕ should be preserved so that at the end both ϕ_0 and ϕ_1 will be available. Then

$$\begin{aligned} \mathrm{sn}(u|m) &= \sin\phi_0 \\ \mathrm{cn}(u|m) &= \cos\phi_0 \\ \mathrm{dn}(u|m) &= \frac{\cos\phi_0}{\cos(\phi_0 - \phi_1)} \; . \end{aligned}$$

Table 7.12 gives typical accuracy measurements for the elliptic function programs. For ellie and ellik, $0 \le \phi \le 2$. For the Jacobian elliptic function $\mathrm{sn}(u|m)$, $0 \le u \le 10$, $0 \le m \le 1$ and the error criterion is absolute. Peak absolute error observed in a consistency check using the addition theorem

$$\mathrm{sn}(u+v) = \frac{\mathrm{sn}(u)\;\mathrm{cn}(v)\;\mathrm{dn}(v) + \mathrm{sn}(v)\;\mathrm{cn}(u)\;\mathrm{dn}(u)}{1 - m\;\mathrm{sn}^2(u)\;\mathrm{sn}^2(v)}$$

was $4 \cdot 10^{-16}$. The program was also tested by the relation given above to the incomplete elliptic integral $F(\phi|m)$. Accuracy of the program deteriorates when u is large.

7.9.10 ellpj.c

Jacobian elliptic functions

```
extern double PIO2, MACHEP;

double ellpj( u, m, sn, cn, dn, ph )
double u, m;
double *sn, *cn, *dn, *ph;
    {
    double ai, b, phi, t, twon;
    double sqrt(), fabs(), sin(), cos(), asin();
    double tanh(), sinh(), cosh(), atan(), exp();
    double a[9], c[9];
    int i;

    if( m < 0.0 || m > 1.0 )
        {
        mtherr( "ellpj", DOMAIN );
        return(0.0);
        }
    if( m < 1.0e-9 )
        {
        t = sin(u);
        b = cos(u);
        ai = 0.25 * m * (u - t*b);
        *sn = t - ai*b;
        *cn = b + ai*t;
        *ph = u - ai;
        *dn = 1.0 - 0.5*m*t*t;
        return;
        }
    if( m >= 0.9999999999 )
        {
        ai = 0.25 * (1.0-m);
        b = cosh(u);
        t = tanh(u);
        phi = 1.0/b;
        twon = b * sinh(u);
        *sn = t + ai * (twon - u)/(b*b);
        *ph = 2.0*atan(exp(u)) - PIO2 + ai*(twon - u)/b;
```

```
          ai *= t * phi;
          *cn = phi - ai * (twon - u);
          *dn = phi + ai * (twon + u);
          return;
          }
A. G. M. scale
     a[0] = 1.0;
     b = sqrt(1.0 - m);
     c[0] = sqrt(m);
     twon = 1.0;
     i = 0;
     while( fabs(c[i]/a[i]) > MACHEP )
          {
          if( i > 7 )
               {
               mtherr( "ellpj", OVERFLOW );
               goto done;
               }
          ai = a[i];
          ++i;
          c[i] = ( ai - b )/2.0;
          t = sqrt( ai * b );
          a[i] = ( ai + b )/2.0;
          b = t;
          twon *= 2.0;
          }
done:
backward recurrence
     phi = twon * a[i] * u;
     do
          {
          t = c[i] * sin(phi) / a[i];
          b = phi;
          phi = (asin(t) + phi)/2.0;
          }
     while( --i );
     *sn = sin(phi);
     t = cos(phi);
     *cn = t;
     *dn = t/cos(phi-b);
     *ph = phi;
     return;
     }
```

Table 7.13: Expansion Coefficients for Zeta Function

j	$(2j)!/B_{2j}$
0	12
1	−720
2	30240
3	−1209600
4	47900160
5	$-1.307674368 \cdot 10^{12}/691$
6	$7.47242496 \cdot 10^{10}$
7	$-1.067062284288 \cdot 10^{16}/3617$
8	$5.109094217170944 \cdot 10^{18}/43867$
9	$-8.028576626982912 \cdot 10^{20}/174611$
10	$1.5511210043330985984 \cdot 10^{23}/854513$
11	$-1.6938241367317436694528 \cdot 10^{27}/236364091$

7.10 Zeta Functions

The zeta function is defined with either one or two arguments. The Hurwitz version with two arguments is

$$\zeta(x, q) = \sum_{k=0}^{\infty}(k+q)^{-x}$$

where $x > 1$ and q is not a negative integer or zero. The Euler-Maclaurin summation formula may be used to derive the expansion

$$\begin{aligned}\zeta(x, q) &= \sum_{k=1}^{n}(k+q)^{-x} + \frac{(n+q)^{1-x}}{x-1} - \frac{1}{2(n+q)^{x}} \\ &\quad + \sum_{j=0}^{\infty} \frac{B_{2j}x(x+1)\cdots(x+2j)}{(2j)!\,(n+q)^{x+2j+1}}\end{aligned}$$

where the B_{2j} are Bernoulli numbers. Exact expressions for the expansion coefficients are given in Table 7.13. A value of $n = 9$ is suitable for double precision calculation.

7.10.1 hurwiz.c

Two argument Hurwitz zeta function program

```
#include "mconf.h"
extern double MAXNUM, MACHEP;
```

Expansion coefficients for Euler-Maclaurin summation formula $(2k)!/B_{2k}$, where B_{2k} are Bernoulli numbers

```
static double A[] = {
    12.0,
    -720.0,
    30240.0,
    -1209600.0,
    47900160.0,
    -1.8924375803183791606e9,
    7.47242496e10,
    -2.950130727918164224e12,
    1.1646782814350067249e14,
    -4.5979787224074726105e15,
    1.8152105401943546773e17,
    -7.1661652561756670113e18
    };

double hurwiz(x,q)
double x,q;
    {
    int i;
    double a, b, k, s, w;
    double apq, t;
    double fabs(), pow();

    if( x == 1.0 )
        return( MAXNUM );
    if( x < 1.0 )
        {
        mtherr("zeta", DOMAIN );
        return(0.0);
        }
```

Euler-Maclaurin summation formula

```
    w = 9.0;
    s = pow( q, -x );
    a = q;
    for( i=0; i<9; i++ )
        {
        a += 1.0;
        b = pow( a, -x );
        s += b;
        if( b/s < MACHEP )
```

```
                goto done;
            }
        w = a;
        s += b*w/(x-1.0);
        s -= 0.5 * b;
        a = 1.0;
        k = 0.0;
        for( i=0; i<12; i++ )
            {
            a *= x + k;
            b /= w;
            t = a*b/A[i];
            s = s + t;
            t = fabs(t/s);
            if( t < MACHEP )
                goto done;
            k += 1.0;
            a *= x + k;
            b /= w;
            k += 1.0;
            }
done:
        return(s);
        }
```

7.10.2 Riemann Zeta Function

The definition of Riemann's zeta function[3] is

$$\zeta(x) = \sum_{k=1}^{\infty} k^{-x} .$$

There is a singularity at $x = 1$. The Euler-Maclaurin summation formula applied to $\zeta(x)$ yields the expansion

$$\begin{aligned}\zeta(x) &= \sum_{k=1}^{n} k^{-x} + \frac{n^{1-x}}{x-1} - \frac{1}{2n^x} \\ &\quad + \sum_{j=0}^{\infty} \frac{B_{2j} x(x+1)\cdots(x+2j)}{(2j)!\; n^{x+2j+1}} .\end{aligned}$$

[3]See Jahnke, E., and F. Emde, *Tables of Functions*, G. E. Stechert & Co., 1941, pp 269-274. For the Euler-Maclaurin formula and Bernoulli numbers, see Fröberg, Carl-Erik, *Introduction to Numerical Analysis*, 2nd ed., Addison-Wesley, 1969, pp 224-233.

A reflection formula is used for negative x:

$$\zeta(x) = (2\pi)^x \sin(\tfrac{1}{2}\pi x)\ \Gamma(1-x)\ \zeta(1-x)\ .$$

An overflow error may occur for large negative x, due to the gamma function in the reflection formula. If $x > 25$ the defining Dirichlet series for $\zeta(x)$ may be used for computation. For certain applications the zeta function is needed only at integer valued arguments. Table 7.14 gives $\zeta(x) - 1$ for x from 0 through 30.

The following rational approximations give double precision accuracy. For $0 \le x < 1$,

$$(1-x)(\zeta(x)-1) \approx \frac{R(x)}{S(x)}$$

where

$R(t) =$	$Q(t) =$
$-3.28717474506562731748 \cdot 10^{-1}t^5$	$1.0t^5$
$+1.55162528742623950834 \cdot 10^{1}t^4$	$+1.95107674914060531512 \cdot 10^{1}t^4$
$-2.48762831680821954401 \cdot 10^{2}t^3$	$+3.17710311750646984099 \cdot 10^{2}t^3$
$+1.01050368053237678329 \cdot 10^{3}t^2$	$+3.03835500874445748734 \cdot 10^{3}t^2$
$+1.26726061410235149405 \cdot 10^{4}t$	$+2.03665876435770579345 \cdot 10^{4}t$
$-1.11578094770515181334 \cdot 10^{5}$,	$+7.43853965136767874343 \cdot 10^{4}$.

For $1 < x \le 10$,

$$(1-1/x)2^x(\zeta(x)-1) \approx \frac{P(1/x)}{Q(1/x)}$$

where

$P(t) =$	$Q(t) =$
$+5.85746514569725319540 \cdot 10^{11}t^8$	$1.0t^8$
$+2.57534127756102572888 \cdot 10^{11}t^7$	$+3.90497676373371157516 \cdot 10^{11}t^7$
$+4.87781159567948256438 \cdot 10^{10}t^6$	$+5.22858235368272161797 \cdot 10^{10}t^6$
$+5.15399538023885770696 \cdot 10^{9}t^5$	$+5.64451517271280543351 \cdot 10^{9}t^5$
$+3.41646073514754094281 \cdot 10^{8}t^4$	$+3.39006746015350418834 \cdot 10^{8}t^4$
$+1.60837006880656492731 \cdot 10^{7}t^3$	$+1.79410371500126453702 \cdot 10^{7}t^3$
$+5.92785467342109522998 \cdot 10^{5}t^2$	$+5.66666825131384797029 \cdot 10^{5}t^2$
$+1.51129169964938823117 \cdot 10^{4}t$	$+1.60382976810944131506 \cdot 10^{4}t$
$+2.01822444485997955865 \cdot 10^{2}$,	$+1.96436237223387314144 \cdot 10^{2}$.

In the above expression the quantity $1 - 1/x$ should be computed as $(x - 1)/x$. This eliminates cancellation error when x is near 1, assuming that x is exact.

For large x, $\zeta(x) - 1$ is close to 2^{-x}, and $\zeta(x) - 1 - 2^{-x}$ is close to 3^{-x}. The logarithm of either of these expressions is nearly proportional to x. This suggests an approximation of the form

$$\zeta(x) - 1 \approx e^{R(x)}\ .$$

Table 7.14: Riemann Zeta Function for Integer Arguments

x	$\zeta(x) - 1$
0	-1.5
1	∞
2	$6.44934066848226436472 \cdot 10^{-1}$
3	$2.02056903159594285400 \cdot 10^{-1}$
4	$8.23232337111381915160 \cdot 10^{-2}$
5	$3.69277551433699263314 \cdot 10^{-2}$
6	$1.73430619844491397145 \cdot 10^{-2}$
7	$8.34927738192282683980 \cdot 10^{-3}$
8	$4.07735619794433937869 \cdot 10^{-3}$
9	$2.00839282608221441785 \cdot 10^{-3}$
10	$9.94575127818085337146 \cdot 10^{-4}$
11	$4.94188604119464558702 \cdot 10^{-4}$
12	$2.46086553308048298638 \cdot 10^{-4}$
13	$1.22713347578489146752 \cdot 10^{-4}$
14	$6.12481350587048292585 \cdot 10^{-5}$
15	$3.05882363070204935517 \cdot 10^{-5}$
16	$1.52822594086518717326 \cdot 10^{-5}$
17	$7.63719763789976227360 \cdot 10^{-6}$
18	$3.81729326499983985646 \cdot 10^{-6}$
19	$1.90821271655393892566 \cdot 10^{-6}$
20	$9.53962033872796113152 \cdot 10^{-7}$
21	$4.76932986787806463117 \cdot 10^{-7}$
22	$2.38450502727732990004 \cdot 10^{-7}$
23	$1.19219925965311073068 \cdot 10^{-7}$
24	$5.96081890512594796124 \cdot 10^{-8}$
25	$2.98035035146522801861 \cdot 10^{-8}$
26	$1.49015548283650412347 \cdot 10^{-8}$
27	$7.45071178983542949198 \cdot 10^{-9}$
28	$3.72533402478845705482 \cdot 10^{-9}$
29	$1.86265972351304900640 \cdot 10^{-9}$
30	$9.31327432419668182872 \cdot 10^{-10}$

Arithmetic	Domain	Trials	Peak	RMS
IEEE	1, 50	10000	$9.8 \cdot 10^{-16}$	$1.3 \cdot 10^{-16}$
DEC	1, 50	2000	$1.1 \cdot 10^{-16}$	$1.9 \cdot 10^{-17}$

Table 7.15: `zetac.c`, relative error

However, $2^{-x} = \mathbf{pow}(2.0, -x)$ can be computed more accurately than the exponential of the imprecise quantity $R(x)$. Therefore the second expression is a more suitable form. The approximation chosen is

$$\zeta(x) - 1 \approx 2^{-x} + e^{P(x)/Q(x)}$$

where, for $10 \leq x \leq 50$,

$P(t) =$	$Q(t) =$
$8.70728567484590192539 \cdot 10^{6} t^{10}$	$1.0 t^{10}$
$+1.76506865670346462757 \cdot 10^{8} t^{9}$	$-7.92625410563741062861 \cdot 10^{6} t^{9}$
$+2.60889506707483264896 \cdot 10^{10} t^{8}$	$-1.60529969932920229676 \cdot 10^{8} t^{8}$
$+5.29806374009894791647 \cdot 10^{11} t^{7}$	$-2.37669260975543221788 \cdot 10^{10} t^{7}$
$+2.26888156119238241487 \cdot 10^{13} t^{6}$	$-4.80319584350455169857 \cdot 10^{11} t^{6}$
$+3.31884402932705083599 \cdot 10^{14} t^{5}$	$-2.07820961754173320170 \cdot 10^{13} t^{5}$
$+5.13778997975868230192 \cdot 10^{15} t^{4}$	$-2.96075404507272223680 \cdot 10^{14} t^{4}$
$-1.98123688133907171455 \cdot 10^{15} t^{3}$	$-4.86299103694609136686 \cdot 10^{15} t^{3}$
$-9.92763810039983572356 \cdot 10^{16} t^{2}$	$+5.34589509675789930199 \cdot 10^{15} t^{2}$
$+7.82905376180870586444 \cdot 10^{16} t$	$+5.71464111092297631292 \cdot 10^{16} t$
$+9.26786275768927717187 \cdot 10^{16}$,	$-1.79915597658676556828 \cdot 10^{16}$.

If $x > 50$, then

$$\zeta(x) - 1 \approx 2^{-x} + 3^{-x} + 4^{-x}$$

to double precision accuracy. Each of the terms will underflow for sufficiently large x, so the program for this should check the argument before calling the exponentiation function. The exact point at which underflow will occur depends on implementation details of the arithmetic. Accuracy of the zeta function program is shown in Table 7.15. Tabulated values at integer x have full machine accuracy, of course.

7.10.3 zetac.c

Riemann zeta function, minus one

```
#include "mconf.h"
```

```
extern double MAXNUM, PI;
```

Tabulation of $\zeta(x)$ for integer argument

```
#ifdef DEC
static short azetac[] = {
0140300,0000000,0000000,0000000,
0077777,0177777,0177777,0177777,
0040045,0015146,0022460,0076462,
0037516,0164001,0036001,0104116,
0037250,0114425,0061754,0022033,
0037027,0040616,0145174,0146670,
0036616,0011411,0100444,0104437,
0036410,0145550,0051474,0161067,
0036205,0115527,0141434,0133506,
0036003,0117475,0100553,0053403,
0035602,0056147,0045567,0027703,
0035401,0106157,0111054,0145242,
0035201,0002455,0113151,0101015,
0035000,0126235,0004273,0157260,
0034600,0071127,0112647,0005261,
0034400,0045736,0057610,0157550,
0034200,0031146,0172621,0074172,
0034000,0020603,0115503,0032007,
0033600,0013114,0124672,0023135,
0033400,0007330,0043715,0151117,
0033200,0004742,0145043,0033514,
0033000,0003225,0152624,0004411,
0032600,0002143,0033166,0035746,
0032400,0001354,0074234,0026143,
0032200,0000762,0147776,0170220,
0032000,0000514,0072452,0130631,
0031600,0000335,0114266,0063315,
0031400,0000223,0132710,0041045,
0031200,0000142,0073202,0153426,
0031000,0000101,0121400,0152065,
0030600,0000053,0140525,0072761
};
```

```
#ifdef IIEEE
static short azetac[] = {
0x0000,0x0000,0x0000,0xbff8,
0xffff,0xffff,0xffff,0x7fef,
0x0fa6,0xc4a6,0xa34c,0x3fe4,
0x310a,0x2780,0xdd00,0x3fc9,
0x8483,0xac7d,0x1322,0x3fb5,
0x99b7,0xd94f,0xe831,0x3fa2,
0x9124,0x3024,0xc261,0x3f91,
0x9c47,0x0a67,0x196d,0x3f81,
0x96e9,0xf863,0xb36a,0x3f70,
0x6ae0,0xb02d,0x73e7,0x3f60,
0xe5f8,0xe96e,0x4b8c,0x3f50,
0x9954,0xf245,0x318d,0x3f40,
0x3042,0xb2cd,0x20a5,0x3f30,
0x7bd6,0xa117,0x1593,0x3f20,
0xe156,0xf2b4,0x0e4a,0x3f10,
0x1bed,0xcbf1,0x097b,0x3f00,
0x2f0f,0xdeb2,0x064c,0x3ef0,
0x6681,0x7368,0x0430,0x3ee0,
0x44cc,0x9537,0x02c9,0x3ed0,
0xba4a,0x08f9,0x01db,0x3ec0,
0x66ea,0x5944,0x013c,0x3eb0,
0x8121,0xbab2,0x00d2,0x3ea0,
0xc77d,0x66ce,0x008c,0x3e90,
0x858c,0x8f13,0x005d,0x3e80,
0xde12,0x59ff,0x003e,0x3e70,
0x5633,0x8ea5,0x0029,0x3e60,
0xccda,0xb316,0x001b,0x3e50,
0x0845,0x76b9,0x0012,0x3e40,
0x5ae3,0x4ed0,0x000c,0x3e30,
0x1a87,0x3460,0x0008,0x3e20,
0xaebe,0x782a,0x0005,0x3e10
};
```

$2^x(1-1/x)(\zeta(x)-1)$

```
static short P[36] = {
0052010,0060466,0101211,0134657,
0051557,0154353,0135060,0064411,
0051065,0133157,0133514,0133633,
0050231,0114735,0035036,0111344,
0047242,0164327,0146036,0033545,
0046165,0065364,0130045,0011005,
0045020,0134427,0075073,0134107,
0043554,0021653,0000440,0177426,
0042111,0151213,0134312,0021402,
```

```
static short P[36] = {
0x3736,0xd051,0x0c26,0x4261,
0x0d21,0x7746,0xfb1d,0x424d,
0x96f3,0xf6e9,0xb6cd,0x4226,
0xd25c,0xa743,0x333b,0x41f3,
0xc6ed,0xf983,0x5d1a,0x41b4,
0xa241,0x9604,0xad5e,0x416e,
0x7709,0xef47,0x1722,0x4122,
0x1fe3,0x6024,0x8475,0x40cd,
0x4460,0x7719,0x3a51,0x4069,
```

```
};
```
Leading 1.0 omitted from Q.
```
static short Q[32] = {
0051665,0153363,0054252,0137010,
0051102,0143645,0121415,0036107,
0050250,0034073,0131133,0036465,
0047241,0123250,0150037,0070012,
0046210,0160426,0111463,0116507,
0045012,0054255,0031674,0173612,
0043572,0114460,0151520,0012221,
0042104,0067655,0037037,0137421,
};
```
$\ln(\zeta(x) - 1 - 2^{-x})$
```
static short A[44] = {
0046004,0156325,0126302,0131567,
0047050,0052177,0015271,0136466,
0050702,0060271,0070727,0171112,
0051766,0132727,0064363,0145042,
0053245,0012466,0056000,0117230,
0054226,0166155,0174275,0170213,
0055222,0003127,0112544,0101322,
0154741,0036625,0010346,0053767,
0156260,0054653,0154052,0031113,
0056213,0011152,0021000,0007111,
0056244,0120534,0040576,0163262,
};
```
Leading 1.0 omitted from B.
```
static short B[40] = {
0145761,0161734,0033026,0015520,
0147031,0013743,0017355,0036703,
0150661,0011720,0061061,0136402,
0151737,0125216,0070274,0164414,
0153227,0032653,0127211,0145250,
0154206,0121666,0123774,0042035,
0155212,0033352,0125154,0132533,
0055227,0170201,0110775,0072132,
0056113,0003133,0127132,0122303,
0155577,0126351,0141462,0171037,
};
```
$(1 - x)(\zeta(x) - 1)$
```
static short R[24] = {
0137650,0046650,0022502,0040316,
0041170,0041222,0057666,0142216,
0142170,0141510,0167741,0075646,
0042574,0120074,0046505,0106053,
0043506,0001154,0130073,0101413,
0144331,0166414,0020560,0131652,
};
```
Leading 1.0 omitted from S.

```
};
static short Q[32] = {
0x57c1,0x6b15,0xbade,0x4256,
0xa789,0xb461,0x58f4,0x4228,
0x67a7,0x764b,0x0707,0x41f5,
0xee01,0x1a03,0x34d5,0x41b4,
0x73a9,0xd266,0x1c22,0x4171,
0x9ef1,0xa677,0x4b15,0x4121,
0x0292,0x1a6a,0x5326,0x40cf,
0xf7e2,0xa7c3,0x8df5,0x4068,
};

static short A[44] = {
0x566f,0xb598,0x9b9a,0x4160,
0x37a7,0xe357,0x0a8f,0x41a5,
0xfe49,0x2e3a,0x4c17,0x4218,
0x7944,0xed1e,0xd6ba,0x425e,
0x13d3,0xcb80,0xa2a6,0x42b4,
0xbe11,0xbf17,0xdd8d,0x42f2,
0x905a,0xf2ac,0x40ca,0x4332,
0xcaff,0xa21c,0x27b2,0xc31c,
0x4649,0x7b05,0x0b35,0xc376,
0x01c9,0x4440,0x624d,0x4371,
0xdcd6,0x882f,0x942b,0x4374,
};

static short B[40] = {
0xc36a,0x86c2,0x3c7b,0xc15e,
0xa7b8,0x63dd,0x22fc,0xc1a3,
0x37a0,0x0c46,0x227a,0xc216,
0x9d22,0xce17,0xf551,0xc25b,
0x3955,0x75d1,0xe6b5,0xc2b2,
0x8884,0xd4ff,0xd476,0xc2f0,
0x96ab,0x554d,0x46dd,0xc331,
0xae8b,0x323f,0xfe10,0x4332,
0x5498,0x75cb,0x60cb,0x4369,
0x5e44,0x3866,0xf59d,0xc34f,
};

static short R[24] = {
0x481a,0x04a8,0x09b5,0xbfd5,
0xd892,0x4bf6,0x0852,0x402f,
0x2f75,0x1dfc,0x1869,0xc06f,
0xb185,0x89a8,0x9407,0x408f,
0x7061,0x9607,0xc04d,0x40c8,
0x1675,0x842e,0x3da1,0xc0fb,
};
```

```
static short S[20] = {
0041234,0013015,0042073,0113570,
0042236,0155353,0077325,0077445,
0043075,0162656,0016646,0031723,
0043637,0016454,0157636,0071126,
0044221,0044262,0140365,0146434,
};
#endif
```

```
static short S[20] = {
0x72ef,0xa887,0x82c1,0x4033,
0xafe5,0x6fda,0xdb5d,0x4073,
0xc67a,0xc3b4,0xbcb5,0x40a7,
0xce4b,0x9bf3,0xe3a5,0x40d3,
0xb9a3,0x581e,0x2916,0x40f2,
};
#endif
```

```
#define MAXL2 127
extern double MACHEP;

double zetac(x)
double x;
    {
    int i;
    double a, b, s, t, w;
    double sin(), floor(), gamma(), pow(), exp();
    double polevl(), p1evl();

    if( x < 0.0 )
        {
        if( x < -30.8148 )
            {
            mtherr( "zetac", OVERFLOW );
            return(0.0);
            }
        s = 1.0 - x;
        w = zetac( s );
        b = sin(0.5*PI*x) * pow(2.0*PI, x)
            * gamma(s) * (1.0 + w) / PI;
        return(b - 1.0);
        }
    if( x >= MAXL2 )
        return(0.0);
```

Tabulated values for integer argument

```
    w = floor(x);
    if( w == x )
        {
        i = x;
        if( i < 31 )
            {
#ifdef UNK
            return( azetac[i] );
#else
```

```
                return( *(double *)&azetac[4*i] );
#endif
                }
        }
    if( x < 1.0 )
        {
        w = 1.0 - x;
        a = polevl( x, R, 5 )/( w * p1evl( x, S, 5 ));
        return( a );
        }
    if( x == 1.0 )
        {
        mtherr( "zetac", SING );
        return( MAXNUM );
        }
    if( x <= 10.0 )
        {
        b = pow( 2.0, x ) * (x - 1.0);
        w = 1.0/x;
        s = (x * polevl( w, P, 8 ))
            / (b * p1evl( w, Q, 8 ));
        return( s );
        }
    if( x <= 50.0 )
        {
        b = pow( 2.0, -x );
        w = polevl( x, A, 10 ) / p1evl( x, B, 10 );
        w = exp(w) + b;
        return(w);
        }
Sum of inverse powers
pseres:
    s = 0.0;
    a = 1.0;
    do
        {
        a += 2.0;
        b = pow( a, -x );
        s += b;
        }
    while( b/s > MACHEP );
    b = pow( 2.0, -x );
    s = (s + b)/(1.0-b);
    return(s);
    }
```

Bibliography

Abramowitz, Milton, and Irene A. Stegun (eds) (1968) *Handbook of Mathematical Functions with Formulas, Graphs, and Mathematical Tables*, National Bureau of Standards Applied Mathematics Series No. 55. U. S. Government Printing Office

AMS55, *see* Abramowitz and Stegun

Bickley, W. G. (1950) "Bessel functions and formulae" in British Association for the Advancement of Science, Bessel functions, Part II. Functions of positive integer order, Mathematical Tables, Volume VI. Cambridge University Press

Blanch, G. (1964) "Numerical evaluation of continued fractions," *SIAM Review 6*, 383-421.

Bleistein, Norman and Richard A. Handelsman (1975) *Asymptotic Expansions of Integrals.* Holt, Rinehart and Winston

Clenshaw, C. W. (1962) *Mathematical Tables, Volume 5, Chebyshev Series for Mathematical Functions.* National Physical Laboratory, Her Majesty's Stationery Office, London

Cody, William J., Jr. and William Waite (1980) *Software Manual for the Elementary Functions.* Prentice-Hall

Exton, H. (1983) *Q Hypergeometric Functions.* Halsted

Feller, William (1968, 1971) *An Introduction to Probability Theory and Its Applications* (two volumes). Wiley

Forsythe, George E. and Cleve B. Moler (1967) *Computer Solution of Linear Algebraic Systems.* Prentice-Hall

Fröberg, Carl-Erik (1969) *Introduction to Numerical Analysis*, 2nd edition. Addison-Wesley

Gradshteyn, I. S. and I. M. Ryzhik (1980) *Table of Integrals, Series, and Products*, edited by Alan Jeffrey. Academic Press

Hamming, Richard W. (1971) *Introduction to Applied Numerical Analysis.* McGraw-Hill

Hart, John F., E. W. Cheney, Charles L. Lawson, Hans J. Maehly, Charles K. Mesztenyi, John R. Rice, Henry C. Thacher, Jr., and Christoph Witzgall (1968) *Computer Approximations.* Wiley

Hastings, Cecil, Jr. (1955) *Approximations for Digital Computers.* Princeton University Press

IEEE Standard for Binary Floating-Point Arithmetic, ANSI/IEEE Std 754-1985. Institute of Electrical and Electronics Engineers, Inc., 345 East 47th Street, New York, NY 10017, USA

IEEE Standard for Radix-Independent Floating-Point Arithmetic, ANSI/IEEE Std 854-1987. Institute of Electrical and Electronics Engineers, Inc., 345 East 47th Street, New York, NY 10017, USA

Jahnke, E., and F. Emde (1941) *Tables of Functions with Formulae and Curves.* G. E. Stechert & Co. There are several editions.

Kahan, W. M. *et al* (1983) PARANOIA (computer program). Obtainable via Richard Karpinski, Computer Center U-76, University of California, San Francisco, CA 94143-0704, USA

Kernighan, Brian W. and Dennis M. Ritchie (1978) *The C Programming Language.* Prentice-Hall

Knuth, Donald E. (1981) *The Art of Computer Programming*, Vol. 2, *Seminumerical Algorithms*, 2nd ed. Addison-Wesley

Lebedev, N. N. (1972) *Special Functions and their Applications*, edited by Richard A. Silverman. Dover

Luke, Yudell L. (1962) *Integrals of Bessel Functions.* McGraw-Hill

Olds, C. D. (1963) *Continued Fractions.* Mathematical Association of America

Ralston, Anthony, and Herbert S. Wilf, (eds) (1960) *Mathematical Methods for Digital Computers.* Wiley

Rivlin, Theodore J. (1969) *An Introduction to the Approximation of Functions.* Blaisdell

Srivastava, J. N. *et al* (1973) *A Survey of Combinatorial Theory.* Elsevier

Titchmarsh, E. C. (1930) *The Zeta-function of Riemann.* Cambridge University Press; also Hafner, 1972

Watson, G. N. (1944) *A Treatise on the Theory of Bessel Functions*, 2nd ed. Cambridge University Press

Index

Mathematics and its Applications

Series Editor: G. M. BELL, Professor of Mathematics, King's College London (KQC), University of London

Gardiner, C.F.	**Algebraic Structures**
Gasson, P.C.	**Geometry of Spatial Forms**
Goodbody, A.M.	**Cartesian Tensors**
Goult, R.J.	**Applied Linear Algebra**
Graham, A.	**Kronecker Products and Matrix Calculus: with Applications**
Graham, A.	**Matrix Theory and Applications for Engineers and Mathematicians**
Graham, A.	**Nonnegative Matrices and Applicable Topics in Linear Algebra**
Griffel, D.H.	**Applied Functional Analysis**
Griffel, D.H.	**Linear Algebra and its Applications: Vol. 1, A First Course; Vol. 2, More Advanced**
Guest, P. B.	**The Laplace Transform and Applications**
Hanyga, A.	**Mathematical Theory of Non-linear Elasticity**
Harris, D.J.	**Mathematics for Business, Management and Economics**
Hart, D. & Croft, A.	**Modelling with Projectiles**
Hoskins, R.F.	**Generalised Functions**
Hoskins, R.F.	**Standard and Non-standard Analysis**
Hunter, S.C.	**Mechanics of Continuous Media, 2nd (Revised) Edition**
Huntley, I. & Johnson, R.M.	**Linear and Nonlinear Differential Equations**
Irons, B. M. & Shrive, N. G.	**Numerical Methods in Engineering and Applied Science**
Ivanov, L. L.	**Algebraic Recursion Theory**
Johnson, R.M.	**Theory and Applications of Linear Differential and Difference Equations**
Johnson, R.M.	**Calculus: Theory and Applications in Technology and the Physical and Life Sciences**
Jones, R.H. & Steele, N.C.	**Mathematics in Communication Theory**
Jordan, D.	**Geometric Topology**
Kelly, J.C.	**Abstract Algebra**
Kim, K.H. & Roush, F.W.	**Applied Abstract Algebra**
Kim, K.H. & Roush, F.W.	**Team Theory**
Kosinski, W.	**Field Singularities and Wave Analysis in Continuum Mechanics**
Krishnamurthy, V.	**Combinatorics: Theory and Applications**
Lindfield, G. & Penny, J.E.T.	**Microcomputers in Numerical Analysis**
Livesley, K.	**Mathematical Methods for Engineers**
Lootsma, F.	**Operational Research in Long Term Planning**
Lord, E.A. & Wilson, C.B.	**The Mathematical Description of Shape and Form**
Malik, M., Riznichenko, G.Y. & Rubin, A.B.	**Biological Electron Transport Processes and their Computer Simulation**
Massey, B.S.	**Measures in Science and Engineering**
Meek, B.L. & Fairthorne, S.	**Using Computers**
Menell, A. & Bazin, M.	**Mathematics for the Biosciences**
Mikolas, M.	**Real Functions and Orthogonal Series**
Moore, R.	**Computational Functional Analysis**
Moshier, S.L.B.	**Methods and Programs for Mathematical Functions**
Murphy, J.A., Ridout, D. & McShane, B.	**Numerical Analysis, Algorithms and Computation**
Nonweiler, T.R.F.	**Computational Mathematics: An Introduction to Numerical Approximation**
Norcliffe, A. & Slater, G.	**Mathematics of Software Construction**
Ogden, R.W.	**Non-linear Elastic Deformations**
Oldknow, A.	**Microcomputers in Geometry**
Oldknow, A. & Smith, D.	**Learning Mathematics with Micros**
O'Neill, M.E. & Chorlton, F.	**Ideal and Incompressible Fluid Dynamics**
O'Neill, M.E. & Chorlton, F.	**Viscous and Compressible Fluid Dynamics**
Page, S. G.	**Mathematics: A Second Start**
Prior, D. & Moscardini, A.O.	**Model Formulation Analysis**
Rankin, R.A.	**Modular Forms**
Scorer, R.S.	**Environmental Aerodynamics**
Shivamoggi, B.K.	**Stability of Parallel Gas Flows**
Smith, D.K.	**Network Optimisation Practice: A Computational Guide**
Srivastava, H.M. & Manocha, L.	**A Treatise on Generating Functions**
Stirling, D.S.G.	**Mathematical Analysis**
Sweet, M.V.	**Algebra, Geometry and Trigonometry in Science, Engineering and Mathematics**
Temperley, H.N.V.	**Graph Theory and Applications**
Temperley, H.N.V.	**Liquids and Their Properties**
Thom, R.	**Mathematical Models of Morphogenesis**
Toth, G.	**Harmonic and Minimal Maps and Applications in Geometry and Physics**
Townend, M. S.	**Mathematics in Sport**
Townend, M.S. & Pountney, D.C.	**Computer-aided Engineering Mathematics**
Trinajstic, N.	**Mathematical and Computational Concepts in Chemistry**
Twizell, E.H.	**Computational Methods for Partial Differential Equations**
Twizell, E.H.	**Numerical Methods, with Applications in the Biomedical Sciences**
Vince, A. and Morris, C.	**Mathematics for Computing and Information Technology**
Walton, K., Marshall, J., Gorecki, H. & Korytowski, A.	**Control Theory for Time Delay Systems**
Warren, M.D.	**Flow Modelling in Industrial Processes**
Wheeler, R.F.	**Rethinking Mathematical Concepts**
Willmore, T.J.	**Total Curvature in Riemannian Geometry**
Willmore, T.J. & Hitchin, N.	**Global Riemannian Geometry**

Numerical Analysis, Statistics and Operational Research

Editor: B. W. CONOLLY, Emeritus Professor of Mathematics (Operational Research), Queen Mary College, University of London

Beaumont, G.P.	**Introductory Applied Probability**
Beaumont, G.P.	**Probability and Random Variables**
Conolly, B.W.	**Techniques in Operational Research: Vol. 1, Queueing Systems**
Conolly, B.W.	**Techniques in Operational Research: Vol. 2, Models, Search, Randomization**
Conolly, B.W.	**Lecture Notes in Queueing Systems**
Conolly, B.W. & Pierce, J.G.	**Information Mechanics: Transformation of Information in Management, Command, Control and Communication**
French, S.	**Sequencing and Scheduling: Mathematics of the Job Shop**
French, S.	**Decision Theory: An Introduction to the Mathematics of Rationality**
Griffiths, P. & Hill, I.D.	**Applied Statistics Algorithms**
Hartley, R.	**Linear and Non-linear Programming**
Jolliffe, F.R.	**Survey Design and Analysis**
Jones, A.J.	**Game Theory**
Kapadia, R. & Andersson, G.	**Statistics Explained: Basic Concepts and Methods**
Moscardini, A.O. & Robson, E.H.	**Mathematical Modelling for Information Technology**
Moshier, S.L.B.	**Mathematical Functions for Computers**
Oliveira-Pinto, F.	**Simulation Concepts in Mathematical Modelling**
Ratschek, J. & Rokne, J.	**New Computer Methods for Global Optimization**
Schendel, U.	**Introduction to Numerical Methods for Parallel Computers**
Schendel, U.	**Sparse Matrices**
Sehmi, N.S.	**Large Order Structural Eigenanalysis Techniques: Algorithms for Finite Element Systems**
Späth, H.	**Mathematical Software for Linear Regression**
Stoodley, K.D.C.	**Applied and Computational Statistics: A First Course**
Stoodley, K.D.C., Lewis, T. & Stainton, C.L.S.	**Applied Statistical Techniques**
Thomas, L.C.	**Games, Theory and Applications**
Whitehead, J.R.	**The Design and Analysis of Sequential Clinical Trials**